ADVANCES IN THE TREATMENT OF ISCHEMIC STROKE

Edited by **Maurizio Balestrino**

Advances in the Treatment of Ischemic Stroke
Edited by Maurizio Balestrino

Published by InTech

Janeza Trdine 9, 51000 Rijeka, Croatia

Publishing Process Manager Maja Bozicevic

Technical Editor Teodora Smiljanic

Cover Designer InTech Design Team

Additional hard copies can be obtained from orders@intechopen.com

Advances in the Treatment of Ischemic Stroke, Edited by Maurizio Balestrino
p. cm.
ISBN 978-953-51-0136-9

For EU product safety concerns:
IN TECH d.o.o., Prolaz Marije Krucifikse Kozulić 3, 51000 Rijeka, Croatia,
info@intechopen.com or visit our website at intechopen.com.

Meet the editor

Dr Maurizio Balestrino was born in Genoa, Italy, the birthplace of Christopher Columbus and of the blue jeans. He received there the Degree in Medicine and the Diploma of Specialist in Neurology. In 1983-1986 he was Research Associate in the Department of Physiology at Duke University, where he started an interest in ischemic brain damage. Back in Italy he has always combined clinical responsibilities as a neurologist and experimental research in brain anoxia or ischemia. He is currently Senior Researcher in the Stroke Unit of the University of Genoa, where he also directs the Laboratory of Experimental Neurophysiology. He has been partner or coordinator in national and international research projects aiming at bridging the gap between experimental research and clinical therapy.

Contents

Preface

The last decade or so has witnessed unprecedented advances in the therapy of ischemic stroke. Intensive preclinical and clinical research in what used to be an almost incurable disease is finally putting at the clinicians' disposal powerful therapeutic tools. While intravenous thrombolysis with recombinant tissue plasminogen activator has been the first and is still the most used tool, other ways of interventions have entered the clinical arena, dramatically improving the therapy of ischemic stroke. As it always happens at times of rapid changes, clinical practice lags behind research findings. While at times it must be so, since clinical practice must wait for evidence confirmation, very often the clinician finds it difficult to receive and process the relevant information in what may appear an overflow of data.

While an all-inclusive review of all available innovations in stroke therapy would probably be impossible in a single book, this one does provide reviews and updated information in several hot issues.

Hypothermia is the first such issue. This powerful therapy is finally coming of age, so far the sole survivor of a host of "neuroprotective" therapies that animal research had developed. While all other neuroprotective therapies have failed in the clinics, hypothermia has grown to become now, basically, recommended practice in the rescue therapy after cardiac arrest, a condition very similar albeit not identical to ischemic stroke. Its application in stroke is currently not routine practice, because of technical difficulties, of still significant side effects (let's not forget that cardiac arrest often occurs in young people, ischemic stroke in elder persons) and because of limited clinical trials. However, there is probably sufficient evidence for considering it on a case-by-case basis in hospitals that have experience in its application. In this book, the contribution by Campbell et al. provide a fine review of both preclinical and clinical issues of hypothermia. Both Nares-López et al. and Shintani and Terao convincingly review the very extensive mechanisms of protection by hypothermia, while Doshi and Hirashima report results from animal research concerning duration and timing of hypothermia, results that are relevant to clinical applications of this technique.

Second, brain regeneration is considered. The last decade of the 20th century has finally rejected the old myth that neurons remain unchanged in number from birth to death and, if damaged, cannot be replaced. On the contrary, we now know that a lively neuronal regeneration is routinely under way in the brain (for example, new neurons are continuously generated in the hippocampal dentate gyrus, a fact that is probably important for memory). Endogenous regeneration is certainly a repair mechanism that occurs after stroke, although still a poorly understood one. In this book, two chapters (by Gu and Wester and by Wappler et al.) provide novel interesting knowledge on the relevance and on the mechanisms of endogenous regeneration after experimental animal ischemic stroke. Moreover, stem cells administration has been extensively investigated, in the hope to replace the neurons that had died after stroke. Unfortunately, human clinical trials in this field have been surprisingly scarce, a fact that still leaves largely unanswered basic questions like: is stem cells administration really useful for stroke, does it really work by replacing dead neurons, or does it rather favor endogenous regeneration and healing, what types of stem cells are better, what stroke types benefit the most from this therapy. Answer to these questions is of paramount importance also because several private hospitals are now offering expensive stem cells transplantation, a legitimate business that however underlines an urgent need to answer the above questions. In this book the two chapters by Afzal and Mocco and by Mitkari et al. help the interested professional navigate this difficult field.

Clot-removal therapies for ischemic stroke (both intravenous thrombolysis and endovascular techniques) entered the neurological armamentarium at the very end of the last century, changing forever the way stroke is treated. As we all recall, they were met with a mix of enthusiasm for their effectiveness and fear for their side effects, chiefly haemorrage. Fear of haemorrhage caused a long list of exclusion criteria that, at least in Europe, prevented intravenous thrombolysis from being administered to many, probably most, patients. In the following years several such criteria were challenged or revised, for example the maximum acceptable time from onset is currently no longer considered 3 hours (as it is still stated in the official approval documents of the therapy) but rather 4.5 hours, following the successful ECASS-III study, and most centers are now administering intravenous thrombolysis off-label between 3 and 4.5 hours from symptoms onset. At the time of this writing (January 2012) we are waiting for the results of the IST-3 trial, that will hopefully clarify other issues in the administration of intravenous thrombolysis (for example, time up to 6 hours from onset, simptoms very mild or too severe, onset with epileptic seizures, and so on). A major issue is patients' age, in fact in Europe r-TPA for intravenous thrombolysis is still officially approved for treatment only in patients younger than 80 y.o. However, this boundary is being strongly challenged, and in this book we (Balestrino et al.) are reporting our so far successful experience with intravenous thrombolysis in patients even older than 90 years.

Endovascular therapy has recently raised great interest both in the therapy of acute stroke and in the therapy of symptomatic or asymptomatic carotid artery stenosis. As for acute stroke, the only randomized, controlled, multicenter, open-label clinical trial with blinded follow-up that has been so far completed is the PROACT-II, that demonstrated superiority of intra-arterial thrombolysis with urokinase plus heparin compared to heparin alone within 6 hours from onset. Such efficacy occurred at the expense of an increased rate of cerebral hemorrhages that was of borderline statistical significance (p=0.06). After that study, several series of patients have been published, that generally support the effectiveness of this procedure. It is noteworthy, and it should not be forgot, that intra-arterial thrombolysis is a complex procedure, whose preliminary activities (angiography, catheterization, etc.) allow it to be started, as an average, one hour later than intravenous thrombolysis. During this hour brain cells continue to die, and nobody has ever demonstrated how intra-arterial and intravenous thrombolysis compare for safety and effectiveness. At the time of this writing (January 2012) a pivotal study ("Synthesis-Expansion", randomizing patients to intravenous vs. intra-arterial thrombolysis within 4.5 hours from onset) is reaching its end and will soon provide much needed answers on whether the two treatments have different efficacy or safety. Waiting for the results of this study, both therapies are largely practiced. In this book, Lacman and Charvát review a popular technique of endoscopic (intra-arterial) treatment, i.e. mechanical thrombectomy (not requiring drugs, only physical disruption of the clot). As we know, this technique is very interesting, not least because it is supposed (lacking the thrombolytic drug) to minimize the bleeding risk. For this reason it is often used also as a "rescue" treatment after failed intravenous thrombolysis.

Another popular application of endovascular techniques is carotid artery stenosis, both asymptomatic (in a person that never had stroke or TIA) and symptomatic. Randomized, controlled clinical trials have underlined how carotid "stenting" is more often loaded with a burden of complications heavier than open-neck conventional surgery, nevertheless this technique still has specific indications (for example, critical patients who could not tolerate open-neck surgery), and it is widely practiced. In this book Tavares and Caldas address relevant technical issues of carotid "stenting". Interestingly, they specifically discuss two consequences of this technique that are often poorly understood. One is the "reperfusion syndrome", a possible harmful consequence of carotid artery recanalization, the other is the effects of carotid artery recanalization on cerebral blood flow and on cognitive defects. Concerning the latter, we should remember that procedures to reopen a clogged carotid artery are usually undertaken to prevent a stroke, but restoration of normal cognitive function could in theory be one more reason to restore carotid artery patency.

Last but not least, research is tackling a fearful complication of ischemic stroke, severe intracranial hypertension in what is called "malignant" infarction of the middle cerebral artery, a condition that in most cases ends with death or severe disability. Hypothermia has been successfully used for this condition, however decompressive surgery is probably the most largely practiced intervention, perhaps due to its larger availability. In this book, Scozzafava et al. review the therapy of cerebral edema, while Valença et al. describe a novel very interesting way (opening the skull "like a window") that they invented to improve this surgery.

In the end, I am grateful to the Authors of the above chapters for providing stimulating and updated reports on many innovative issues in treating ischemic stroke. I am also indebted to the InTech publisher for having stimulated me to edit this book and having provided me, in this process, with powerful online tools and with professional human assistance. In particular, I would like to thank Ms. Ana Pantar for her skill and determination in starting this project and making it possible, and Ms. Maja Bozicevic for her kind, extensive and continuous assistance in streamlining the publishing procedure.

Maurizio Balestrino, MD
Department of Neuroscience,
Ophthalmology and Genetics, University of Genova,
Italy

Part 1

Hypothermia in the Acute Phase

1

Cerebral Ischemia and Post-Ischemic Treatment with Hypothermia

Kym Campbell, Neville W. Knuckey and Bruno P. Meloni
Centre for Neuromuscular and Neurological Disorders, University of Western Australia, Australian Neuromuscular Research Institute, Department of Neurosurgery, Sir Charles Gairdner Hospital, Nedlands, WA, Australia

1. Introduction

With its strict dependence on a continuous supply of oxygen and glucose to meet its energy needs and its high metabolic rate, the brain is particularly sensitive to any compromise of blood supply. Brain ischemia, as occurs in a number of disease states but most importantly in ischemic stroke and during cardiac arrest, rapidly results in exhaustion of ATP, triggering an energy crisis. Within minutes, the failure of ion pumps sees the depolarisation of neuronal cell membranes and the consequent release of stored presynaptic glutamate, leading, by way of overstimulation of glutamate receptors (=excitotoxicity), to many-fold increases in intracellular calcium and zinc concentrations. Severely affected cells die within only a few minutes. In those cells that are less severely injured, ongoing cellular and tissue damage occurs due to activation of proteolytic enzymes, oxidative and nitrosative stress (Forder & Tymianski, 2009), altered calcium homeostasis, initiation of active cell death pathways (apoptosis, necrosis, autophagy and necroptosis), inflammation (microglia and astrocyte activation, neutrophil infiltration within 4 - 6 hours), cortical spreading depressions, disruption of the blood brain barrier (BBB; starting at 2 hours, followed by a second phase from 24 - 72 hours; Brouns & De Deyn, 2009), microvascular injury (which promotes BBB disruption, inflammation and impairs vascular control of blood flow), hemostatic activation (platelet activation and the intrinsic pathway) and edema. Additionally, though reperfusion is the cornerstone of treatment, when/if it is established, many of these damaging events can be exacerbated.

These processes are interconnected, rather than sequential, and presenting them as a list might be misleading if it were to be taken that counteracting one event would therefore prevent those occurring later in the list, even if they do in fact occur later in time. The list only acts as a summary of the damaging events, against which can be checked the likelihood of a potential therapy to do some good. The importance of each process waxes and wanes at different times during and after the ischemic episode, so an important principle of effective therapy is that it will need to be applied at the time of the injurious events to counteract its effect. In many respects hypothermia is, in theory, the ideal therapy, with multiple mechanisms of action in opposition to the consequences of ischemia.

2. Background of therapeutic hypothermia

Cooling of the body for therapeutic purposes is not a new concept in medicine. For example, in 1941 the British Medical Journal noted that generalised therapeutic hypothermia was under investigation for the treatment of various cancers, such as bladder carcinoma (Anonymous, 1941). It was also suggested that whole body cooling might find use in patients with intractable pain, morphine addiction, leukaemia, and schizophrenia, though no mention was made at this time of stroke or cardiac arrest. In the early 1950s, however, studies were being performed in which animals were cooled to very low temperatures (16 - 19ºC in macacus rhesus monkeys, 2.5 - 5ºC in groundhogs) to permit cardiac surgery (Bigelow & McBirnie, 1953). Cardiac output was completely stopped in these experiments for long periods (15 - 24 minutes in the monkeys, 1 - 2 hours in the groundhogs) with few deaths, and no apparent neurological deficits when the animals were recovered.

Thus, it has been known for several decades at least that a state of hypothermia decreases central neurological injury in the face of ischemia. This has led to many animal studies, of various designs, which have tended to confirm the potential for hypothermia to reduce ischemic brain damage (for review see Meloni et al., 2008). In recent years, clinical trials have proven that moderate hypothermia, using a target body temperature of 33ºC, improves outcomes for cardiac arrest survivors (Bernard et al., 2002; Hypothermia after cardiac arrest study group, 2002; Meloni et al., 2008), and its use is, at the time of writing, under investigation in several ongoing or planned trials following ischemic and hemorrhagic stroke (Table 1; Meloni et al., 2008). Between them, these studies will answer several of the important questions regarding the best use of therapeutic hypothermia.

Study*	**Method of cooling**	**Number of subjects**	**Delay from stroke onset**	**Tempe-rature**	**Duration**
Cerebral hypothermia in ischemic lesion (CHIL). Stroke Trials Directory	Cold infusion induction + endovascular cooling or local head cooling	80	Within 6h	33°C	24h
Cooling in acute stroke-II (COAST-II). Stroke Trials Directory	Cold infusion induction + endovascular cooling#	50	Within 3h; 30 - 90min after tPA	35°C	24h
Mild hypothermia in acute ischemic stroke (MHAIS) Stroke Trials Directory	Surface cooling#	36	Within 6h	35°C	12h
Mild hypothermia in acute ischemic stroke trial – Edinburgh (HAIST-E). EUROHYP	Cold infusion induction + surface cooling#	24	Within 4.5h	35 or 33°C	12 or 24h

Study*	Method of cooling	Number of subjects	Delay from stroke onset	Tempe-rature	Duration
Cooling for ischemic stroke trial (COOLIST); NTR2616. Nederlands Trial Register	Cold infusion induction + surface cooling#	84	Within 4.5h	35, 34.5 or 34°C	24h
Mild hypothermia in acute ischemic stroke (MASCOT; Pilot). EUROHYP	Surface cooling	40	Within 24	33°C	24h
Mild hypothermia in acute ischemic stroke: surface. vs endovascular cooling (HAIS-SE). Stroke Trials Directory	Cold infusion induction; surface or endovascular cooling #	60	Within 4.5h; 30min after tPA	34°C	12, 18 or 24h
Cooling in intracerebral hemorrhage (CINCH). EUROHYP	Endovascular cooling	50	Within 6 - 18h	35°C	8 d
Hypothermia for intracerebral hemorrhage. Clinical Trials	Surface cooling	20	Within 6h	34°C	24h
European stroke research network for hypothermia (EuroHYP). EUROHYP	Cold infusion induction + surface or endovascular cooling#	1500	Within 6h; within 90min after tPA	34 - 35°C	24h

* For more detail see Stroke Trials Directory, EUROHYP Nederlands Trial Register and Clinical Trials web sites (details provided in reference list). # Patient awake and treated with pethidine and/or buspirone to control shivering and improve comfort.

Table 1. Current clinical trials of hypothermia in stroke (ischemic and hemorrhagic).

3. Neuroprotective mechanisms of hypothermia

There is evidence that therapeutic hypothermia has beneficial effects by numerous mechanisms including reduction of metabolic rate, promotion of energy recovery after ischemia, inhibition of glutamate release, inhibition of cell death pathways, inhibition of free radical formation, inhibition of inflammation, preservation of the BBB, stimulation of neurotrophin expression, and numerous effects on molecular responses to ischemia (e.g., inhibition of AMPK and MAPK activation, inhibition of SMAC/Diablo, p53).

Intuitively, a reduced metabolic rate might be expected to be one of the more important means by which hypothermia could protect against ischemia, and it does make a contribution, though the effect is easily overestimated. It has been calculated that, on the measure of reduced metabolic oxygen consumption, 5 minutes of ischemia at 37°C would cause approximately equivalent damage to 15 minutes of ischemia at 27°C (Schaller & Graf, 2003). Thus, the benefit is only moderate even at substantially lower temperatures than are usually considered suitable for most therapeutic purposes. There is evidence, however, that hypothermia also expedites the recovery of ATP stores after a period of ischemia, as well as improving the return of energy metabolism to pre-ischemic levels (Erecinska et al., 2003; Zhao et al., 2007). The combination of reduced demand and improved recovery both during and after what is a state of failed energy supply is perhaps enough to explain the outstanding neuroprotection afforded by intraischemic hypothermia. That is, without taking into account any of its other actions, hypothermia reduces the severity of any one incidence of cerebral ischemia. Its influence does not end there, however, though clearly hypothermia will necessarily be less effective when delayed.

While excitotoxicity, principally attributable to overstimulation of the NMDA subtype of glutamate receptor, is a critical component of the ischemic cascade, its very early occurrence means it is likely to remain a frustrating target for therapeutic intervention. That said, there is fair evidence that hypothermia reduces, or at least delays, the release of glutamate from ischemic neurons, probably by delaying the onset of anoxic depolarisation (Zhao et al., 2007). Using a cardiac arrest model, it was shown that hypothermia (31°C/20min) either during ischemia or initiated at the time of reperfusion reduced extracellular glutamate concentrations (measured at the hippocampus), but not when the initiation of hypothermia was delayed by as little as 5 minutes after reperfusion (Takata et al., 2005). Hachimi-Idrissi et al. (2004) found a long lasting (more than 2 hours) inhibition of both glutamate and dopamine release in hypothermia (34°C/1h) treated animals when commenced after resuscitation in an asphyxiation/cardiac arrest model.

Importantly, the activity of hypothermia in reducing oxidative damage after ischemia or ischemia-like insults is well-supported. Shin et al. (2010) found that, in rats, the death of neurons induced by hypoglycemia could be reduced by maintaining brain temperature at 33 - 34°C for 1 hour, and that this was associated with reductions in zinc ion release/translocation, generation of ROS, and activation of microglia. Maier et al. (2002) reported that intra-ischemic hypothermia (33°C/2h) reduced production of the superoxide anion after transient focal cerebral ischemia, and Van Hemelrijck et al. (2005) showed that hypothermia (34°C/2h during ischemia) reduced hydroxyl radical formation, by inhibition of neuronal NOS, during the resuscitative phase after focal cerebral ischemia.

There is general agreement from several studies that hypothermia improves membrane stability, reducing disruption of the BBB as well as protecting neuronal cell membranes. Kiyatkin and Sharma, 2009, found that hypothermia at 34 - 35°C in normal brains slightly increased BBB permeability to albumin (compared to normothermia), but this did not result in edema. The mechanism appears to be related to temperature dependent variations in the sodium and chloride content of brain tissue, such that hypothermia reduces these ions by an undescribed mechanism, preventing osmotic draw into this tissue and producing a relative dehydration. In any case, the effect was mild, certainly compared to hyperthermia, which markedly increased both BBB permeability and edema. Baumann et al. (2009) found, on the other hand, that after global ischemia hypothermia (32°C/6h commencing after reperfusion) stabilised blood vessels and decreased BBB permeability, probably by preservation of the basement membrane. Nagel et al. (2008) measured extravasation of MRI contrast agent after transient focal ischemia (tMCAO), and found that hypothermia (33°C/4h starting 60min into 90min MCAO) greatly reduced BBB disruption. Huang et al. (1999) showed that hypothermia (29°C/6h commencing after reperfusion) particularly reduces the second phase of BBB disruption that occurs around 24 hours after transient focal ischemia.

Besides these non-specific harmful processes, there are in the penumbra a variety of pro-apoptotic responses to ischemia that are mediated by particular molecular pathways. Identifying these pathways and investigating interventions to counteract them is a field of substantial current activity. There is too much to summarise here, and any such attempt would be likely to be out of date very soon. Nevertheless, a couple of examples are useful to give the flavour of the work being done, but for more additional information see recent reviews by Zhao et al. (2007) and González-Ibarra et al. (2011).

AMP-activated protein kinase (AMPK) is responsive to energy stress, and, when phosphorylated, suppresses anabolic and promotes catabolic activity, evidently in order to maintain ATP supplies. Perhaps paradoxically, AMPK inhibition reduces ischemic brain damage, and there is evidence that hypothermia (32°C/6h commencing after reperfusion) inhibits AMPK activation after transient focal cerebral ischemia in mice (Li et al., 2011). Li & Wang (2011) demonstrated that hypothermia (33°C during transient focal ischemia) reduced expression of the protein complex second mitochondrion-derived activator of caspases (SMAC), an important molecule in the activation of apoptosis, which is upregulated in response to a variety of insults. The reduction in SMAC expression was also associated with reduced neurological impairment in rats after focal ischemia.

4. Hypothermia and glial cells

The best neurological outcomes will be achieved by measures taken to address the consequences of ischemia not just in neurons and the BBB, but in glial cells as well. Studies of the effects of hypothermia on glia are relatively few, but those there are suggest that hypothermia promotes survival and inhibits pathological responses in microglia and astrocytes. Hypothermia does reduce activation and proliferation of microglia, thus reducing oxidative and nitrosative stress (Si et al., 1997). Reduced activation of microglia associated with hypothermia has been demonstrated by several studies using different animal models, for example, during and after global cerebral ischemia (Kumar and Evans, 1997; Webster et al., 2009), after transient focal cerebral ischemia (Inamasu et al., 2000), and after hypoxia/ischemia (Fukui et al., 2006). Hachimi-Idrissi et al. (2004) found that astrocyte proliferation is also inhibited by hypothermia after asphyxiation/cardiac arrest and resuscitation. Haun et al. (1993) found that astrocyte cultures were made relatively resistant to an *in vitro* glucose-oxygen deprivation injury by hypothermia.

5. Depth of hypothermia

The efficacy of hypothermia increases as the depth of hypothermia increases, though the response is not linear. In a large meta-analysis of animal studies, van der Worp et al. (2007) found that the greatest therapeutic response (reducing mean infarct volumes by approximately 55%) was achieved by cooling to below 30ºC, though cooling to even 35ºC, the highest level of hypothermia included, still resulted in a considerable positive response (infarct volume reduction of 30%). What's more, the adverse effects of the treatment (specifically cardiac arrhythmias, coagulopathies and immunosuppression) also increase with increasing depth, as do the technical difficulties involved in bringing patients to deeper body temperatures in the first place. Therefore, the optimum target will be the best balance between therapeutic effect versus detrimental effect versus practicality. While the optimum target temperature is still to be determined, based on preclinical and clinical studies it will probably be in the range 33 to 35ºC, that is, what is usually referred to as moderate or mild hypothermia. As mentioned earlier hypothermia at 33°C is being used following cardiac arrest in comatose survivors admitted to intensive care wards. However, from a clinical standpoint, hypothermia of 35°C offers the advantage of being achievable in awake subjects outside of intensive care units, which would comprise the majority of stroke patients. Furthermore several of the ongoing and planned stroke trials listed in Table 1 will provide data that aims to specifically address the question of the relative efficacy of hypothermia at 33°C versus 35°C.

It is worth considering here the use of hypothermia during (when it is most effective) cardiothoracic and neurosurgical procedures, in which body temperatures are lowered from anywhere from 26 - 35°C, specifically to protect tissues, including the brain, during an anticipated period of compromised blood supply. For example during cardiac surgery, there are two distinct levels of hypothermia that are commonly used; a target body temperature of 34 - 35ºC is now becoming accepted as the standard for Cardiopulmonary Bypass (CPB), while in especially critical cases surgeons may opt to use Deep Hypothermic Circulatory Arrest (DHCA) in which patients are cooled to a rather extreme 15 - 26ºC (Choi et al., 2009; Cook, 2009; Mackensen et al., 2009).

6. Timing and duration of hypothermia

As noted earlier, the pathophysiology of cerebral ischemia is dynamic and multifaceted, with numerous damaging mechanisms occurring, becoming important at different times, and lasting for different durations, many of which interact to exacerbate the effect of another. It is an oversimplification to say that hypothermia reduces the impact of all of these damaging processes, but it's fair to say that that is the trend. Consequently, the earlier hypothermia is commenced and the longer it is maintained while the ischemic damaging processes are occurring will permit the greatest neuroprotective effect. This contention is, however, not particularly well borne out by the evidence from animal trials (van der Worp et al., 2007), though there are some possible reasons for this finding (van der Worp et al.,

2010). In the bulk of animal trials, hypothermic treatment is used during or very soon after ischemia, when it is most effective. Prolonging hypothermia in this situation allows little opportunity to improve on the highly effective neuroprotection afforded by early treatment, while permitting the adverse effects of hypothermia treatment (eg. coagulopathies, immunosuppression, pneumonia) to become significant. Furthermore, there will be a time after ischemia where delayed hypothermia will not be effective at inhibiting neuroregenerative processes. There is evidence, however, that the longer treatment is delayed, short periods of hypothermia have little or no effect, while prolonged hypothermia (>24h) can be very effective (Clark et al., 2008; Colbourne et al., 1999ab; Zhu et al., 2005).

Again, however longer treatment consumes more resources and also increases the health risks, particularly in patients who require sedation or anaesthesia to maintain the hypothermic state. The optimum duration of hypothermia will most likely be in the range 12 - 48 hours, and is likely also to be dependent on factors such as the specific cause of ischemia (stroke, cardiac arrest), severity of ischemia, age of patient and the time delay to commencing hypothermia after ischemia. In terms of therapeutic window, this will vary depending on the type of ischemia (focal vs global) and severity, but could be up to 6 hours following stroke (focal ischemia; Ohta et al., 2007) and up to 12 hours following global ischemia (Colbourne et al., 1999b; Coimbra & Walsh, 1994). With respect to rewarming it is becoming increasingly accepted that slow rewarming at the rate of 0.2 - 0.3°C/hour is most desirable (Bardutzky & Schwab, 2007; Bernard & Buist, 2003).

7. Cooling methods

One of the most significant barriers to therapeutic hypothermia is the technical difficulty involved in inducing the target temperature in the target tissue in a timely and safe manner. Large mammals such as humans are very efficient at maintaining a normal body temperature in the face of attempts to cool the body. Available techniques are surface cooling by refrigerative blankets, cooling helmets, cold air blowers, intravascular heat exchangers, intravascular cold fluids and, which is currently under investigation, intranasal evaporative cooling (Castrén et al., 2010; Jordan & Carhuapoma, 2007). An alternative approach is the use of pharmaceutical agents, such as the neurotensin analogue NT77 to alter the body's temperature set-point as monitored and controlled by the hypothalamus, thus allowing an effectively physiological induction of hypothermia (Katz et al., 2004). Each of these has advantages and disadvantages in cost, accuracy, degree of control, rate of cooling and ease of application. Mild hypothermia can be induced in awake patients, as long as steps are taken to manage the associated discomfort (see below), but moderate to deep hypothermia requires sedation or anaesthesia with intubation, ventilation and intensive care measures.

At present the intravenous infusion of cold salt solutions (4°C) at a rate of 20 - 30ml/kg over 20 - 30 minutes is gaining acceptance as the method of choice to induce hypothermia (Bernard et al., 2003; Polderman et al., 2005; Moore et al., 2008). The cold saline infusion procedure has several attractions as it is: i) inexpensive ii) safe; iii) relatively straight forward; iv) fast at inducing mild to moderate hypothermia (33 - 35°C); v) applicable in the field allowing early hypothermia induction; vi) suited for use in both comatose and awake subjects; and vii) often indicated anyway as a means of improving physiological parameters (blood pressure, renal function, acid-base homeostasis; Bernard et al., 2003). Following hypothermia induction by cold saline infusion one or more of the cooling procedures outlined above would then be implemented to provide a more precise control of body temperature.

To further aid the induction and maintenance of hypothermia the use of pethidine (meperidine) alone or with other agents, such as, the anxiolytic buspirone or magnesium are being used (Kliegel et al., 2007; Mokhtarani et al., 2001; Martin-Schild et al., 2009; Zweifler et al., 2004; Table 1). The use of these agents, along with simple measures such as warm gloves and socks is especially useful when inducing hypothermia in awake patients to minimize discomfort and shivering (Mahmood & Zweifler, 2007).

8. Combination with other treatments

A substantial advantage of hypothermia is that it presents little or no obstacle to the application of other treatments, and in fact has been shown to enhance or act synergistically with some other neuroprotective approaches (Campbell et al., 2008; Zhu et al., 2005). In reviewing the literature, we have found that hypothermia in combination treatments generally has additive or synergistic effects, and in several instances medications which were thought to be neuroprotective were later found to induce hypothermia and in fact were not neuroprotective at all when normal body temperatures were maintained (Campbell et al., 2007; Nurse & Corbett, 1996). It is especially important that any potential stroke treatment should be compatible with tPA thrombolysis, and in this respect it appears based on *in vitro* data that at least for mild hypothermia (i.e. 35°C), it will not significantly reduce the effectiveness of tPA (Schwarzenberg et al., 1998; Shaw et al., 2007; Yenari et al., 1995).

9. Concluding remarks

There is compelling experimental and clinical evidence that mild to moderate hypothermia is effective following global and focal (ischemic stroke) cerebral ischemia. However, it is likely that the depth and duration of hypothermia that provides the best benefit to patients will vary depending on the type (global vs focal) and severity of brain ischemia, the time that hypothermia is commenced, and patient age and presence of co-morbidities (diabetes, hypertension). Therefore further experimental and clinical trials will be required to determine hypothermia protocols that best suit individual patients. Moreover, based upon the available human studies, it appears that the use of hypothermia, in particular mild hypothermia (35°C) is feasible and safe to implement in clinical situations. In addition, based on current information therapeutic hypothermia should be commenced as soon as possible after the ischemic event, and maintained for durations of 12 - 48 hours to achieve a sustained benefit in terms of neuronal recovery and survival and functional benefits. To this end, future experimental studies in global and focal ischemia models and the results of the clinical stroke trials, will no doubt, help address further refinement of therapeutic hypothermia protocols to better suit individual cases. In addition, evaluation of the effectiveness of hypothermia in combination with other potential neuroprotective agents such as magnesium, caffeinol, glutamate antagonists and anti-oxidants could further improve efficacy.

10. References

Anonymous. (1941). Treatment by hypothermia. *British Medical Journal*, 2 (4206), pp. 231-232

Bardutzky, J & Schwab, S. (2007). Antiedema therapy in ischemic stroke. *Stroke*, 38 (11), pp. 3084-3094

Baumann, E, Preston, E, Slinn, J & Stanimirovic, D. (2009). Post-ischemic hypothermia attenuates loss of the vascular basement membrane proteins, agrin and SPARC, and the blood-brain barrier disruption after global cerebral ischemia. *Brain Research*, 1269 (1), pp. 185-197

Bernard, SA & Buist, M. (2003). Induced hypothermia in critical care medicine: a review. *Critical Care Medicine*, 31 (7), pp. 2041-2051

Bernard, S, Buist, M, Monteiro, O & Smith K. (2003). Induced hypothermia using large volume, ice-cold intravenous fluid in comatose survivors of out-of-hospital cardiac arrest: a preliminary report. *Resuscitation*, 56 (1), pp. 9-13

Bernard, SA, Gray, TW, Buist MD, Jones BM, Silvester W, Gutteridge G & Smith, K. (2002). Treatment of comatose survivors of out-of-hospital cardiac arrest with induced hypothermia. *New England Journal of Medicine*, 346 (8), pp. 557-563

Bigelow, WG & McBirnie, JE. (1953). Further experiences with hypothermia for intracardiac surgery in monkeys and groundhogs. *Annals of Surgery*, 137 (3), pp. 361-365

Brouns, R & De Deyn, PP. (2009). The complexity of neurobiological processes in acute ischemic stroke. *Clinical Neurology & Neurosurgery*, 111 (6), pp. 483-495

Campbell K, Meloni, BP & Knuckey, NW. (2008). Combined magnesium and mild hypothermia (35 degrees C) treatment reduces infarct volumes after permanent middle cerebral artery occlusion in the rat at 2 and 4, but not 6 h. *Brain Research*, 1230 (1), pp. 258-264

Campbell, K, Meloni, BP & Knuckey, NW. (2007). Hypothermia in combination therapy for cerebral ischaemia: a review of animal trials. *International Journal of Neuroprotection and Neuroregeneration*, 3 (3), pp. 210-224

Castrén, M, Nordberg, P, Svensson, L, Taccone, F, Vincent, JL, Desruelles, D, Eichwede, F, Mols, P, Schwab, T, Vergnion, M, Storm, C, Pesenti, A, Pachl, J, Guérisse, F, Elste, T, Roessler, M, Fritz, H, Durnez, P. Busch, HJ, Inderbitzen, B & Barbut, D. (2010). Intra-arrest transnasal evaporative cooling: a randomized, prehospital, multicenter study (PRINCE: Pre-ROSC IntraNasal Cooling Effectiveness). *Circulation*, 122 (7), pp. 729-736

Choi, R, Andres, RH, Steinberg, GK & Guzman, R. (2009). Intraoperative hypothermia during vascular neurosurgical procedures. *Neurosurgical Focus*, 26 (5), pp. 1-6

Clark, DL, Penner, M, Orellana-Jordan, IM & Colbourne, F. (2008). Comparison of 12, 24 and 48 h of systemic hypothermia on outcome after permanent focal ischemia in rat. *Experimental Neurology*, 212 (2), pp. 386-392

Clinical Trials. www.clinicaltrials.gov; Trial identifier number: NCT0122114

Coimbra C & Wieloch T. (1994). Moderate hypothermia mitigates neuronal damage in the rat brain when initiated several hours following transient cerebral ischemia. *Acta Neuropathologic*, 87(4), pp. 325-331

Colbourne, F, Li, H & Buchan, AM. (1999a). Indefatigable CA1 sector neuroprotection with mild hypothermia induced 6 hours after severe forebrain ischemia in rats. *Journal of Cerebral Blood Flow & Metabolism*, 19 (7), pp. 742-749

Colbourne F, Sutherland GR & Auer RN. (1999b). Electron microscopic evidence against apoptosis as the mechanism of neuronal death in global ischemia. *Journal of Neuroscience*, 19(11), pp. 4200-4210

Cook, DJ. (2009). CON: Temperature regimens and neuroprotection during cardiopulmonary bypass: does rewarming rate matter? *Anesthesia & Analgesia*, 109 (6), pp. 1733-1737

Erecinska, M, Thoresen, M & Silver, IA. (2003). Effects of hypothermia on energy metabolism in mammalian central nervous system. *Journal of Cerebral Blood Flow & Metabolism*, 23 (5), pp. 513–530

EUROHYP. www.eurohyp.org

Forder, JP & Tymianski, M. (2009). Postsynaptic mechanisms of excitotoxicity: involvement of postsynaptic density proteins, radicals, and oxidant molecules. *Neuroscience*, 158 (1), pp. 293-300

Fukui, O, Kinugasa, Y, Fukuda, A, Fukuda, H, Tskitishvili, E, Hayashi, S, Song, M, Kanagawa, T, Hosono, T, Shimoya, K & Murata, Y. (2006). Post-ischemic hypothermia reduced IL-18 expression and suppressed microglial activation in the immature brain. *Brain Research*, 1121 (1), pp. 35-45

González-Ibarra, FP, Varon, J & López-Meza, EG. (2011). Therapeutic hypothermia: critical review of the molecular mechanisms of action. *Frontiers in Neurology*, 2 (4), pp. 1-8

Hachimi-Idrissi, S, Van Hemelrijck, A, Michotte, A, Smolders, I, Sarre, S, Ebinger, G, Huyghens, L & Michotte, Y. (2004). Postischemic mild hypothermia reduces neurotransmitter release and astroglial cell proliferation during reperfusion after asphyxial cardiac arrest in rats. *Brain Research*, 1019 (1-2), pp. 217-225

Haun, S, Trapp, VL & Horrocks, LA. (1993). Hypothermia decreases astroglial injury and arachidonate release during combined glucose-oxygen deprivation. *Brain Research*, 631 (1), pp. 352-356

Huang, ZG, Xue, D, Preston, E, Karbalai, H & Buchan, AM. (1999). Biphasic opening of the blood-brain barrier following transient focal ischemia: effects of hypothermia. *Canadian Journal of Neurological Sciences*, 26 (4), pp. 298-30

Hypothermia after cardiac arrest study group. (2002). Mild therapeutic hypothermia to improve the neurologic outcome after cardiac arrest. *New England Journal of Medicine*, 346 (8), pp. 549-556

Inamasu, J, Suga, S, Sato, S, Horiguchi, T, Akaji, K, Mayanagi, K & Kawase, T. (2000). Post-ischemic hypothermia delayed neutrophil accumulation and microglial activation following transient focal ischemia in rats. *Journal of Neuroimmunology*, 109 (2), pp. 66-74

Jordan, JD & Carhuapoma, JR. (2007). Hypothermia: comparing technology. *Journal of the Neurological Sciences*, 261 (1-2), pp. 35-38

Katz, LM, Young, A, Frank, JE, Wang, Y & Park, K. (2004). Neurotensin-induced hypothermia improves neurologic outcome after hypoxic-ischemia. *Critical Care Medicine*, 32 (3), pp. 806-810

Kiyatkin, EA & Sharma, HS. (2009). Permeability of the blood-brain barrier depends on brain temperature. *Neuroscience*, 161 (3), pp. 926-939

Kliegel, A, Janata, A, Wandaller, C, Uray, T, Spiel, A, Losert, H, Kliegel, M, Holzer, M, Haugk, M, Sterz, F & Laggner, AN. (2007). Cold infusions alone are effective for induction of therapeutic hypothermia but do not keep patients cool after cardiac arrest. *Resuscitation*, 73 (1), pp. 46-53

Kumar, K, & Evans, AT. (1997). Effect of hypothermia on microglial reaction in ischemic brain. *Neuroreport*, 8 (4), pp. 947-950

Li, J, Benashski, S & McCullough, LD. (2011). Post-stroke hypothermia provides neuroprotection through inhibition of amp-activated protein kinase. *Journal of Neurotrauma*, 28 (7), pp. 1281-1288

Li, H & Wang, D. (2011). Mild hypothermia improves ischemic brain function via attenuating neuronal apoptosis. *Brain Research*, 1368 (1), pp. 59-64

Mackensen, GB, McDonagh, DL & Warner, DS. (2009). Perioperative hypothermia: use and therapeutic implications. *Journal of Neurotrauma*, 26 (3), pp. 342-358

Mahmood, MA & Zweifler RM. (2007). Progress in shivering control. *Journal of the Neurological Sciences*, 261 (1-2), pp. 47-54

Maier, CM, Sun, GH, Cheng, D, Yenari, MA, Chan, PH & Steinberg, GK. (2002). Effects of mild hypothermia on superoxide anion production, superoxide dismutase expression, and activity following transient focal cerebral ischemia. *Neurobiology of Disease*, 11 (1), pp. 28-42

Martin-Schild, S, Hallevi, H, Shaltoni, H, Barreto, AD, Gonzales, NR, Aronowski, J, Savitz, SI & Grotta, JC. (2009). Combined neuroprotective modalities coupled with thrombolysis in acute ischemic stroke: a pilot study of caffeinol and mild hypothermia. *Journal of Stroke & Cerebrovascular Diseases*, 18 (2), pp. 86-96

Meloni, BP, Mastaglia, FL & Knuckey, NW. (2008). Therapeutic applications of hypothermia in cerebral ischaemia. *Therapeutic Advances in Neurological Disorders*, 1 (2), pp. 75–98

Mokhtarani, M, Mahgoub, AN, Morioka, N, Doufas, AG, Dae, M, Shaughnessy, TE & Bjorksten AR. (2001). Buspirone and meperidine synergistically reduce the shivering threshold. *Anesthesia & Analgesia*, 93 (5), pp. 1233-1239

Nagel, S, Su, Y, Horstmann, S, Heiland, S, Gardner, H, Koziol, J, Martinez-Torres, FJ & Wagner, S. (2008). Minocycline and hypothermia for reperfusion injury after focal cerebral ischemia in the rat: effects on BBB breakdown and MMP expression in the acute and subacute phase. *Brain Research*, 1188 (1), pp. 198-206

Nederlands Trial Register. www.trialregister.nl

Nurse, S & Corbett, D. (1996). Neuroprotection after several days of mild, drug-induced hypothermia. *Journal of Cerebral Blood Flow & Metabolism* 16 (3), pp. 474-480

Ohta, H, Terao, Y, Shintani, Y & Kiyota Y. (2007). Therapeutic time window of post-ischemic mild hypothermia and the gene expression associated with the neuroprotection in rat focal cerebral ischemia. Neuroscience Research, 57 (3), pp. 424-433

Polderman, KH, Rijnsburger, ER, Peerdeman, SM & Girbes, AR. (2005). Induction of hypothermia in patients with various types of neurologic injury with use of large volumes of ice-cold intravenous fluid. *Critical Care Medicine*, 33 (12), pp. 2744-2751

Schaller, B & Graf, R. (2003). Hypothermia and stroke: the pathophysiological background. *Pathophysiology*, 10 (1), pp. 7-35

Schwarzenberg, H, Müller-Hülsbeck, S, Brossman, J, Glüer, CC, Bruhn, HD & Heller M. (1998). Hyperthermic fibrinolysis with rt-PA: in vitro results. *Cardiovascular and Interventional Radiology*, 21 (2), pp. 142-145

Shaw, GJ, Dhamija, A, Bavani, N, Wagner, KR & Holland, CK. (2007). Arrhenius temperature dependence of in vitro tissue plasminogen activator thrombolysis. *Physics in Medicine and Biology*, 52 (11), pp. 2953-2967

Shin, BS, Won, SJ, Yoo, BH, Kauppinen, TM & Suh, SW. (2010). Prevention of hypoglycemia-induced neuronal death by hypothermia. *Journal of Cerebral Blood Flow & Metabolism*, 30 (2), pp. 390–402

Si, Q-S, Nakamura Y, & Kataoka, K. (1997). Hypothermic suppression of microglial activation in culture: inhibition of cell proliferation and production of nitric oxide and superoxide. *Neuroscience*, 81 (1), pp. 223–229

Stroke Trials Directory. www.strokecenter.org/trials

Takata, K, Takeda, Y, Sato, T, Nakatsuka, H, Yokoyama, M & Morita, K. (2005). Effects of hypothermia for a short period on histologic outcome and extracellular glutamate concentration during and after cardiac arrest in rats. *Critical Care Medicine*, 33 (6), pp. 1340-1345

van der Worp, HB, Sena, ES, Donnan, GA, Howells, DW & Macleod, MR. (2007). Hypothermia in animal models of acute ischaemic stroke: a systematic review and meta-analysis. *Brain*, 130 (12), pp. 3063-3074

van der Worp, HB, Macleod, MR & Kollmar, R; for the European Stroke Research Network for Hypothermia (EuroHYP). (2010). Therapeutic hypothermia for acute ischemic stroke: ready to start large randomized trials? *Journal of Cerebral Blood Flow & Metabolism*, 30 (6), pp. 1079-1093

Van Hemelrijck, A, Hachimi-Idrissi, S, Sarre, S, Ebinger, G & Michotte, Y. (2005). Post-ischaemic mild hypothermia inhibits apoptosis in the penumbral region by reducing neuronal nitric oxide synthase activity and thereby preventing endothelin-1-induced hydroxyl radical formation. *European Journal of Neuroscience*, 22 (6), pp. 1327-1337

Webster, CM, Kelly, S, Koike, MA, Chock, VY, Giffard, RG & Yenari, MA. (2009). Inflammation and NFKB activation is decreased by hypothermia following global cerebral ischemia. *Neurobiology of Disease*, 33 (2), pp. 301–312

Yenari, MA, Palmer, JT, Bracci, PM & Steinberg GK. (1995). Thrombolysis with tissue plasminogen activator (tPA) is temperature dependent. *Thrombosis Research*, 77(5), pp. 475-481

Zhao, H, Steinberg, GK & Sapolsky, RM. (2007). General versus specific actions of mild-moderate hypothermia in attenuating cerebral ischemic damage. *Journal of Cerebral Blood Flow & Metabolism*, 27 (12), pp. 1879–1894

Zhu, H, Meloni, BP, Bojarski, C, Knuckey, MW & Knuckey, NW. (2005). Post-ischemic modest hypothermia (35C) combined with intravenous magnesium is more effective at reducing CA1 neuronal death than either treatment used alone following global cerebral ischemia in rats. *Experimental Neurology*, 193 (2), pp. 361-368

Zweifler, RM, Voorhees, ME, Mahmood, MA & Parnell, M. (2004). Magnesium sulfate increases the rate of hypothermia via surface cooling and improves comfort. *Stroke*, 35 (10), pp. 2331-2334

2

Hypothermia as an Alternative for the Management of Cerebral Ischemia

Felipe Eduardo Nares-López, Gabriela Leticia González-Rivera and María Elena Chánez-Cárdenas
Laboratorio de Patología Vascular Cerebral, Instituto Nacional de Neurología y Neurocirugía "Manuel Velasco Suárez" México

1. Introduction

Cerebral ischemia results from the decrease in oxygen and glucose supply by the transient or permanent reduction of cerebral blood flow, triggering excitotoxic, oxidative, inflammatory and apoptotic events which end up in brain tissue death. Cerebral ischemia is one of the leading causes of death in industrialized countries, a medical emergency with few specific treatments available to minimize the acute injury and provide neuroprotection and brain repair. In fact, current therapies are limited to clot removal, aspirin, and decompressive hemicraniectomy for ischemic stroke. To date, alteplase, recombinant tissue-type plasminogen activator (rt-PA) is the only approved therapy for acute ischemic stroke. A relevant concern in stroke research is that despite the increase in pharmacological studies, these treatments have shown to be ineffective or to cause adverse effects. Among more than 700 drugs which have been studied and found to be effective in animal stroke models, yet none has been proved efficacious in clinical studies.

The reduction of cerebral blood flow as a consequence of a thrombus or embolus occlusion results in brain injury with metabolic and functional deficits. The extent of damage depends on the severity and duration of cerebral blood flow decrease; and according to the remaining blood supply, an ischemic core and a penumbra area can be identified.

The core is defined by almost complete energetic failure that ends up in necrotic cell death. Cells in the hypoperfused penumbra are non-functional, however, structural integrity and viability are retained. Experimental and medical evidence indicates that if the blood flow is not restored throughout reperfusion within hours, the penumbral region becomes part of the core. Hence, penumbra is the target to rescue since brain tissue at this region remains potentially viable for 16 to 48 hours, enabling clinicians to intervene and reduce post-stroke disability.

In addition to medication, the development of novel and rational strategies directed to reduce impairments after stroke have been improved. Ischemic preconditioning, electroacupuncture, hypothermia and stem cell therapy are the most relevant non-pharmacological strategies for the management of the patient who suffers an ischemic stroke, which is at present, one of the most frequent diseases at adult age and the principal cause of disability in many countries.

Hypothermia is considered as one of the most effective options to treat stroke patients in the management of the adverse events taking place in the brain. Hypothermia alters different events in cerebral injury including reduction in metabolic and enzymatic activity, release and re-uptake of glutamate, inflammation, production of reactive oxygen species, blood-brain barrier breakdown and shift of cell death and survival pathways. Although stroke models vary in methodology, several laboratories have consistently shown that hypothermia reduces the extent of neurologic damage and improves neurologic function.

The aim of this chapter is to provide a recent review of basic research in hypothermia treatment. Beside the clinical studies that incorporate hypothermia, numerous efforts have been performed in recent years to understand the mechanisms underlying protection by hypothermia. The clinical and basic research concurrence will allow a better understanding of hypothermia mechanisms in the near future, making its incorporation more efficient as a co-adyuvant in stroke treatment.

2. Protective hypothermia

The reduction of body temperature or hypothermia during an adverse event such as cardiac arrest, cardiopulmonary resuscitation (Hassani, 2010), neonatal hypoxia, hepatic encephalopathy (Barba et al., 2008) and ischemic stroke has been applied in humans as well as in animal models. In all cases, hypothermia has shown to preserve cerebral function. Clinical trials have proved that hypothermia is an effective protector of brain injury (Jacobs et al., 2007), and laboratory animal studies provided a considerable amount of evidence supporting hypothermia protection after focal, global, transient and permanent cerebral ischemia models as well as *in vitro* approaches with ischemia or hypoxia treatments (van der Worp et al., 2010; Yenari & Hemmen, 2010).

The amelioration of ischemia/reperfusion-induced oxidative stress, inflammatory and apoptotic responses are the most promising mechanisms to understand the biological action of hypothermia protection. There is relevant evidence that suggests that hypothermic protection occurs mainly by reducing cerebral metabolism, supporting the protective effects of hypothermia in the different steps of the ischemic cascade as we explain below.

2.1 Hypothermia Induction

Mild (>32°C) to moderate (28-32°C) systemic hypothermia has been studied widely. It has been reported that mild hypothermia improves neurological function suppressing apoptosis pathogenesis, and moderate hypothermia limits some of the metabolic responses by altering neurotransmitter release, attenuating energy depletion, decreasing radical oxygen species production and reducing neuronal death and apoptosis.

In clinical stroke, hypothermia is an effective neuroprotective strategy when applied for a long period after the ischemic event, since it has been observed that the optimum conditions for hypothermic neuroprotection are mostly affected by the duration and timing of cooling. Several works have been performed to determine the timing, duration and deepness of experimental hypothermia. In gerbils subjected to a global ischemia model, the immediate induction or the intensification of hypothermia improved survival rate, suppressed post-ischemic hypoperfusion and prevented vasoconstriction. However, the therapeutic hypothermia time window was narrow, suggesting that it should be induced immediately after the onset of ischemia in order to improve survival (Noguchi et al., 2011).

There are different methods for hypothermia induction in stroke patients and in basic research with *in vivo* models of ischemia. In clinical practice, the main methods to perform hypothermia are surface cooling and endovascular cooling. Both methods have advantages and disadvantages. Surface cooling can be induced with inexpensive methods such as air blankets, alcohol bathing and even fans to decrease temperature. Ice packs, neck bands and head caps are more sophisticated and practical techniques. In addition, surface cooling can be performed in awake patients with ischemic stroke. It is non-expensive, non invasive and allows the use of hypothermia in combination with thrombolytics. In fact, the combination of hypothermia with intravenous tissue plasminogen activator in patients treated within 6 h after ischemic stroke has shown hopeful results (Hemmen et al., 2010). However, complete control of body temperature is not possible with surface cooling, shivering and discomfort of the patient may occur (reviewed in Yenari et al., 2008).

Endovascular cooling seems to be a more efficient way to generate and control hypothermia. Its invasive nature leads to time loss and also requires trained personal in endovascular techniques (Polderman & Herold, 2009). A pilot study with patients with acute ischemic stroke included within 3 h after symptom onset suggests that ice cold saline infusion combined with pethidine and buspirone (to prevent shivering), lowered body temperature to 35.4±0.7 °C in a fast manner and without major side effects. The results of this small uncontrolled case series work, suggest that the induction of hypothermia with an infusion represents a fast approach for induction of hypothermia that could ameliorate the damage caused by the delayed induction observed in the majority of clinical cases (Kollmar et al., 2009). The authors suggest that this rapid induction of hypothermia by ice cold saline infusion is an effective and rapid way to induce mild to moderate hypothermia for stroke treatment in an ambulance car.

The procedure to induce hypothermia also has an important effect in its neuroprotective effect in animal models. Recently, Wang et al (2010) reported the use of systemic, head or local vascular ischemia in rats with middle cerebral artery occlusion. Their results showed that the use of vascular cooling is the most effective procedure to reduce infarct volume as well as in the functional outcome rather than the other two methods (Wang et al., 2010). However in animal models many methods to induce hypothermia are used, including the removal of heating blanket with a spontaneously decrease in temperature (Doshi et al., 2009).

2.2 Alternative agents to induce hypothermia

2.2.1 Helium

Inert gases such as xenon and helium have also been used to produce hypothermia. Helium is considered a "cost-efficient" inert gas with no anesthetic properties, in contrast to the availability and cost of xenon. David et al (2009) have shown that rats subjected to transient middle cerebral artery occlusion and hypothermia generated by helium administered after reperfusion showed an improvement in neuroprotection. Helium produces cortical protection evaluated by infarct size and reduction of behavioral motor deficits at 25 °C hypothermia but not at 33 °C. The post-ischemic helium hypothermia administration is important as a possible clinical application.

3. Hypothermia amelioration of ischemic damage

3.1 The ischemic cascade

The decrease in oxygen and glucose supply by the transient or permanent reduction of cerebral blood flow in cerebral ischemia triggers a series of excitotoxic, oxidative, inflammatory and apoptotic events known as the "ischemic cascade" which ends up in brain tissue death.

Brain cells are dependent almost exclusively on oxygen and glucose supply for energy production through oxidative phosphorylation. The oxygen and glucose reduction causes accumulation of lactate increasing acidosis. ATP depletion triggers a series of pathologic events including the loss of membrane potential, peri-infarct depolarizations, glutamate and aspartate excitotoxicity, the increase in Ca^{2+} concentration, oxidative stress and free radical generation, protein synthesis inhibition, inflammation and apoptosis. The disruption of ion homeostasis originated by the disturbance of Na^{+}/K^{+}-ATPase and Ca^{2+}/H-ATPase pumps, and the reversed Na^{+}-Ca^{2+} transporter, triggers an increase in intracellular Na^{+}, Cl^{-} and Ca^{2+} concentrations, as well as extracellular K^{+}. Besides this biochemical response, within minutes after the onset of ischemia, there is an increase in gene expression. Cells respond to stress by adjusting the gene expression program in order to deal with the stress condition, to trigger a recovery process or to lead to signaling for additional tissue injury (Dirnalg, 1999; Durukan & Tatlisumak, 2007).

3.1.1 Brain edema and blood brain barrier breakdown

The effects of hypothermia on the disruption of the blood brain barrier have been implicated in many studies. The role of temperature in blood brain barrier function has been studied in cortex, thalamus, hippocampus and hypothalamus of rats subjected to hyperthermia. Astrocytic activation, a larger content of brain water, Na^{+}, K^{+} and Cl^{-} as well as structural abnormalities that suggest brain edema were observed, demonstrating that brain temperature is an important factor in regulating blood brain barrier integrity, permeability and brain edema (Kiyatkin & Sharma, 2009). The effect of temperature in blood brain barrier integrity has also been studied using hypoxia and high ambient temperature to follow the permeability to Na^{+} and the expression of the endothelial barrier antigen, a protein associated with blood brain barrier. A clear effect in the increase of Na^{+} and a reduction in the endothelial barrier antigen were observed, as well as an exacerbation with hyperthermia (Natah et al., 2009).

The dependence of blood brain barrier integrity and brain edema with temperature has important implications, since even the thrombolytic therapy using rTPA is able to cause hemorrhagic damage. In a work by Hamann et al (2004), it has been proposed that hypothermia could be used as a protection to basal lamina, a component along with the interendothelial tight junctions and perivascular astrocytes of the blood brain barrier. Basal lamina has the main function of preventing extravasation of cellular blood elements, and the loss of its integrity results in hemorrhage. In order to determine whether hypothermia could maintain microvascular integrity in ischemic stroke, the loss of collagen type IV component of the basal lamina, the non-cellular proteolytic system that degrades basal lamina matrix metalloproteinase (MMP)-2 and MMP-9, plasminogen-plasmin system urokinase-type plasminogen activator (uPA) and tissue-type plasminogen activator (tPA) were determined. Rats were subjected to 3 h ischemia and 24 h reperfusion with the suture model and hypothermia between 32-34°C was applied 30 min before reperfusion. Results were compared with a normothermic group. This work shows that infarct size was considerably reduced in hypothermia treated rats; collagen type IV loss from basal lamina of cerebral microvessels was considerable reduced, as well as MMP-2, MMP-9, tPA and uPA activities, showing that hypothermia preserves microvascular integrity and reduces hemorrhage and the activities of MMP-2, MMP-9, uPA, and tPA (Hamman et al., 2004).

Even that hypothermia has been successful in the protection against neuronal death in several models of ischemia, a recent work using C57BL/6J mice subjected to occlusion of bilateral common carotid arteries , demonstrated that hypothermia induced by the removal of heating blanket with a spontaneously decrease in temperature was an effective protection against neuronal death detected by histological damage and terminal deoxynucleotidyl transferase (TdT)-mediated dUTP-biotin nick end labeling (TUNEL). However, brain edema was not prevented by hypothermia treatment (Doshi et al., 2009).

3.1.2 Metabolic downregulation

As mentioned before, in cerebral ischemia two main regions of damage can be defined according to the severity and duration of the cerebral blood flow reduction: 1) The *core*, where complete abolishment of blood supply occurs (less than 12 ml/100g/min) and 2) the *penumbra*, in which collateral blood supply from surrounding arteries assures a flow of approximately 30 ml/100g/min. In normal non pathological conditions cerebral blood flow is decreased under hypothermic conditions, however, the effect of hypothermia in cerebral blood flow during ischemia events is not completely clear since some works support that hypothermia reduces or has no effect in cerebral blood flow and other reports show that it is increased during ischemia. Metabolic suppression has been proposed as one of the most relevant mechanisms underlying the hypothermic treatment. Hypothermia (in the range of 22 to 37 °C) reduces the rate of oxygen consumption fall in body temperature by approximately 5% per degree Celsius. It also decreases glucose consumption and lactate levels (Yenari et al., 2008).

The deepness and extent of hypothermia stimulus has been considered of importance to the outcome and success of therapeutic hypothermia, and the developing methods to monitor and control hypothermia treatments is significant. Single voxel proton magnetic resonance spectroscopy (H´-MRS) is an important tool to detect metabolites and mechanisms that could be changing during hypothermia. Recently, using H´-MRS 7 Tesla MRI scanner, Chan et al (2010) determined the levels of metabolites in response to normothermia and hypothermia. Cortex and thalamus changes of metabolites involved in osmolality, brain temperature and energy metabolism were detected with important implications in the understanding of hypothermia protective mechanisms. For example, it was observed that lactate, the substrate of energy responsible for anaerobic metabolism and anaerobic glycolysis increased 43% in the cortex during hypothermia. This observation has been considered as a change in energy metabolism associated with neuroprotection which results from an increase in glycolysis and depression of tricarboxilic acid cycle.

Myo-inositol, a metabolite involved in diverse cellular processes such as signal transduction, membrane structure, vesicular trafficking and as an osmolite modulating cell volume, was increased 21 % in cortex during normothermia. They showed that this technique is able to detect metabolic changes in specific regions of the brain in alive animals in a noninvasively manner. In thalamus, taurine was 16% increased during hypothermia suggesting its role as a regulator of temperature and protecting neurons via its agonistic gamma-aminobutyric acid effect. In the same region choline decreased 29%, and authors suggest that this decrease could imply thermoregulation via muscarinic receptors which act against hypothermia. The main contribution of this work is the demonstration that this noninvasive method can detect changes *in vivo*, of significant metabolites involved in neuroprotective hypothermia (Chan et al., 2010).

Using acid-base related parameters as well as the antioxidant-oxidant effects of deep (21-22°C) hypothermia before acute hypoxic insults in rats, it was observed that during hypothermia mild metabolic acidosis appeared in arterial blood. It suggests that hypothermia induced acidosis contributes to a reduction of potential in liver (Alva et al., 2010). The determination of lactate levels showed that blood lactate increased in normothermia, and hypothermia prevents this increase contributing to the prevention of tissue damage.

3.1.3 Glutamate release and peri-infarct depolarizations

One of the critical steps in cerebral ischemia damage is excitotoxicity by glutamate and aspartate release as a result of membrane depolarization. The activation of glutamate receptors and increase in Ca^{2+}, Na^+ and Cl^- levels initiate molecular events that end in cell death by excitotoxicity. The release of glutamate into the extracellular compartment is one of the early and most intense events of the ischemic cascade. The reduction of glutamate release in hypothermia supports the idea of protection through metabolism downregulation.

It has been observed that the increase in glutamate levels is delayed in the ischemic core as a consequence of hypothermia treatment in permanent focal cerebral ischemia (Baker et al., 1995) and the extracellular glutamate concentration is reduced in the penumbra when analyzed by microdialysis after permanent middle cerebral artery occlusion. This last study suggests that the protection observed by hypothermia probably involves a reduction in the pool of diffusible glutamate in the core but has little effect on glutamate release in the penumbra (Winfree et al., 1996). The increase in glutamate is related to the initiation of peri-infarct depolarizations. Results using the N-methyl-D-aspartate receptor antagonist MK 801 and moderate hypothermia (32-34 °C) have shown neuroprotective effects alone and in combination supporting these observations (Alkan et al., 2001). Several studies have shown that hypothermia decreased glutamate efflux by attenuating the initial rise of extracelluar K^+ and preventing Ca^{2+} accumulation (reviewed in Yenari et al., 2008).

As a result of glutamate excitotoxicity, peri-infarct depolarizations contribute to the increase in infarct volume. The use of temporal NADH fluorescence images to obtain temporal and spatial resolutions to follow the propagation of peri-infarct depolarizations was performed with spontaneously hypertensive rats subjected to permanent focal ischemia by occlusion of the middle cerebral and left common carotid arteries. Hypothermia (30°C) maintained 2 h and applied before ischemia, showed that hypothermia delays the appearance, however does not modify the dynamics of propagation of peri-infarct depolarizations. The authors suggest that peri-infarct depolarizations could have a greater effect on the infarct area in hypothermic rats (Sasaki et al., 2009) and that the inefficacy to suppress peri-infarct depolarizations is the cause of the absence of hypothermia protection in several models of cerebral ischemia.

3.1.4 Oxidative stress

The generation of reactive oxygen species (ROS), reactive nitrogen species and free radicals is increased as a consequence of ischemic damage and particularly by the restoration of blood supply. Xantine oxidase, cyclooxygenase, NADPH oxidase, the mitochondrial respiratory chain and the inflammatory response are the major sources of free radicals in cerebral ischemia (Margaill et al., 2005). Reactive species play an important role in both necrotic and apoptotic cell death in cerebral ischemia and reperfusion. ROS generate oxidative stress and triggers tissue inflammation, cause damage to the cellular membrane by lipid peroxidation, DNA damage and disruption of cellular processes. Particularly, superoxide radical has an important responsibility of ischemia damage, since a number of ROS are derived from superoxide.

Using an *in vivo* real-time quantitative superoxide analysis system with an electrochemical sensor previously developed, it has been demonstrated the increase in superoxide in the jugular veins of rats during ischemia/reperfusion in a forebrain ischemia model (Aki et al., 2009). The effect of pre-ischemic hypothermia (32°C) in the generation of superoxide or as a post ischemia treatment (immediately after reperfusion) was determined in the same model. Both pre and post ischemic hypothermia successfully decreased the superoxide generated by ischemic rats. Hypothermia also decreased oxidative stress, early inflammation and endothelial injury markers in both treatments (Koda et al., 2010).

It has been observed that hypothermia maintains the glutation potential of the liver in an *in vivo* acute hypoxia model, it also avoids the increase in malondialdehide and prevents tissue damage induced by hypoxia (Alva et al., 2010).

The inhibition of superoxide generation using hypothermia was also evaluated in an insulin-induced hypoglycemia model. They showed that intracellular accumulation of zinc promotes the production of ROS through NADPH oxidase activation after hypoglycemia (Shin et al., 2010). This work also provides evidence that hypothermia could affect other mechanisms such as vesicular Zn^{2+} release and translocation, which affects part of the excitotoxic neuronal death. Hypothermia prevents massive Zn^{2+} release which ends in cell death, and hyperthermia aggravates it, showing that Zn^{2+} release is dependent on temperature (Suh et al., 2004, Shin et al., 2010). However, despite all the evidences of ROS reduction and hypothermia protection, the determination of ROS during the first 60 minutes of ischemia in normothermic and hypothermic conditions using a global cerebral ischemia-reperfusion model, showed that the widely hypothermia protection effect observed does not correlate with the oxidative stress induced by ROS, as observed with electron spin resonance system (Kunimatsu et al., 2001). The use of different ischemia models, time and temperature are probably the explanation to contradictory results with respect to ROS generation.

3.1.5 Inflammation

Hypothermia attenuates inflammation by suppressing activating kinases of nuclear factor-kappa B (NFkappaB). In a global cerebral ischemia model by bilateral carotid artery occlusion, microglial activation was observed and hypothermia decreased this activation as well as nuclear NFkappaB translocation and activation (Webster et al., 2009). NFkappaB is activated in cerebral ischemia, controlling the expression of inflammatory genes. In a study using middle cerebral artery occlusion by 2 h and hypothermia at 33 °C, it was observed that the decrease in temperature decreased NFkappaB translocation and binding activity. Regulatory proteins such as IkappaB kinase were also affected decreasing its activity, suggesting that hypothermia exerts its protective effect by NFkappaB inhibition (Han et al., 2003).

3.2 Hypothermia and apoptosis signaling pathways

3.2.1 Protein kinases activated by hypoxia

In order to understand the protective mechanisms of hypothermia several efforts have been performed along years. It has been proposed that the mitochondria and the phosphatidylinositol 3-kinase (PI3-K)/Akt (protein kinase B) signaling pathway are determinant for neuronal survival controlling proapoptosis and antiapoptosis in ischemic neurons during stroke. Akt activity has been implicated in the endogenous neuroprotection observed by preconditioning (Miyawaki et al., 2008) and as part of the neuroprotective response to cerebral ischemia (Kamada et al., 2007). Several pharmacological efforts have been performed to target PI3-K/Akt pathway, since it is known that PI3-K/Akt downstream phosphorylated Bad and proline-rich Akt substrate survival signaling cascades are upregulated in surviving neurons in the ischemic brain (Chan et al., 2004).

One of the most relevant efforts is the demonstration that PI3/Akt pathways are involved in neuroprotection by hypothermia. After the distal middle cerebral artery occlusion of rats using intra-ischemic hypothermia (30ºC), Zhao et al (2005) observed a reduction in infarct size and the improvement of neurological outcome up to two months. Relevant information was obtained from this work besides observed tissue protection and functional response: 1) decrease of Akt activity observed in normothermic animals after stroke was attenuated by hypothermia; 2) hypothermia improved phosphorylation and attenuates dephosphorylation of phosphatase and tensin homolog deleted on chromosome 10 (PTEN) and phosphoinositide-dependent protein kinase 1 (PDK1); 3) consequent to the observed tissue protection and the involvement of the Akt pathway, the inhibition of PI3K (an upstream activator of Akt) increases infarct size of hypothermic ischemic rats; 4) phosphorylation of forkhead transcription factor (FKHR) was improved by hypothermia attenuating its apoptotic effects, since dephosphorylated FKHR acts as a transcription factor increasing Bcl-2 interacting mediator of cell death (Bim) and Fas ligand; 5) the nuclear translocation of the transcription factor P-β-catenin, observed in stroke normothermic rats was blocked by hypothermia in the penumbra, but not in the ischemic core, suggesting an important role of β-catenin in stroke excitotoxicity (Zhao et al., 2005).

The same group evaluated the effect of hypothermia in the activation of a kinase implicated in neuroprotection *in vitro,* the epsilon protein kinase C (εPKC), demonstrating that εPKC preservation is an important component in the protective effect of hypothermia. Using the permanent distal middle cerebral artery occlusion plus 1 h of transient bilateral common carotid artery occlusion in normothermic (37°C) and hypothermic (30°C) rats, the neuronal full length εPKC expression and localization was evaluated with inmmunofluorescence by confocal microscopy and western blot. Normothermic ischemic rats showed a decrease of εPKC in the ischemic core at 4 h after common carotid artery release, hypothermia blocked this decrease. Hypothermia also affects cellular distribution. In non ischemic rats, εPKC is in the cytoplasm of neurons. When cellular εPKC distribution was determined, normothermic ischemic rats showed an εPKC decrease in cytosol as well as in membranal fractions of the ischemic core, blocked by hypothermia. In the penumbra, the membranal εPKC was decreased after the ischemic damage in normothermic rats and hypothermia blocked this decrease. Even hypothermia blocks εPKC cleavage, inhibition of caspase-3 assays showed that this caspase is not involved in this process, suggesting the action of other proteases (Shimohata et al., 2007a).

In contrast to εPKC protective effect, delta protein kinase C (δPKC) is involved in the tissue damage by ischemia. δPKC activation depends on catalytic cleavage, phosphorylation and translocation to membranes. The inhibition of δPKC activity using the specific inhibitor δV1-1, decreased infarct size after transient cerebral ischemia (Brigth et al., 2004). The translocation to the nucleus and mitochondria, the caspase-3 dependent proteolytic cleavage to generate the δPKC catalytic fragment (which increased in the membrane fraction, mitochondria, and nuclei); and the release of cytochrome c as earlier as 10 min after reperfusion, are processes triggered by common carotid artery occlusion. Hypothermia blocked all these processes; in fact, the subcellular translocation of the activated δPKC was attenuated in the penumbra but not in the ischemic core (Shimohata et al., 2007b). All these protective effects were corroborated with the use of a specific δPKC activator, ψδRACK, in the hypothermic rats.

Primary targets of ischemia-reperfusion injury are vascular endothelial cells through the stimulation of calcium overload, ROS generation and the triggering of inflammatory process, which in turn begin apoptotic programs. In order to determine mechanisms involved in hypothermia protection to ischemia and reperfusion damage, human umbilical endothelial cells were used. Using hypothermia (33°C), a clear reduction in cell apoptosis induced by ischemia/reperfusion was observed. Characterization of this process showed that hypothermia reduces ischemia/reperfusion-induced apoptosis as observed by TUNEL, expression of activated caspase-3 and poly-ADP ribose polymerase (PARP). Hypothermia also reversed Fas/caspase 8 activation pathway and attenuated the Bax/Bcl-2 ratio compared with normothermic cells. Since JNK1/2 and p38 MAPK signaling pathways play an important role in oxidative stress-induced apoptosis, the effect of hypothermia in this pathway was studied, showing that hypothermia inhibits both extrinsic- and intrinsic-dependent apoptotic pathways and activation of JNK1/2 activation via MKP-1 induction (Yang et al., 2009).

3.2.2 Apoptotic proteins

The decrease in apoptosis contributes in a significant manner to hypothermia protection. Ischemia and reperfusion have shown to increase the number of active caspase-3 immunoreactive nuclei, and hypothermia clearly reduced this induction (Kunimatsu et al., 2001).

Recently, Li & Wang (2011) used mild hypothermia (33 °C) in rats subjected to middle cerebral artery occlusion and determined neurological impairment and the expression of Second Mitochondrion-derived Activator of Caspases (SMAC) as an index of cellular apoptosis. Mild hypothermia significantly improved the neurological deficit scores while results from protein and transcript expression of SMAC showed a significantly decrease, suggesting that mild hypothermia could be protecting the functions of cells by attenuating apoptotic death (Li & Wang, 2011).

It has been previously shown that the pro-apoptotic protein SMAC/DIABLO expression is increased in cortex and hippocampus in transient cerebral ischemia as well as in ischemia-reperfusion injury (Saito et al., 2003; Scarabelli & Stephanou, 2004; Siegelin et al., 2005). SMAC increases 3 h after cerebral ischemia with a peak at 24 h. Apparently, hypothermia could be down-regulating SMAC production attenuating caspases activation. Cell apoptosis via SMAC involve mitochondrial and death receptor pathways, inducing changes in mitochondrial membrane permeability and subsequently release membrane proteins, such as SMAC, into the cytoplasm. SMAC leads cells toward apoptosis through apoptosis-related protein. The prevention of loss of mitochondrial transmembrane potential, release of apoptotic proteins (citochrome c and apoptosis inducing factor [AIF]) and the activation of apoptotic proteins such as caspase 3, as well as the attenuation in the elevation of oxidative stress markers have been observed also in *in vitro* cells with hypothermia in a model of iron and ascorbic acid neurotoxicity (Hasegawa et al., 2009).

Rats subjected to global cerebral ischemia with the four-vessel occlusion model with hypothermia (31-32 °C) and hyperthermia (41-42°C) confirmed the protective effects of hypothermia in the decrease of mortality rate (at 72 and 168 h post reperfusion), and in the increase of surviving neurons in hippocampus under hypothermic conditions. Hypothermia clearly reduced p53 and increased bcl-2 proteins reducing neuronal death. In addition hyperthermia had the opposite effect in the expression of both proteins (Zhang et al., 2010).

3.3 Regulation of gene and protein expression

Hypothermia induces changes in inflammatory, apoptotic and metabolic genes as the result of gene expression regulation.

In general, it is considered that hypothermia downregulates gene expression. However, there are reports that show the upregulation of certain genes, particularly those involved in cell survival. The understanding of gene and protein expression as result of hypothermia in cerebral ischemia could lead to the possible therapeutic target genes or pathways regulated as a result of decreasing body temperature.

Recently, an analysis of gene and protein expression using DNA microarrays and proteomics approach in a 2 h middle cerebral artery occlusion model and mild hypothermia (35°C) has shown that it is possible to determine target molecules. The authors proposed that suppression of neuroinflammatory cascades MIP-3α-CCR could contribute to the neuroprotective effects of hypothermia and also identified Hsp 70 as a neuroprotective factor stimulated by hypothermia (Shintani et al., 2010; Terao et al., 2009,).

3.3.1 Effect of hypothermia in proteins involved in gene expression

The decrease in temperature and/or the decrease in oxygen concentration involve a series of events that modulate transcription and translation. Novel proteins have been discovered in the last decades. Here we show the role of cold inducible proteins and hypoxia inducible factor-1 as proteins that regulate the efficient transcription and translation of proteins in ischemia and ischemia/hypothermia events.

3.3.1.1 Cold-inducible RNA-binding proteins

Cold inducible proteins have the function to ensure the efficient translation of specific mRNAs at temperatures below the physiological standard. Hypothermia induces the synthesis of amino terminal consensus sequence RNA-binding domain proteins (CS-RBD). The "cold-inducible RNA-binding protein" (CIRP) is one of these proteins, and has been involved in protection as the result of hypothermia treatment in *in vitro* studies. CIRP regulates gene expression at translational level. It binds to the 5´-untranslated region (5´-UTR) or 3´-UTR of specific transcripts, affecting translation and transcript stability (Lleonart, 2010).

CIRP has been proposed as a therapeutic target in cerebral ischemia by Liu et al (2010). They determined mRNA expression in hippocampus and cortex of rat brains subjected to hypothermia (30°), cerebral ischemia (by four vessel occlusion model of forebrain ischemia) and hypothermia plus cerebral ischemia by real time quantitative PCR analysis. mRNA CIRP expression was followed at 2, 6 and 24 h showing an increase in cortex after cerebral ischemia with a previous hypothermia treatment. In order to clarify the relationship between CIRP and energy metabolism, they determined lactate and piruvate concentrations, showing that CIRP has a neuroprotective effect in hypothermia; however, it is not related to energy metabolism.

The contribution of CIRP to the neuroprotection observed by hypothermia has also been studied using MEMB5 cells, a neural stem cell line from mouse forebrain (Saito et al., 2010). These cells proliferate in the presence of epidermal growth factor (EGF). EGF deprivation at 37 °C results in apoptosis induction, as well as a decrease of the nestin neural stem cell marker and an increase of the astrocyte marker glial fibrillary acidic protein (GFAP). In contrast, MEMB5 cells at moderate hypothermia prevented apoptosis and decreased the observed GFAP expression of normothermic cells. This observation is important because it suggests that hypothermia prevents neural stem cells differentiation, and it has been hypothesized that the preservation of neural stem cells is one of the neuroprotective mechanisms of therapeutic hypothermia, since it could maintain the capability of cells to differentiate and proliferate after an ischemic event. CIRP mRNA and protein was increased in the MEB5 in hypothermic cells, a response according to previous observations showing that ischemia/reperfusion decreases CIRP mRNA (Xue et al., 1999) whereas hypothermia increases it in an *in vivo* cerebral ischemia model (Liu et al., 2010).

The relevance of CIRP expression was confirmed using CIRP iRNA, increasing apoptosis in hypothermic cells without EGF. This result suggests that the induced CIRP plays the role of a survival factor in neural stem cells. The prevention of apoptosis observed with induced CIRP at hypothermia has been suggested to be the result of the activation of extracellular signal-regulated kinase ERK (Artero-Castro et al., 2009; Sakurai et al., 2006; Schmitt et al., 2007).

3.3.1.2 Hypoxia inducible factor

Hypoxia inducible factor is a transcription factor which binds to hypoxic response elements-driven promoters of genes that mediate adaptive reactions to reduction in oxygen availability. Hypoxia inducible factor is regulated by oxygen accessibility and has been considered as a therapeutic target in cerebral ischemia (Aguilera et al., 2009). Recently it has been reported that persisting low temperature affects its stabilization and protein accumulation. Since the normal regulatory degradation processes of HIF are not affected by hypothermia, it has been hypothesized that probably hypothermia elevates intracellular oxygen tension by decreasing oxygen consumption, suppressing in turn HIF-1 alpha subunit induction. These results were obtained with different cell lines (T98G cells from human glioblastoma multiform, HeLa cells (derived from human cervical carcinoma) and Hep3B cells (derived from human hepatoma) as well as mice subjected to hypoxia and hypothermia (18°C in a mouse incubator). The translation of HIF-1 alpha protein showed to be dependent on time exposure to hypothermia. The down-regulation of HIF protein expression observed with hypothermia has relevant implications in ischemia and hypothermia studies, since this transcription factor is a master regulator of the hypoxic response to oxygen decrease (Tanaka et al., 2010).

3.3.2 Gene and protein expression in CA1 neurons as result of hypothermia

Hippocampal CA1 layer is a region that presents a typical apoptosis cell death after ischemic damage. As a matter of fact , accumulating evidence has indicated that the postischemic DNA fragmentation in the hippocampal CA1 area in experimental ischemic models is a key phenomenon for the delayed neuronal death and is considered as apoptosis. Hypothermia has shown to protect CA1 neurons attenuating the down-regulation of GluR2 mRNA in a model of forebrain ischemia using two days of mild hypothermia induced after 1 h cerebral ischemia, suggesting that the observed attenuation and CA1 neurons protection responds to cooling (Colbourne et al., 2003). Another interesting protein is the β-galactosidase-binding lectin Galectin-3, which has been observed expressed in experimental models of stroke (Walther et al., 2000; Yan et al., 2009) and increased in microgial cells in the hippocampal CA1 layer after a transient ischemic insult. Galectin-3 is a protein involved in apoptotic regulation, inflammation and cell differentiation and used as a marker of activated microglia. After 5 min of bilateral common carotid arteries of gerbils, galectin-3 expression was observed in microglial cells in CA1 region. Hypothermia (31°C) prevents galectin-3 expression suggesting that hypothermia protection occurs through the inhibition of microglial activation and probably by preventing neuronal death (Satoh et al., 2011). Even when galectin-3 has been considered apoptotic, its role as an inflammatory mediator in neonatal hypoxia ischemia injury through the modulation of the inflammatory response has been reported (Doverhag et al., 2010).

4. Combined therapies

The preservation of tissue and reduction of brain damage observed during hypothermia and the easiness to achieve and maintain low temperatures, constitute an attractive alternative against stroke. However, because of the complexity of the pathophysiological mechanisms involved in the ischemic cascade, it is common to observe the use of one or two drugs besides hypothermia in the search of a neuroprotective compound capable of blocking the metabolic insult. Combination therapy is based in the extension of the therapeutic window of hypothermia using pharmacotherapy and establishes an interesting approach in hypothermia research.

4.1 Hypothermia, tirilazad and magnesium

The combined administration of the antioxidant tirilazad, magnesium and hypothermia has shown that the hypothermia protection increases with the use of pharmacotherapy, extending the therapeutic window of hypothermia treatment. Zausinger et al (2003) used hypothermia (2 h, 33°C) at 0, 1, 3 and 5 h after transient focal ischemia induction combined with two administrations of tirilazad (3 mg/kg) and magnesium (1 mmol·L^{-1}·kg^{-1}) in 1 h intervals. Infarct volume was reduced by 74%, 49% and 45% when hypothermia was applied at 0, 1 and 3 h. No improvements were observed at 5 h. The same combination therapy was performed in a permanent ischemia with middle cerebral artery occlusion. Two h at 33 °C hypothermia and two times of drug administration (30 min before and 1 h after middle cerebral artery occlusion) showed a 52 % infarct size reduction after 6 h occlusion. Interestingly, a separate group with 7 days of permanent middle cerebral artery occlusion was followed daily with neurological tests and body weight. Even high mortality ocurred in this group, neurological recovery was observed in survivor rats, as well as a decrease in infarct size (Schöller et al., 2004).

Moderate hypothermia (30°C) has shown higher protection than mild hypothermia (33°C), however moderate hypothermia is associated with severe side effects. Nonetheless, the combined effect observed with mild hypothermia, magnesium and tirilazad showed a comparable protection with that observed using moderate hypothermia, suggesting that combination therapy could be a promising approach in clinical applications.

4.2 Hypothermia and magnesium

Combination therapy with magnesium has generated controversy, because magnesium treatment alone has shown to be ineffective in normothermic rats subjected to ischemia (Zhu et al., 2004, a and b). The quantification of infarct volumes at magnesium 360 or 720 umol/kg in rats subjected to middle cerebral artery occlusion showed that the 360 umol/kg dose reduces the striatal infarct volume by 32 %. The authors observed that mild spontaneous hypothermia was responsible of the observed neuroprotective effect of magnesium (Campbell et al 2008a).

The therapeutic time window of combined magnesium and mild hypothermia treatment was determined after permanent middle cerebral artery occlusion. The administration of a magnesium sulfate infusion (360 µmol/kg, then 120 µmol/kg/h) and mild hypothermia (35°C) after 2, 4 or 6 h ischemia showed that combination therapy considerably reduced infarct volumes at 2 and 4 h but not at 6 h, supporting the use of combined therapy even at delayed hours after ischemia onset (Campbell et al 2008b).

Although the controversy in the use of magnesium and hypothermia in cerebral ischemia exists, a lot of experimental evidence demonstrates its efficacy in laboratory studies and it is necessary to perform more clinical trials to translate this combinatorial therapy to clinical use. Actually, combined mild hypothermia (35ºC) and magnesium is recognized as a neuroprotective treatment able to minimize ischemic damage, even in delayed treatments (Meloni et al., 2009).

4.3 Hypothermia and citicoline

Hypothermia has been combined with other compounds. One of them is citicoline, an endogenous compound that has shown to stabilize membrane function and reduce free radical generation during ischemia (Rao et al., 2000).

Mild hypothermia combined with citicoline resulted in an additive effect in attenuating apoptosis in focal cerebral ischemia/reperfusion injury. Hypothermia (34±1ºC) during middle cerebral artery occlusion (2 h) followed by 24 h reperfusion was used in combination with 4000 mg/kg *i.p.* citicoline. Combined therapy with citicoline and hypothermia resulted in reduced apoptotic cell death before the development of the apoptosome, thus preventing apoptosis and neuronal damage. Bcl-2, caspases 3 and 9 as well as Bax proteins were immunohistochemistry tested, showing that the use of citicoline with hypothermia is more effective than citicoline or hypothermia used alone. A decrease in cerebral injury was observed and apparently, the tissue protection is due to the suppression of apoptotic processes (Sahin et al., 2010).

4.4 Hypothermia and ginkgolides

Despite the neuroprotective role of hypothermia, the combined use of hypothermia with protective compounds not always has a synergistic effect. This is the case of hypothermia and ginkgolides in astrocytes subjected to ischemia and reperfusion in which hypothermia attenuates, rather than enhances, the protective effect of ginkgolides on astrocytes from ischemia and reperfusion-induced injury. Co-treatment with different doses of ginkgolides at 32 and 28 ºC hypothermia during 24, 48 and 72 h before 24 h ischemia followed by 24 h reperfusion showed that the use of ginkgolides without hypothermia treatment had an improvement in cell viabilities and in anti-apoptotic properties, and this protective effects were not observed in the co-treatment (Fang et al., 2009).

4.5 Hypothermia and xenon

Xenon has been considered a great promise used as a neuroprotectant in *in vivo* and *in vitro* studies. Combined therapy with hypothermia in a neonatal rat hypoxia-ischemia model has shown that the combination of 50% xenon and hypothermia of 32°C has a protective effect in the restoration of long-term functional outcomes and global histopathology, showing that the combined xenon/hypothermia has a greater protection (Hobbs et al., 2008).

5. Conclusion

Despite all the efforts to develop pharmacological treatments to contend with cerebral ischemic damage, nowadays, there are no efficient treatments to deal with this pathology. Non-pharmacological treatments emerge as reasonable approaches to compete against ischemic damage.

Hypothermia has shown positive results in cardiac arrest and neonatal hypoxic ischemia and in patients with acute brain injury. However, there are no large clinical stroke studies that could assist in the comprehension of the hypothermia role in stroke. In laboratory studies, hypothermia is a consistent protective agent. The present chapter shows the contribution of hypothermia research in the last years to understand its participation during the different stages of the ischemic damage cascade. Blood brain barrier integrity, metabolic rate decrease, redox state changes, inflammatory, apoptotic, signalling and gene regulation are targets where hypothermia protection participates. Hypothermia is affected by diverse factors such as timing, duration and deepness. All these factors contribute to disparity in obtained results. However, it is clear that this research is necessary in order to determine the exact targets, the time when hypothermia begins and duration of hypothermia. Additionally, since brain ischemia is a multifactorial problem and hypothermia has demonstrated to decrease damage in several stages of the process, combination therapy is an alternative to improve treatments by the extension of the therapeutic time window of hypothermia protection.

6. Acknowledgments

Nares-López F. E. received a scholarship from BECANET, number 053103.

7. References

Aguilera, P., Vázquez-Contreras, E., Gómez-Martínez, C. D., & Chánez-Cárdenas, M. E. (2009). Hypoxia Inducible Factor-1 as a Therapeutic Target in Cerebral Ischemia. *Current Signal Transduction Therapy*, Vol. 4, No. 3, (September, 2009). pp 162-173, ISSN 1574-3624

Aki, S.H., Fujita, M., Yamashita, S., Fujimoto, K., Kumagai, K., Tsuruta, R., Kasaoka, S., Aoki, T., Nanba, M., Murata, H., Yuasa, M., Maruyama, I., Maekawa, T. (2009). Elevation of jugular venous superoxide anion radical is associated with early inflammation, oxidative stress, and endothelial injury in forebrain ischemia-reperfusion rats. *Brain Research*, Vol. 1292, (October 2009), pp. 180-90, ISSN 0006-8993

Alkan, T., Kahveci, N., Buyukuysal, L., Korfali, E., & Ozluk, K. (2001). Neuroprotective effects of MK 801 and hypothermia used alone and in combination in hypoxic-ischemic brain injury in neonatal rats. *Archives of Physiology and Biochemistry*, Vol. 109, No. 2, (May 2010), pp. 135-44, ISSN 1381-3455

Alva, N., Carbonell, T., & Palomeque, J. (2010). Hypothermic protection in an acute hypoxia model in rats: Acid-base and oxidant/antioxidant profiles. *Resuscitation*, Vol. 81, No. 5, (May 2010), pp. 609-16, ISSN 0300-9572

Artero-Castro, A., Callejas, F. B., Castellvi, J., Kondoh, H., Carnero, A., Fernández-Marcos, P.J., Serrano, M., Ramón y Cajal, S., & Lleonart, M. E. (2009). Cold-inducible RNA-binding protein bypasses replicative senescence in primary cells through extracellular signal-regulated kinase 1 and 2 activation. *Molecular and Cellular Biology*, Vol. 29, No. 7, (April 2009), pp.1855-1868, ISSN 0270-7306

Baker, C. J., Fiore, A. J., Frazzini, V. I., Choudhri, T. F., Zubay, G. P., & Solomon, R. A. (1995). Intraischemic hypothermia decreases the release of glutamate in the cores of permanent focal cerebral infarcts. *Neurosurgery*, Vol. 36, No. 5, (May 1995), pp. 994-1001, ISSN 0148-396X

Barba, I., Chatauret, N., Garcia-Dorado, D., & Cordoba, J. (2008). A 1H nuclear magnetic resonance-based metabonomic approach for grading hepatic encephalopathy and monitoring the effects of therapeutic hypothermia in rats. *Liver International*, Vol. 28, No. 8, (September 2008), pp. 1141-1148, ISSN 1478-3223

Bright, R., Raval, A. P., Dembner, J. M., Perez-Pinzon, M. A., Steinberg, G. K., Yenari, M. A., & Mochly-Rosen, D. (2004). Protein kinase C delta mediates cerebral reperfusion injury in vivo. *Journal of Neuroscience*, Vol. 24, No. 31, (August 2004), pp. 6880-8, ISSN 0270-6474

Campbell, K., Meloni, B.P., & Knuckey, N.W. (2008a). Combined magnesium and mild hypothermia (35 degrees C) treatment reduces infarct volumes after permanent middle cerebral artery occlusion in the rat at 2 and 4, but not 6 h. *Brain Research*, Vol. 1230, (September 2008), pp. 258-64, ISSN 0006-8993

Campbell, K., Meloni, B.P., Zhu, H., & Knuckey, N.W. (2008b). Magnesium treatment and spontaneous mild hypothermia after transient focal cerebral ischemia in the rat. *Brain Research Bulletin*, Vol. 77, No. 5, (September 2008), pp. 320-2, ISSN 0361-9230

Chan, P.H. (2004). Future targets and cascades for neuroprotective strategies. *Stroke*, Vol. 35, No. 11 (Suppl 1), (November 2004), pp. 2748-50, ISSN 0039-2499

Chan, K.W., Chow, A.M., Chan, K.C., Yang. J., & Wu, E. X. (2010). Magnetic resonance spectroscopy of the brain under mild hypothermia indicates changes in neuroprotection-related metabolites. *Neurosci Lett.*, Vol. 475, No. 3, (May 2010), pp.150-155, ISSN 0304-3940

Colbourne, F., Grooms, S. Y., Zukin, R. S., Buchan, A. M., & Bennett, M. V. (2003). Hypothermia rescues hippocampal CA1 neurons and attenuates down-regulation of the AMPA receptor GluR2 subunit after forebrain ischemia. *Proc Natl Acad Sci U S A*, Vol. 100, No. 5, (March 2003), pp. 2906-10, ISSN 0027-8424

David, H. N., Haelewyn, B., Chazalviel, L., Lecocq, M., Degoulet, M., Risso, J. J., &Abraini, J. H. (2009). Post-ischemic helium provides neuroprotection in rats subjected to middle cerebral artery occlusion-induced ischemia by producing hypothermia. *Journal of Cerebral Blood Flow & Metabolism*, Vol. 29, No. 6, (June 2009), pp. 1159-65, ISSN 0271-678X

Dirnagl, U., Iadecola, C., & Moskowitz, M.A. (1999). Pathobiology of ischaemic stroke: an integrated view. *Trends in Neuroscience.*, Vol. 22, No. 9, (September 1999), pp. 391-7, ISSN 0166-2236

Doshi, M., Kuwatori, Y., Ishii, Y., Sasahara, M., & Hiroshima, Y. (2009). Hypothermia during ischemia protects against neuronal death but not acute brain edema following transient forebrain ischemia in mice. *Biological & Pharmaceutical Bulletin*, Vol. 32, No. 12, (December 2009), pp. 1957-61, ISSN 0918-6158

Doverhag, C., Hedtjärn, M., Poirier, F., Mallard, C., Hagberg, H., Karlsson, A., & Sävman, K. (2010). Galectin-3 contributes to neonatal hypoxic-ischemic brain injury. *Neurobiology of Disease*, Vol. 38, No. 1, (April 2010), pp. 36-46, ISSN 0969-9961

Durukan, A., & Tatlisumak, T. (2007). Acute ischemic stroke: overview of major experimental rodent models, pathophysiology, and therapy of focal cerebral ischemia. *Pharmacology, Biochemistry & Behavior*, Vol. 87, No. 1, (May 2007), pp. 179-97, ISSN 0091-3057

Fang, D., Ming, Q. Z., Li, Z., Mei, W. X., & Ya, K. (2009). Hypothermia attenuates protective effects of ginkgolides on astrocytes from ischemia/reperfusion injury. *Neurochemistry International*, Vol. 55, No. 4, (December 2000), pp. 181-6, ISSN 0197-0186

Hobbs, C., Thoresen, M., Tucker, A., Aquilina, K., Chakkarapani, E., & Dingley, J. (2008). Xenon and hypothermia combine additively, offering long-term functional and histopathologic neuroprotection after neonatal hypoxia/ischemia. *Stroke*, Vol. 39, No. 4, (April 2008), pp. 1307-13, ISSN 0039-2499

Koda, Y., Tsuruta, R., Fujita, M., Miyauchi, T., Kaneda, K., Todani, M., Aoki, T., Shitara, M., Izumi, T., Kasaoka, S., Yuasa, M., & Maekawa, T. (2010). Moderate hypothermia suppresses jugular venous superoxide anion radical, oxidative stress, early inflammation, and endothelial injury in forebrain ischemia/reperfusion rats. *Brain Research*, Vol. 1311, (January 2010), pp. 197-205, ISSN 0006-8993

Hamann, G.F., Burggraf, D., Martens, H.K., Liebetrau, M., Jäger, G., Wunderlich, N., DeGeorgia, M., & Krieger, D.W. (2004). Mild to moderate hypothermia prevents microvascular basal lamina antigen loss in experimental focal cerebral ischemia. *Stroke*, Vol. 35, No. 3, (March 2004), pp. 764-769, ISSN 0039-2499

Han, H.S., Karabiyikoglu, M., Kelly, S., Sobel, R.A., & Yenari, M.A.(2003). Mild hypothermia inhibits nuclear factor-kappaB translocation in experimental stroke. *Journal of Cerebral Blood Flow & Metabolism*, Vol. 23, No. 5, (May 2003), pp. 589-98, ISSN 0271-678X

Hasegawa, M., Ogihara, T., Tamai, H., & Hiroi, M. (2009). Hypothermic inhibition of apoptotic pathways for combined neurotoxicity of iron and ascorbic acid in differentiated PC12 cells: reduction of oxidative stress and maintenance of the glutathione redox state. *Brain Research*, Vol. 1283, (August 2009), pp. 1-13, ISSN 0006-8993

Hassani, H., & Meyer, S. (2010). Hypothermia for neuroprotection in adults after cardiopulmonary resuscitation. *American Family Physician*, Vol. 82, No. 5, (September 2010), pp. 477-478, ISSN 0002-838X

Hemmen, T. M., Raman, R., Guluma, K. Z., Meyer, B. C., Gomes, J. A., Cruz-Flores, S., Wijman, C. A., Rapp, K. S., Grotta, J. C., Lyden, P. D.; ICTuS-L Investigators. (2010). Intravenous thrombolysis plus hypothermia for acute treatment of ischemic stroke (ICTuS-L): final results. *Stroke*, Vol. 41, No. 10, (October 2010), pp. 2265-2270, ISSN 0039-2499

Jacobs, S., Hunt, R., Tarnow-Mordi, W., Inder, T., & Davis, P. (2007). Cooling for newborns with hypoxic ischaemic encephalopathy. *Cochrane Database of Systematic Reviews*, Vol. 4, (October 2007), pp. CD003311, ISSN 1469-493X

Kamada, H., Nito, C., Endo, H., & Chan, P.H. (2007). Bad as a converging signaling molecule between survival PI3-K/Akt and death JNK in neurons after transient focal cerebral ischemia in rats. *Journal of Cerebral Blood Flow & Metabolism*, Vol. 27, No. 3, (March 2007), pp. 521-33, ISSN 0271-678X

Kiyatkin, E. A., & Sharma, H. S. (2009). Permeability of the blood-brain barrier depends on brain temperature. *Neuroscience*, Vol. 161, No. 3, (July 2009), pp. 926-39, ISSN 0306-4522

Kollmar, R., Schellinger, P. D., Steigleder, T., Köhrmann, M., Schwab, S. (2009). Ice-cold saline for the induction of mild hypothermia in patients with acute ischemic stroke: a pilot study. *Stroke*, Vol. 40, No.5, (May 2009), pp. 1907-1909, ISSN 0039-2499

Kunimatsu, T., Kobayashi, K., Yamashita, A., Yamamoto, T.,& Lee, M.C. (2011). Cerebral reactive oxygen species assessed by electron spin resonance spectroscopy in the initial stage of ischemia-reperfusion are not associated with hypothermic neuroprotection. *Journal of Clinical Neuroscience*, Vol. 18, No. 4, (April 2011), pp. 545-8, ISSN 0967-5868

Li, H., & Wang, D. (2011). Mild hypothermia improves ischemic brain function via attenuating neuronal apoptosis. *Brain Research*, Vol. 1368, (January 2011), pp. 59-64, ISSN 0006-899

Liu, A., Zhang, Z., Li, A., & Xue, J. (2010). Effects of hypothermia and cerebral ischemia on cold-inducible RNA-binding protein mRNA expression in rat brain. *Brain Research*, Vol. 1347, (August 2010), pp.104-110, ISSN 0006-8993

Lleonart, M.E. (2010). A new generation of proto-oncogenes: cold-inducible RNA binding proteins. *Biochim. Biophys. Acta*, Vol. 1805, No. 1, (January 2010), pp.43-52, ISSN 0006-3002

Margaill, I., Plotkine, M., & Lerouet, D. (2005). Antioxidant strategies in the treatment of stroke. *Free radical Biology and Medicine*, Vol. 39, No. 4, (August 2005), pp. 429-43 ISSN 0891-5849

Meloni, B. P., Campbell, K., Zhu, H., & Knuckey, N. W. (2009). In search of clinical neuroprotection after brain ischemia: the case for mild hypothermia (35 degrees C) and magnesium. *Stroke*, Vol. 40, No. 6, (June 2009), pp. 2236-40, ISSN 0039-2499

Miyawaki, T., Mashiko, T., Ofengeim, D., Flannery, R.J., Noh, K.M., Fujisawa, S., Bonanni, L., Bennett, M.V., Zukin, R.S., & Jonas, E. A. (2008). Ischemic preconditioning blocks BAD translocation, Bcl-xL cleavage, and large channel activity in mitochondria of postischemic hippocampal neurons. *Proc Natl Acad Sci U S A*, Vol. 105, No. 12, (March 2008), pp. 4892-7, ISSN 0027-8424

Natah, S. S., Srinivasan, S., Pittman, Q., Zhao, Z., Dunn, J.F. (2009). Effects of acute hypoxia and hyperthermia on the permeability of the blood-brain barrier in adult rats. *Journal of Applied Physiology*, Vol. 107, No. 4, (October 2009), pp. 1348-56, ISSN 8750-7587

Noguchi, K., Matsumoto, N., Shiozaki, T., Tasaki, O., Ogura, H., Kuwagata, Y., Sugimoto, H., & Seiyama, A. (2011). Effects of timing and duration of hypothermia on survival in an experimental gerbil model of global ischaemia. *Resuscitation*, Vol. 82, No. 4, (April 2011), pp. 481-6, ISSN 0300-9572

Polderman, K. H., & Herold, I. (2009). Therapeutic hypothermia and controlled normothermia in the intensive care unit: practical considerations, side effects, and cooling methods. *Critical Care Medicine* Vol. 37, No. 3, (March 2009), pp. 1101-1120, ISSN 0090-3493

Rao, A. M., Hatcher, J. F., & Dempsey, R J. (2000). Lipid alterations in transient forebrain ischemia: possible new mechanisms of CDP-choline neuroprotection. *Journal of Neurochemistry*, Vol. 75, No. 6, (December 2000), pp. 2528-35, ISSN 0022-3042

Saito. K., Fukuda, N., Matsumoto, T., Iribe, Y., Tsunemi, A., Kazama, T., Yoshida-Noro, C., & Hayashi, N. (2010). Moderate low temperature preserves the stemness of neural stem cells and suppresses apoptosis of the cells via activation of the cold-inducible RNA binding protein. *Brain Research*, Vol. 1358, (October 2010), pp.20-29, ISSN 0006-8993

Saito, A., Hayashi, T., Okuno, S., Ferrand-Drake, M., & Chan, P.H. (2003). Interaction between XIAP and Smac/DIABLO in the mouse brain after transient focal cerebral ischemia. *Journal of Cerebral Blood Flow & Metabolism*, Vol. 23, No. 9, (September 2003), pp.1010-1019, ISSN 0271-678X

Sahin, S., Alkan, T., Temel, S. G., Tureyen, K., Tolunay, S., & Korfali, E. (2010). Effects of citicoline used alone and in combination with mild hypothermia on apoptosis induced by focal cerebral ischemia in rats. *Journal of Clinical Neuroscience*, Vol. 17, No. 2, (February 2010), pp. 227-31, ISSN 0967-5868

Sakurai, T., Itoh, K., Higashitsuji, H., Nonoguchi, K., Liu, Y., Watanabe,H., Nakano, T., Fukumoto,M., Chiba, T., & Fujita, J. (2006). Cirp protects against tumor necrosis factor-alpha-induced apoptosis via activation of extracellular signal-regulated kinase. *Biochim. Biophys. Acta*, Vol. 1763, No. 3, (March 2006), pp.290-295, ISSN: 0006-3002

Sasaki, T., Takeda, Y., Taninishi, H., Arai, M., Shiraishi, K., & Morita, K. (2009). Dynamic changes in cortical NADH fluorescence in rat focal ischemia: evaluation of the effects of hypothermia on propagation of peri-infarct depolarization by temporal and spatial analysis. *Neurosci Lett.* Vol. 449, No. 1, (January 2009), pp. 61-5, ISSN 0304-3940

Satoh, K., Niwa, M., Goda, W., Binh, N. H., Nakashima, M., Takamatsu, M., & Hara, A. (2011). Galectin-3 expression in delayed neuronal death of hippocampal CA1 following transient forebrain ischemia, and its inhibition by hypothermia. *Brain Research*, Vol. 1382, (March 2011), pp. 266-74, ISSN 0006-8993

Scarabelli, T. M., & Stephanou, A., (2004). Minocycline inhibits caspase activation and reactivation, increases the ratio of XIAP to smac/DIABLO and reduces the mitochondrial leakage of cytochrome C and smac/DIABLO. *Journal of the American College of Cardiology*, Vol. 43, No. 5, (March 2004), pp. 865-874, ISSN 0735-1097

Schmitt, K. R., Diestel, A., Lehnardt, S., Schwartlander, R., Lange, P.E., Berger, F., Ullrich, O., & Abdul-Khaliq, H. (2007). Hypothermia suppresses inflammation via ERK signaling pathway in stimulated microglial cells. *Journal of Neurimmunology*, Vol. 189, No. 1-2, (September 2007), pp.7-16, ISSN 0165-5728

Schöller, K., Zausinger, S., Baethmann, A., & Schmid-Elsaesser, R. (2004). Neuroprotection in ischemic stroke--combination drug therapy and mild hypothermia in a rat model of permanent focal cerebral ischemia. *Brain Research*, Vol. 1023, No. 2, (October 2004), pp. 272-8, ISSN 0006-8993

Siegelin, M. D., Kossatz, L. S., Winckler. J., & Rami, A. (2005). Regulation of XIAP and Smac/DIABLO in the rat hippocampus following transient forebrain ischemia. *Neurochemistry International*, Vol. 46, No. 1, (January 2005), pp.41-51, ISSN 0197-0186

Shimohata, T., Zhao, H., & Steinberg, G. K.(2007). Epsilon PKC may contribute to the protective effect of hypothermia in a rat focal cerebral ischemia model. *Stroke*, Vol. 38, No.2, (February 2007), pp. 375-380, ISSN 0039-2499

Shimohata, T., Zhao, H., Sung, J.H., Sun, G., Mochly-Rosen, D., & Steinberg, G.K. (2007). Suppression of deltaPKC activation after focal cerebral ischemia contributes to the protective effect of hypothermia. *Journal of Cerebral Blood Flow & Metabolism*, Vol. 27, No. 8, (August 2007), pp. 1463-75, ISSN 0271-678X

Shin, B. S., Won, S. J., Yoo, B. H., Kauppinen, T. M., & Suh, S. W. (2010). Prevention of hypoglycemia-induced neuronal death by hypothermia. *Journal of Cerebral Blood Flow & Metabolism*, Vol. 30, No. 2, (February 2010), pp. 390-402, ISSN 0271-678X

Shintani, Y., Terao, Y., & Ohta, H. (2011). Molecular mechanisms underlying hypothermia-induced neuroprotection. *Stroke Research and Treatment*, Vol. 1, (December 2010), 809874, ISSN 2042-0056

Suh, S. W., Garnier, P., Aoyama, K., Chen, Y., & Swanson, R.A. (2004). Zinc release contributes to hypoglycemia-induced neuronal death. *Neurobiology of Disease*, Vol. 16, No. 3, (August 2004), pp. 538–45, ISSN 0969-9961

Tanaka, T., Wakamatsu, T., Daijo, H., Oda, S., Kai, S., Adachi, T., Kizaka-Kondoh, S., Fukuda, K., & Hirota, K. (2010). Persisting mild hypothermia suppresses hypoxia-inducible factor-1alpha protein synthesis and hypoxia-inducible factor-1-mediated gene expression. *American Journal of Physiology*. Regulatory, integrative and comparative physiology, Vol. 298, No. 3, (March 2010), pp. R661-71, ISSN 0363-6119

Terao, Y., Miyamoto, S., Hirai, K., Kamiguchi, H., Ohta, H., Shimojo, M., Kiyota, Y., Asahi, S., Sakura, Y., & Shintani, Y. (2009). Hypothermia enhances heat-shock protein 70 production in ischemic brains. *Neuroreport*, Vol. 20, No. 8, (May 2009), pp. 745-9, ISSN 0959-4965

van der Worp, H. B., Macleod, M. R., Kollmar, R.; European Stroke Research Network for Hypothermia (EuroHYP). (2010). Therapeutic hypothermia for acute ischemic stroke: ready to start large randomized trials? *Journal of Cerebral Blood Flow & Metabolism*, Vol. 30, No. 6, (June 2010), pp. 1079-93, ISSN 0271-678X

Walther, M., Kuklinski, S., Pesheva, P., Guntinas-Lichius, O., Angelov, D.N., Neiss, W.F., Asou, H., & Probstmeier, R. (2000). Galectin-3 is upregulated in microglial cells in response to ischemic brain lesions, but not to facial nerve axotomy. *Journal of Neuroscience Research*, Vol. 61, No. 4, (August 2000), pp. 430-5, ISSN 0360-4012

Wang, F., Luo, Y., Ling, F., Wu, H., Chen, J., Yan, F., He, Z., Goel, G., Ji, X., & Ding, Y. (2010). Comparison of neuroprotective effects in ischemic rats with different hypothermia procedures. *Neurological Research*, Vol. 32, No. 4, (May 2010), pp. 378-383, ISSN 0161-6412

Webster, C. M., Kelly, S., Koike, M.A., Chock, V.Y., Giffard, R.G., & Yenari, M.A. (2009). Inflammation and NFkappaB activation is decreased by hypothermia following global cerebral ischemia. *Neurobiology of Disease*, Vol. 33, No. 2, (February 2009), pp. 301-12, ISSN 0969-9961

Winfree, C. J., Baker, C. J., Connolly, E. S., Jr., Fiore, A. J., & Solomon, R.A. (1996). Mild hypothermia reduces penumbral glutamate levels in the rat permanent focal cerebral ischemia model. *Neurosurgery*, Vol. 38, No. 6, (June 1996), pp. 1216-22, ISSN 0148-396X

Xue, J. H., Nonoguchi, K., Fukumoto, M., Sato, T., Nishiyama, H., Higashitsuji, H., Itoh, K., & Fujita, J., (1999). Effects of ischemia and H_2O_2 on the cold stress protein CIRP expression in rat neuronal cells. *Free Radical. Biology. & Medicine.* Vol. 27, No. 11-12, (December 1999), pp.1238-1244, ISSN: 0891-5849

Yan, Y. P., Lang, B. T., Vemuganti, R., & Dempsey, R.J. (2009). Galectin-3 mediates post-ischemic tissue remodeling. *Brain Research*, Vol. 1288, (September 2009), pp. 116–124, ISSN 0006-8993

Yang, D., Guo, S., Zhang, T., & Li, H. (2009) Hypothermia attenuates ischemia/reperfusion-induced endothelial cell apoptosis via alterations in apoptotic pathways and JNK signaling. *FEBS Lett.* Vol. 583, No. 15, (August 2009), pp. 2500-2506, ISSN: 0014-5793

Yenari, M., Kitagawa, K., Lyden, P., & Perez-Pinzon, M. (2008). Metabolic downregulation: a key to successful neuroprotection? *Stroke*, Vol. 39, No. 10, (October 2008), pp. 2910-2927 ISSN 0039-2499

Yenari, M. A., & Hemmen, T. M. (2010). Therapeutic hypothermia for brain ischemia: where have we come and where do we go? *Stroke*, Vol. 41, No. 10 Suppl, (October 2010), pp. S72-4 ISSN 0039-2499

Zhang, H., Xu, G., Zhang, J., Murong, S., Mei, Y., & Tong, E. (2010). Mild hypothermia reduces ischemic neuron death via altering the expression of p53 and bcl-2. *Neurological Research*, Vol. 32, No 4, (May 2010), pp. 384-9, ISSN 0161-6412

Zhao, H., Shimohata, T., Wang, J.Q., Sun, G., Schaal, D.W., Sapolsky, R.M., & Steinberg, G.K. (2005). Akt contributes to neuroprotection by hypothermia against cerebral ischemia in rats. *Journal of Neuroscience*, Vol. 25, No. 42, (October 2005), pp. 9794-9806, ISSN 0270-6474

Zhu, H. D., Martin, R., Meloni, B., Oltvolgyi, C., Moore, S., Majda, B., & Knuckey N. (2004). Magnesium sulfate fails to reduce infarct volume following transient focal cerebral ischemia in rats. *Neuroscience Research*, Vol. 49, No. 3, (July 2004), pp. 347-53, ISSN 0168-0102

Zhu, H., Meloni, B. P., Moore, S. R., Majda, B. T., & Knuckey, N. W. (2004). Intravenous administration of magnesium is only neuroprotective following transient global ischemia when present with post-ischemic mild hypothermia. *Brain Research*, Vol. 1014, No. (1-2) (July 2004), pp. 53-60, ISSN 0006-8993

Zausinger, S., Schöller, K., Plesnila, N., & Schmid-Elsaesser, R. (2003). Combination drug therapy and mild hypothermia after transient focal cerebral ischemia in rats. *Stroke*, Vol. 34, No. 9, (September 2003), pp. 2246-51, ISSN 0039-2499

3

Timing of Hypothermia (During or After Global Cerebral Ischemia) Differentially Affects Acute Brain Edema and Delayed Neuronal Death

Masaru Doshi[1] and Yutaka Hirashima[2]
[1]Faculty of Pharmaceutical Sciences, Teikyo University
[2]Department of Neurosurgery, Imizu City Hospital
Japan

1. Introduction

Hypothermia (HT) is one of the most effective neuroprotective therapies for brain injury caused by cardiac arrest in humans (Benard et al., 2002), although there is as yet no evidence of such an effect of HT on cerebrovascular diseases from a large-scale clinical trial. We recently reported the experimental studies on the importance of timing of HT using global cerebral ischemia in mice (Doshi et al., 2009, 2011). we summarized our findings in this chapter because our results may be relevant to clinical studies.

2. Experimental methods

2.1 Global cerebral ischemia using C57BL/6J mice

Transient forebrain ischemia (global cerebral ischemia), which is induced by occlusion of the bilateral common carotid arteries (BCCA) in mice, causes delayed neuronal death in the hippocampus and is known as a model of brain injury following transient cardiac arrest (Kawase et al., 1999). C57BL/6J mice are widely used as a background strain for genetic alterations and have been valuable for investigating the molecular mechanism of delayed neuronal death following transient forebrain ischemia (Yang et al., 1997; Tajiri et al., 2004). We recently demonstrated that acute brain edema, one of the most important disorders following cerebral ischemia, occurred in the forebrain in this C57BL/6J mouse model (Doshi et al., 2009).

Forebrain ischemia was induced by BCCA occlusion with clips for 15 min under 1 % halothane anesthesia in air using a face mask. Rectal temperature was monitored using a digital thermometer and maintained at 37±0.5℃ with a heating blanket (normothermia, NT). The control mice underwent a sham operation without BCCA occlusion under halothane anesthesia for 15 min.

2.2 Evaluation of delayed neuronal death and acute brain edema

Delayed neuronal death in the hippocampus 7 days after reperfusion was determined by both hematoxylin-eosin (HE) staining and TdT-mediated dUTP-biotin nick end labeling

(TUNEL) assay. To evaluate the acute brain edema, the water content in the forebrain 1 hour after reperfusion was determined by the weight differences between wet and dry samples. The percentage of water in the forebrain was calculated as follows: ((wet weight-dry weight) / wet weight)×100.

2.3 Induction of hypothermia

HT during ischemia was spontaneously induced by removing the heating blanket, and the mice were allowed to recover from anesthesia at room temperature (23-25℃) until 1 hr after reperfusion. In contrast, artificial HT after reperfusion (rHT) was induced by placing the mice on a refrigerant, which was maintained at 18-19℃ until 1 hr after reperfusion in a styrofoam box. In NT mice, rectal temperature was maintained at 37±0.5℃ during ischemia. In HT mice, rectal temperature significantly decreased by 28-30℃ during ischemia. On the other hand, the patterns of the change in rectal temperature after reperfusion in rHT mice were significantly lower than that in HT mice (Fig. 1).

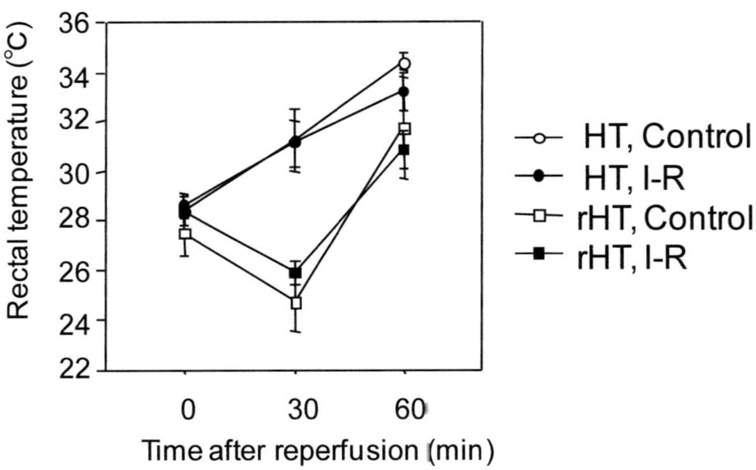

Fig. 1. Changes in rectal temperature after reperfusion in both HT and rHT mice.

Forebrain ischemia was induced by BCCA occlusion for 15 min under 1 % halothane anesthesia in both HT and rHT mice, as described in the Materials and Methods. The control mice (Control) underwent a sham operation without BCCA occlusion under halothane anesthesia for 15 min. The rectal temperature was monitored 0, 30, and 60 min after reperfusion. Data are expressed as mean ± SE (control : n=3, I-R : n=7). Statistical analysis was performed by analysis of variance (ANOVA) of repeated measures for the comparison of the changes in rectal temperature between the HT and rHT groups.

3. Effect of hypothermia during ischemia on occurrence of delayed neuronal death

HT during ischemia has been shown to protect against delayed neuronal death in several animal models of cerebral ischemia including the global cerebral ischemia in C57BL/6J mice (Yang et al., 1997). We confirmed the protective effect of HT during ischemia against delayed neuronal death in the hippocampus under our experimental conditions. As shown in Table 1, there were no histological changes in the hippocampus 7 d after reperfusion in all HT mice, whereas delayed neuronal death occurred in three of the four NT mice. In this model, controlled NT during cerebral ischemia is important for the induction of neuronal death following cerebral ischemia (Ohtaki et al., 2006), supported by data revealing that HT during ischemia protects against neuronal death in the BCCA occlusion C57BL/6J mouse model (Yang et al., 1997). Our data in this study were consistent with previous data. However, the neuroprotective effect of HT after ischemia-reperfusion still remains to be solved in NT mice.

Mice	The number of delayed neuronal death-positive mice / total mice
NT (ischemia 15 min)	3 / 4
HT (ischemia 15 min)	0 / 5*
HT (ischemia 45 min)	0 / 5*

*p<0.05 vs NT (Fisher's exact test)

Table 1. Effect of HT during ischemia on occurrence of delayed neuronal death in hippocampus of C57BL/6J mice.

4. Effect of timing of hypothermia on occurrence of acute brain edema

Brain edema, defined as an increase in brain water content, is one of the major reperfusion pathologies following cerebral ischemia along with delayed neuronal death and infarction. The increase in brain water content results in serious pathologic situations such as elevation of intracranial pressure and reduction of cerebral blood flow, and subsequently causes cerebral herniation and death (Kempski, 2001). Even now, the treatment options for brain edema are limited to the use of hyperosmotic agents, such as glycerol or mannitol, and surgical decompression. That is, despite the clinical significance of brain edema, the mechanisms of brain water transport and edema formation in ischemic injuries remain unclear.

Several studies have shown that HT, which was monitored by measuring rectal temperature, protects against delayed neuronal death and infarction (Maier et al., 1998; Tsuchiya et al., 2002), but few studies have investigated the effects of HT on brain edema in mouse models of cerebral ischemia. Therefore, we investigated the effect of HT on the occurrence of acute brain edema following the global cerebral ischemia in C57BL/6J mice.

4.1 Effect of hypothermia during ischemia on occurrence of acute brain edema

We first investigated the effect of HT during global cerebral ischemia on the occurrence of acute brain edema in the C57BL/6J mouse model. In both NT and HT mice, the water content 1 h after reperfusion was significantly higher than that of the control mice, but no significant differences were detected between the NT and HT mice (Fig. 2). This data indicated the ineffectiveness of HT during ischemia against acute brain edema, unlike its effect on neuronal death, suggesting that HT during ischemia may not necessarily be an effective therapy for all reperfusion pathologies following cerebral ischemia. However, we found that the rectal temperature of the HT mice recovers from HT during ischemia to NT mice levels within 1 hr after reperfusion. Therefore, we speculated that the ineffectiveness of

HT during ischemia against acute brain edema is due to the immediate recovery of rectal temperature after reperfusion at room temperature in the C57BL/6J mouse model.

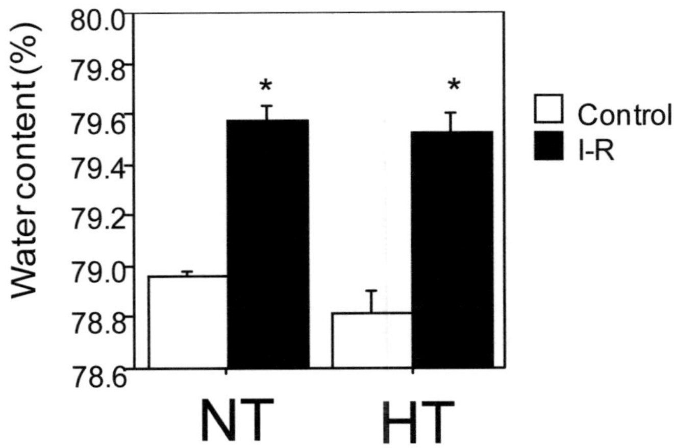

Fig. 2. Effect of HT during ischemia on occurrence of brain edema in C57BL/6J mice.

Forebrain ischemia was induced by BCCA occlusion for 15 min under 1 % halothane anesthesia in both normothermia (NT) and hypothermia (HT). The control mice (Control) underwent a sham operation without BCCA occlusion under halothane anesthesia for 15 min. The water content in the forebrain treated with BCCA occlusion for 15 min and 1 h after reperfusion (I-R) was measured. Data are expressed as the mean ± SE (control : n=3, I-R : n=5). Statistical analysis was performed by two-way ANOVA and an unpaired Student's *t*-test for comparison between control and I-R mice groups (*p<0.01 vs control).

4.2 Effect of hypothermia after reperfusion on occurrence of acute brain edema

We next investigated the effect of rHT on acute brain edema in the C57BL/6J mouse model. The water content 1 hr after reperfusion was significantly higher in both the HT and rHT mice than in the control mice ($p<0.05$). However, the water content 1 hr after reperfusion was significantly lower in the rHT mice than in the HT mice (Fig. 3). Already, we have shown an increase in brain water content during BCCA occlusion in a previous study (Doshi et al., 2009). Therefore, these results indicated that rHT suppresses the additional increase in brain water content after reperfusion, but not the increase in brain water content during BCCA occlusion.

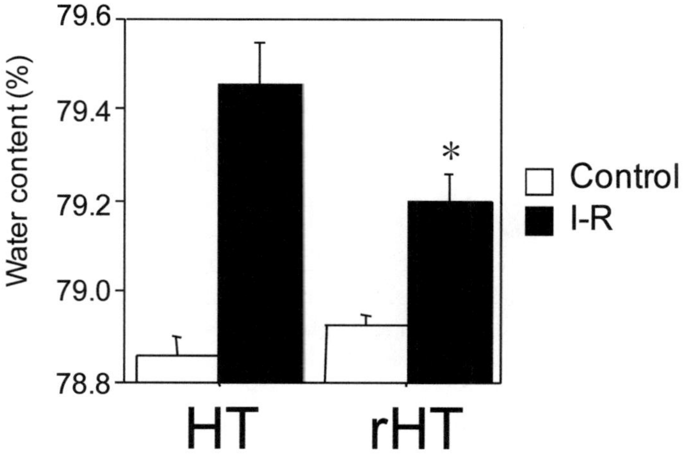

Fig. 3. Effect of HT after reperfusion on occurrence of brain edema in the BCCA occlusion C57BL/6J mouse model.
Forebrain ischemia was induced by BCCA occlusion for 15 min under 1 % halothane anesthesia in both HT and rHT mice. The control mice (Control) underwent a sham operation without BCCA occlusion under halothane anesthesia for 15 min. The water content in the forebrain treated with BCCA occlusion for 15 min and 1 hr after reperfusion (I-R) was measured. Data are expressed as mean ± SE (control : n=3, I-R : n=7). Statistical analysis was performed by two-way ANOVA and the unpaired Student's *t*-test (*$p<0.05$ vs HT).

5. Conclusion

HT during ischemia protects against delayed neuronal death in the hippocampus. In addition, rHT suppresses aggravation of acute brain edema following global cerebral ischemia in mice. Therefore, these findings indicate that the timing of therapeutic HT differs depending on the pathology following global cerebral ischemia. Further experimental studies should be performed to establish a standard method for the clinical application of therapeutic HT because there are still several controversial issues in the development of cooling techniques and in the determination of optimal duration and temperature.

6. References

Benard, SA., Gray, TW., Buist, MD., Jones, BM., Silvester, W., Gutteridge, G., & Smith, K. (2002). Treatment of comatose survivors of out-of-hospital cardiac arrest with induced hypothermia, *N Engl J Med*, 346, 557-563.

Doshi, M., Kuwatori, Y., Ishii, Y., Sasahara, M., & Hirashima, Y. (2009). Hypothermia during ischemia protects against neuronal death but not acute brain edema following transient forebrain ischemia in mice, *Biol Pharm Bull*, 32, 1957-1961.

Doshi, M., Higuchi, E., & Hirashima, Y. (2011). Hypothermia after reperfusion suppresses aggravation of acute brain edema following transient forebrain ischemia in mice, *J Health Sci*, 57, 82-85.

Kawase, M., Murakami, K., Fujimura, M., Morita-Fujimura, Y., Gasche, Y., Kondo, T., Scott, RW., & Chan, PH. (1999). Exacerbation of delayed cell injury after transient global ischemia in mutant mice with CuZn superoxide dismutase deficiency, *Stroke*, 30, 1962-1968.

Kempski, O. (2001). Cerebral edema, *Seminars in Nephrology*, 21, 303-307.

Maier, CM., Ahern, K., Cheng, ML., Lee, JE., Yenari, MA., & Steinberg, GK. (1998). Optimal depth and duration of mild hypothermia in a focal model of transient cerebral ischemia, *Stroke*, 29, 2171-2180.

Ohtaki, H., Nakamachi, T., Dohi, K., Yofu, S., Hodoyama, K., Matsunaga, M., Aruga, T., & Shioda, S. (2006). Controlled normothermia during ischemia is important for the induction of neuronal cell death after global ischemia in mouse, *Acta Neurochir*, 96, 249-253.

Tajiri, S., Oyadomari, S., Yano, S., Morioka, M., Gotoh, T., Hamada, J-I., Ushio, Y., & Mori, M. (2004). Ischemia-induced neuronal cell death is mediated by the endoplasmic reticulum stress pathway involving CHOP, *Cell Death Differ*, 11, 403-415.

Tsuchiya, D., Hong, S., Suh, SW., Kayama, T., Panter, SS., & Weinstein, PR. (2002). Mild hypothermia reduces zinc translocation, neuronal cell death, and mortality after transient global ischemia in mice, *J Cereb Blood Flow Metab*, 22, 1231-1238.

Yang, G., Kitagawa, K., Matsushita, K., Mabuchi, T., Yagita, Y., Yanagihara, T., & Matsumoto, M. (1997). C57BL/6 strain is most susceptible to cerebral ischemia following bilateral common carotid occlusion among seven mouse strains: selective neuronal death in the murine transient forebrain ischemia, *Brain Res*, 752, 209-218.

4

Molecular Mechanisms Underlying the Neuroprotective Effect of Hypothermia in Cerebral Ischemia

Yasushi Shintani and Yasuko Terao[1]
Biology Research Laboratories
[1]CNS Drug Discovery Unit,
Takeda Pharmaceutical Company Ltd.,
Japan

1. Introduction

Cerebral ischemia is a serious dynamic event in the brain involving heterogeneous cell types. Neuroprotective agents represent a potential approach for the treatment of acute stroke. Presently, recombinant tissue plasminogen activator (rtPA) is the only drug that is approved for the management of acute ischaemic stroke, except for the antioxidant Edaravone in Japan (Yoshida et al., 2006). Although stroke patients can receive rtPA therapy within the initial 3 h therapeutic window, there is an increased risk of intracranial haemorrhage, disruption of the blood brain barrier, seizures, or the progression of neuronal damage (Laloux, 2001). Thus, there is a continued need to explore novel neuroprotective strategies for the management of ischemic stroke. A large number of therapeutic agents have been tested, including N-methyl-D-aspartate receptor antagonists, calcium channel blockers, and antioxidants, for the management of stroke, but none has provided significant neuroprotection in clinical trials (Green et al., 2003; Kidwell et al., 2001).

Therapeutic hypothermia lowers a patient's body temperature in order to reduce the risk of the ischemic injury to the brain following a period of insufficient blood flow (Lampe and Becker, 2011; Yenari and Hemmen, 2010). The normal human adult body temperature is between 34.4–37.8°C and is maintained at a constant level through homeostasis or thermoregulation. Therapeutic hypothermia is defined as the artificial maintenance of the body temperature at <35°C and is subdivided into 4 different categories: mild (32–35°C), moderate (28–32°C), severe (20–28°C), and profound (<20°C). Nowadays, mild-to-moderate hypothermia (31–33°C) is usually applied for neuroprotection. The practical usage of hypothermia for clinical purposes was begun by the ancient Egyptians, Greeks, and Romans (Polderman, 2004). Ancient people observed the clinical usefulness of hypothermia for accidents and applied it to various diseases/symptoms. Modern clinical interest in hypothermia began in the 1930s with the description of the successful rescue of a drowned person with hypothermia after a prolonged period of asphyxia. After the first scientific report in 1945, which described the clinical application of hypothermia to patients with a severe head injury, hypothermia was subsequently applied to intracerebral aneurysm surgery and cerebral protection during complete circulatory arrest.

In the past few decades, the neuroprotective effects of hypothermia have been well established in experimental animals (Kawai et al., 2000; Miyazawa et al., 2003; Yanamoto et al., 2001) and in patients with cardiac arrest (THCASG, 2002 and Bernard et al., 2002). The initiation of moderate hypothermia within a few hours of severe ischemia can reduce the subsequent neuronal death and profoundly improve behavioural recovery. Although hypothermia is the only clinical intervention that appears to be neuroprotective after the initial injury, its key mechanisms have not been clarified. In other words, the neuroprotective effects of hypothermia provide many insights into the pathology of stroke, and thus may reveal clues for novel drug targets.

2. Neuroprotection against cerebral ischemia by hypothermia

Ischemia and cerebral hemorrhage are the two main causes of strokes. Ischemia accounts for ~85% of all reported incidents of stroke, and occurs when a thrombus or embolus blocks cerebral blood flow, resulting in cerebral ischemia and consequently neuronal damage and cell death. Conversely, hemorrhage occurs following the rupture of any blood vessel in the brain, resulting in rapid cerebral damage, and accounts for the remaining 15% of stroke cases. In each situation, the interruption of blood flow to the brain results in the reduced supply of oxygen and nutrients to the neurons. At the molecular level, the pathophysiology of ischemia is complicated and involves multiple sequential steps: progressive neural injury beginning with the activation of glutamate receptors, followed by the production and release of proinflammatory cytokines, nitric oxide (NO), free oxygen radicals, and proteases. As a result, neurons in an ischemic brain suffer irreversible and fatal damage. From the electrophysiological viewpoint, neurons depolarize massively giving rise to anoxic depolarization (Bureš et al., 1974) or to peri-infarct depolarizations (Gyngell et al., 1995). They are both characterized by swelling of neurons, massive influx of Na^+ and Ca^{2+} into neurons, massive release of K^+ into the interstitial space, release of glutamate, acidification of the tissue (Somjen, 2001; Dreier, 2011; Balestrino, 1995).

The lack of blood supply results in two identifiable areas, namely the core and the penumbra. The core is a neuronal dead area that is not therapeutically accessible, whereas the penumbra is a still salvageable zone (Bandera et al., 2006). As a consequence of the reduced blood supply inside the core, adenosine triphosphate (ATP) levels are reduced, leading to the depression of cellular metabolism. Energy loss results in impaired ion homeostasis, which leads to rapid depolarization and a large influx of calcium and potassium ions. The increased levels of intracellular calcium induces the activation of excitotoxic glutamatergic transmission, NO synthase, caspase, xanthine oxidase, and the release of reactive oxygen species. Glutamate release activates phospholipases, phospholipid hydrolysis, and the release of arachidonic acid. The generation of free radicals and lipid peroxidation and the activation of immediate early genes, such as *c-fos, c-jun*, and the inflammatory cascade, lead to progressive ischemic damage, resulting in necrotic as well as apoptotic cell death. Conversely, the penumbra represents viable tissue surrounding the core and receives a trivial amount of blood from collateral arteries; therefore, the penumbra is the target for drug intervention and has the potential for recovery.

Precisely controlled mild hypothermia has been proven to have neuroprotective properties and reduces the risk of the detrimental effects that often occur during profound hypothermia. Large numbers of the phenomena observed in ischemic brains can be ameliorated by hypothermia, including the reduction of oxygen radical production, with the subsequent reduction in peroxidase damage to lipids, proteins, and DNA, thereby supporting the behavioral recovery of patients. Hypothermia also decreases microglial activation, ischemic depolarization, cerebral metabolic demand for oxygen, and the release of glycerin and excitatory amino acids. We and others have demonstrated that inflammation potentiates cerebral ischemic injury and that hypothermia can reduce inflammation by suppressing the infiltration of neutrophils into ischemic regions (Ohta et al., 2007; Shintani et al., 2011; Terao et al., 2009; Zheng and Yonari, 2004). Furthermore, the inhibition of reactive oxygen species production by leukocytes and microglia in the ischemia brain (Kil et al., 1996), NF-κB activation (Han et al., 2003), neutrophil infiltration (Wang et al., 2002), and cytochrome c release (Yenari et al., 2002) has been also reported in hypothermia-treated ischemic rat brains.

Recently, Lin et al., (2011) reported that whole-body hypothermia broadens the therapeutic window of intranasally administered recombinant human insulin-like growth factor 1 (IGF-1) in a neonatal rat cerebral hypoxia–ischemia model. They ligated the right common carotid artery of postnatal day 7 rat pups, followed by 8% oxygen inhalation for 2 h. After the hypoxia-ischemia treatment, the pups were divided into 2 groups and maintained under different temperatures, room temperature (24.5 ± 0.2°C) and a cool environment (21.5 ± 0.3°C), for 2 or 4 h before being returned to room temperature. IGF-1 was administered intranasally at 1 h intervals starting at 0, 2, or 4 h after hypothermia. Although the administration of hypothermia or IGF-1 alone at 2 h after hypoxia-ischemia treatment did not provide neuroprotection, the combined treatment of hypothermia and IGF-1 significantly protected the neonatal rat brain from hypoxia-ischemia injury. Hypothermia extended the therapeutic window of IGF-1 to 6 h after hypoxia-ischemia. It was observed that the combination therapy decreased the infiltration of polymorphonuclear leukocytes, the activation of microglia/macrophages, and the attenuation of nuclear factor kappa-B (NF-κB) activation. These findings broaden the potential application of hypothermia, i.e., not only for its neuroprotective effects but also for its synergistic effects by the combined use of hypothermia with already existing therapeutic drugs.

3. Molecular mechanisms of hypothermia-induced neuroprotection

3.1 Proteins influenced by hypothermia

The detailed molecular mechanisms underlying the neuroprotection induced by hypothermia against ischemia have been approached by degrees. In most cases, molecules, whose expression is known to be affected by ischemia, have been identified and the effects of hypothermia on the molecule have been explored retrospectively. Table 1 lists the molecules whose expression is affected by hypothermia during ischemia and/or reperfusion. The majority of these proteins are apoptosis, inflammation-related, and signalling molecules, such as kinases and transcription factors. For example, apoptosis-related molecules, such as B-cell lymphoma 2 (Bcl2), Bcl-associated X protein (Bax), caspases, calpains, cytochrome c, and Fas/Fas ligand (FasL), are up- or down-regulated by hypothermia in accordance with its neuroprotective effect, thereby preventing neuronal cell death. Inflammatory molecules, such as tumor necrosis factor alpha (TNFα), interleukin (IL)-1β, IL-6, monocyte chemotactic protein-1 (MCP-1), macrophage inflammatory protein-3

Molecule	Hypothermia	Expression		Ischemia/observation	Reference
		Ischemia	Hypothermia		
IL-1α	33°C culture	N.D.	Increase	In vitro cell culture	Yanagawa et al., 2002
IL-1β, TNFα	2 h hypothermia, started 20 min after ET-1 injection	Increase	Decrease	Rat endothelin (ET)-1-induced transient focal cerebral ischemia; hypothermia reduced astrogliosis at 1 and 3 d after stroke onset.	Ceulemans et al., 2011
IL-1β, TNFα	33°C culture	Increase	Decrease	In vitro cell culture; 30 h oxygen – glucose deprivation (OGD)	Webster et al., 2009
IL-6	33°C culture	N.D.	Decrease	In vitro cell culture	Yanagawa et al., 2002
IL-18	32°C for 24 h after hypoxia-ischemia	Increase	Decrease	Hypoxia-ischemia	Fukui et al., 2006
MIP-3α	34°C after MCAO	Increase	Decrease	2 h MCAO	Terao et al., 2009
IL-1β, IL-6, IL-10, TNFα, ICAM-1	33°C for 24 h	Increase	Decrease	Pigs were subjected to cardiac arrest following temporary coronary artery occlusion.support.	Meybohm et al., 2010
MCP-1	34 °C after MCAO	Increase	Decrease	2 h MCAO	Ohta et al., 2007
Intercellular adhesion molecule-1 (ICAM-1)	33°C etc.	Increase	Decrease	2 h MCAO; extracellular signal-regulated kinase-1/2 (ERK-1/2) activation, and induction of leukocyte infiltration and inflammatory reaction by ischemia were inhibited by hypothermia.	Choi et al., 2011; Inamasu et al., 2001; Koda et al., 2010
NF-κB	33°C for 2 h after MCAO	No change	No change	2 h MCAO; NF-kB was translocated from cytoplasm to nucleus by hypothermia.	Han et al., 2003
NF-κB's inhibitory protein (IkB-α)	33°C for 2 h after MCAO	Increase	Decrease	2 h MCAO	Han et al., 2003
IkB-α, NOS, TNFα	30-34°C	Increase	Decrease	Phosphorylation of IkB-α was suppressed by hypothermia.	Yenari et al., 2006

Molecule	Hypothermia	Expression		Ischemia/observation	Reference
		Ischemia	Hypothermia		
Bcl-2	30-34°C	Down	Increase	1 h hypoxia-ischemia etc.	Eberspacher et al., 2005; Jieyong et al., 2006; Zhao et al., 2004; Yenari et al., 2002; Zhang et al., 2001
Bcl-2	31-32°C hypothermia for 60 min after MCAO	N.D.	Increase	Rat global cerebral ischemia (20 min); the mortality in rats was evaluated at 72 h and 168 h reperfusion.	Zhang et al., 2010
Bcl2	33°C culture	Increase	Decrease	In vitro cell culture; 5 h OGD at 37°C; human umbilical vein endothelial cells were used in this study.	Yang et al., 2009
Bax	33°C culture	Decrease	Increase	In vitro cell culture; 5 h OGD at 37°C; human umbilical vein endothelial cells were used in this study.	Yang et al., 2009
Bax	34°C during ischemia	Increase	Decrease	Bcl-2, p53, and Mdm-2; no change	Eberspacher et al, 2003, 2005
Fas, FasL	33°C during ischemia etc.	Increase	Decrease	2h MCAO; soluble FasL (sFasL) was decreased by hypothermia, while membrane-bound FasL (mFasL) increased.	Liu et al., 2008; Phanithi et al., 2000
Akt	34°C after ischemia	Increase	Decrease	Hypoxia-ischemia for 1 h	Tomimatsu et al., 2001
Caspase-3	30-34°C	Increase	Decrease	1 h hypoxia-ischemia etc.	Fukuda et al., 2001; Pabello et al., 2005; Phanithi et al., 2000; Tomimatsu et al., 2001
Caspase 8	33°C during ischemia	Increase	Decrease	2h MCAO	Liu et al., 2008
Caspase-3/9	33°C for 10 min before ischemia and maintained 3 h after reperfusion	Increase	Decrease	10 min MCAO and hypotention	Zhao et al., 2004, 2005
Calpain	32°C for 10 min before and during reperfusion	Increase	Decrease	3 h MCAO and 24 h reperfusion	Liebetrau et al., 2004

Molecule	Hypothermia	Expression		Ischemia/observation	Reference
		Ischemia	Hypothermia		
Cytochrome c, AIF	30 or 33°C for 2 h during and/or after MCAO	N.D.	N.D.	2 hr MCAO; cytochrome c release and AIF translocation from mitochondria to nuclei were stimulated by hypothermia.	Zhao et al., 2007
Cytochrome c, AIF	34 or 36°C	Increase	Decrease	50 min MCAO; cytochrome c release and AIF translocation from mitochondria to nuclei were stimulated by hypothermia.	Zhu et al., 2006
Akt	30°C for 10 min before ischemia and maintained for 1 h after ischemia	N.D.	N.D.	1 h MCAO; Akt activity was inhibited by ischemia and stimulated by hypothermia.	Zhao et al., 2005
c-Fos, AP-1	Cold room (1°C) during MCAO	Increase	Decrease	1 h MCAO; c-Jun expression was not affected by hypothermia.	Akaji et al., 2003
AMPK (phospho-rylation)		N.D.	N.D.	Ischemia-induced phosphorylation of AMPK was inhibited by hypothermia.	Li et al., 2011
cold-inducible RNA-binding protein (CIRP)	Moderate (30 ± 2°C) hypothermia for 2 h	Increase	Increase	20 min MCAO	Liu et al., 2010
p53	31-32°C	N.D.	Decrease	Rat global cerebral ischemia (20 min); the mortality in rats was evaluated at 72 h and 168 h reperfusion.	Zhang et al., 2010
p53, PUMA, NOXA	33°C	Increase	Decrease	2 h MCAO; DNA damage-dependent signaling events, including NAD depletion, p53 activation, and mitochondrial translocation of PUMA and NOXA	Ji et al., 2007
hypoxia-inducible factor-1 (HIF-1)	28 - 32°C culture	Increase	Decrease	In vitro cell culture	Tanaka et al., 2010
high-mobility group box 1 (HMGB1)	32°C	N.T.	Decrease		Koda et al., 2010
Hsp70		Increase	Increase	2 h MCAO	Terao et al., 2009

Molecule	Hypothermia	Expression		Ischemia/observation	Reference
		Ischemia	Hypothermia		
Agrin, SPARC (BM-40, osteonectin)	32° C	Decrease	Increase	Transient 20 min forebrain ischemia	Baumann et al., 2009
GSK3β	33°C	N.D.	Decrease	Global cerebral ischemia	Kelly et al., 2005
GSK 3β	Moderate hypothermia (30°C) blocked degradation of total GSK 3β.	N.D.	N.D.	Dephosphorylated after stroke in normothermia	Zhang et al., 2008
β-catenin		N.D.	N.D.	Phosphorylation of b-catenin was increased and degraded by ischemia. Hypothermia did not inhibit the phosphorylation, but it blocked degradation in the ischemic penumbra.	Zhang et al., 2008
MLK3, MKK4/7, JNK3, c-Jun, FasL	32°C hypothermia for 10 min before ischemia and maintained for 3 h after ischemia	Increase	Decrease	Alternation of the assembly of the GluR6-PSD95-MLK3 signaling module	Hu et al., 2008
Galanin	33°C	Decrease	Increase	60 min transient MCAO	Theodorsson et al., 2008
COX-2	33-34°C	Increase	Decrease	Global ischemic insult	Xiang et al., 2007
MMP-2, MMP-9	33°C etc.	Increase	Decrease	90 min MCAO	Lee et al., 2005; Nagel et al., 2008
TIMP-2	33°C during ischemia	N.D.	Increase	2 h MCAO	Lee et al., 2005
BDNF	33-34°C for 24 h	Decrease	Increase	Permanent MCAO	Xie et al. 2007
α, β, γ-PKCs, CaM kinase II	30 min of ischemia followed by 60 min of reperfusion	N.D.	N.D.	30 min ischemia followed by 60 min reperfusion; hypothermia inhibited translocation of CaM kinase II and α, β, γ-PKC	Harada et al., 2002
PKCε	30°C during MCAO	Decrease	Increase	1 h MCAO	Shimohata et al., 2007
PKCδ	30°C during MCAO	Increase	Decrease	1 h MCAO	Shimohata et al., 2007

Molecule	Hypothermia	Expression		Ischemia/observation	Reference
		Ischemia	Hypothermia		
hypoxanthine phosphoribosyl transferase (HPRT)	33-34˚C throughout identical injury and reperfusion periods	No change	No change	8 min hypoxia or ischemia, and 30 min or 4 h of cerebral reperfusion	Cherin et al., 2006
calcium sensing receptor (CaSR)	33˚C for 3 h after reperfusion	Increase	Decrease	10 min ischemia followed by 1-3 d reperfusion	Kim et al., 2011
Inhibitory gamma-aminobutyricacid-B receptor 1 (GABA-B-R1)	33˚C for 3 h after reperfusion	Decrease	Increase	10 min ischemia followed by 1-3 d reperfusion	Kim et al., 2011
AMPA receptor subunit GluR2	1 h after reperfusion at 32˚C, then warmed to 34˚C and 36˚C in a step-by-step manner.	Decrease	Increase	1 h MCAO	Colbourne et al., 2003
SOD	33˚C during ischemia	Increase	Decrease	2 h MCAO	Maier et al., 2002
iNOS	33˚C during or after ischemia	Increase	Decrease	1 h MCAO	Han et al., 2002
GPR78	34±0.5˚C	Decrease	Increase	ischemic for 15 min and then reperfused for 3 h under	Aoki et al., 2001
74 proteins including glycolysis, plasticity, and redox-related proteins.	33˚C during ischemia	N.D.	Glycolysis and plasticity-related proteins were preserved but redox-related proteins were lowered by hypothermia.	15 min ischemia	Teilum et al., 2007

N.D. = not determined.

Table 1 Proteins of which expression and/or modification are altered by ischemia and hypothermia

alpha (MIP-3α), and NF-κB, are also up-regulated after middle cerebral artery occlusion (MCAO) and down-regulated by hypothermia. These molecules are considered to enhance cell damage through the activation of microglia and/or astrocytes. Recently, more direct and dynamic inflammatory changes have been speculated to be induced by hypothermia during ischemia. And now it has become clear that the neuroinflammatory response could be detrimental and that even peripheral immune responses can be regulated by the brain (Ceulemans et al., 2011).

Another type of change evoked by hypothermia in ischemic brains is protein modification. For instance, Li et al. (2011) reported that hypothermia reduced the activation of 5′ adenosine monophosphate-activated protein kinase (AMPK), a ubiquitously distributed kinase, by the dephosphorylation of its regulatory residues in ischemic mice. They showed that hypothermic neuroprotection was ameliorated by compound C, an AMPK inhibitor, and that genetic deletion of one of the catalytic isoforms of AMPK completely reversed the effects of hypothermia on stroke outcome after acute and chronic survival. Their study provides evidence that hypothermia exerts its protective effects, in part, by inhibiting AMPK activation, at least in experimental focal stroke. AMPK is known to participate in an energy sensing cascade and to serve as a master regulator of metabolism in response to ATP depletion. Recently, additional roles of AMPK in a variety of other cellular processes have been revealed, in the cytoplasm and nucleus, as a controllor of cell polarity and a transcriptional regulator. A more interesting function of AMPK in signaling pathways is its role as a responder to cellular stress and damage, and the relevance of AMPK signaling in various diseases is becoming a hot topic in the investigation of ischemic physiology.

3.2 Inflammatory proteins

3.2.1 Cytokines

Neuroinflammation is involved in the pathogenesis of many central nervous system (CNS) diseases. In stroke, excess inflammatory activation results in brain injury and ultimately causes severe neuronal apoptosis (Zheng and Yonari, 2004). Anti-inflammatory therapies using immunosuppressants (Furuichi et al., 2004) or biogenetics, e.g., an anti-ICAM-1-neutralizing antibody (Matsuo et al., 1994), have been applied in preclinical and clinical trials. In this context, the effect of hypothermia on neuroinflammation has been vigorously explored.

Webster et al. (2009) measured the levels of proinflammatory cytokines, such as TNF-α and IL-β, in microglial culture supernatants after stimulation with lipopolysaccharide (LPS) or 2 h oxygen–glucose deprivation (OGD) exposure followed by 24 h reperfusion. There was a marked increase in the production and release of inflammatory cytokines by microglia following LPS stimulation and OGD, which was attenuated by hypothermia. Glutamate, a major neurotransmitter, was also released from microglia stimulated with LPS at 37°C, but reduced levels were observed when they were stimulated with LPS at 33°C. IL-18 is another proinflammatory cytokine that may contribute to brain injury. Fukui et al. (2006) showed that the effects of hypothermia treatment after hypoxia-ischemia (rectal temperature of 32°C for 24 h) on IL-18 expression. IL-18 expression in the ipsilateral hemispheres of the normothermia group significantly increased at 72 h after hypoxia-ischemia compared with controls; however, IL-18 expression was significantly decreased in the hypothermia group.

NF-kB is a transcription factor that is activated after cerebral ischemia. The activation of NF-kB leads to the expression of many inflammatory genes involved in the pathogenesis of stroke. Yenari and Han (2006) showed that hypothermia decreases the translocation of NF-kB from the cytoplasm to nucleus and its binding activity to NF-kB regulatory proteins. Mild hypothermia appears to suppress the phosphorylation of IkB-α, an NF-kB inhibitory protein, by decreasing the expression and activity of IkB kinase-γ (IKKγ). As a consequence, hypothermia suppressed the expression of 2 NF-kB target genes, inducible nitric oxide synthase (iNOS) and TNFα.

We previously determined that the therapeutic time window of post-ischemia mild hypothermia was 4 h after reperfusion (Ohta et al., 2007). Thus, we considered that the gene expression changes that occur before the 4 h time point might be important for the neuroprotective effect induced by mild hypothermia, even though the neuroprotection afforded by hypothermia alone during this period was found to be insufficient. After a series of investigations with hypothermia, we hypothesized that the gene expression changes that were observed after the 4 h timepoint, following the discontinuation of hypothermia, are related to the ischemic damage detected at 2 d after MCAO. Genes that were upregulated after the 4 hr timepoint, including *early growth response-2 (Egr-2), neurotransmitter-induced early genes-1* (*Ania-1*), and *macrophage inflammatory protein-3α* (*MIP-3α*), were found to be important for the neuroprotection afforded by hypothermia. We selected the following 12 genes that might exert substantial neuroprotective activity: *c-Fos, Egr-1, Egr-4, neuron-derived orphan receptor-1 (Nor1), MAP kinase phosphatase-1* (*MKP-1*), *MKP-CPG21, MIP-3α, monocyte chemotactic protein-1* (*MCP-1*), *brain-derived neurotrophic factor* (*BDNF*), *IL-1β, Ania-1,* and *Ania-7*. Egr-1 is known as a master switch that is activated by ischemia to trigger the expression of pivotal regulators of inflammation, e.g., IL-1α, MCP-1, and MIP-2, in addition to coagulation and vascular hyperpermeability.

3.2.2 Chemokines

Chemokines are also well known to be detrimental factors in the brain. For instance, MCP-1 is considered to be a promising drug target due to its possible role in exacerbating ischemic injury, controlling blood-brain barrier permeability, and driving leukocyte infiltration into the brain parenchyma in stroke (Dimitrijevic et al., 2006, 2007; Schilling et al., 2009). We have observed that *MCP-1* gene expression was upregulated by ischemia and that the expression stimulated by ischemia was supressed by hypothermia (Ohta et al., 2007).

We further conducted comprehensive gene expression analyses of ischemic rat brains with or without hypothermia by using a rat ischemia-reperfusion model in order to elucidate the underlying mechanisms and discover novel target molecules. In this study, we revealed that cerebral *MIP-3a* and *CC-chemokine receptor 6 (CCR6)* genes were significantly induced in the core and penumbra regions of MCAO rat brains, and hypothermia suppressed the expression of both genes (Terao et al., 2009). MIP-3α is expressed in macrophages, dendritic cells, and lymphocytes. Depending on the conditions, MIP-3α can act constitutively or inducibly and serves as a chemoattractant, especially in epithelial immunological systems such as those of the skin and mucosa (Charbonnier et al., 1999; Cook et al., 2000). In the CNS, MIP-3α expression has been reported in autoimmune encephalomyelitis (Ambrosini et al., 2003) and stroke patients (Lu et al., 2004; Utans-Schneitz et al., 1998), but its full role has not yet been determined. CCR6, the sole receptor for MIP-3α, is expressed in multiple leukocyte subsets, and is implicated in diverse inflammatory responses in animal models, such as allergic airway disorders, inflammatory bowel disease, and autoimmune encephalitis (Schutyser et al., 2003). Strikingly, the intracerebral administration of an anti-rat MIP-3α-neutralizing antibody significantly reduced infarct volumes in MCAO rats compared with those of vehicle- and control mouse IgG-treated rats, suggesting that MIP-3α-CCR6 signaling is dominant in the neuroinflammatory cascades of brain ischemia. Interestingly, the administration of MIP-3α into the striatum induced *CCR6* gene expression in a dose-dependent manner, but not *CCR1* or *CCR2* expression. The intrastriatal injection of IL-1β and TNF-α into control rats upregulated *MIP-3α* and *CCR6* mRNA expression levels in a sequential fashion. Taken together with the robust induction of IL-1β and TNF-α in ischemic brains during an acute phase of MCAO prior to MIP-3α expression, these cytokines may directly evoke MIP-3α production in the CNS. Furthermore, *MIP-3α* mRNA expression was markedly induced by IL-1β and TNF-α in rat astrocytes, but not in microglia or neurons. Astrocytes stimulated by ischemic stress turn into their active form, expressing glial fibrillary acidic protein (GFAP), and appear around the damaged area after ischemic injury (Zoli et al., 1997). Rat primary microglia constitutively express the *CCR6* gene under normal culture conditions, while astrocytes and neurons do not. Interestingly, we found that the expression of *iNOS* and *IL-1β* was induced in MIP-3α-treated microglia. Microglia are activated and accumulate around the injured area following ischemia (Wood, 1995). We observed that MIP-3α was produced by rat primary cultured astrocytes in response to IL-1β and TNF-α treatment, while hypothermia significantly suppressed the expression of both cytokines. Therefore, the activation of astrocytes and microglia may accelerate brain injury-induced neuroinflammation via MIP-3α-CCR6 signaling, whereas hypothermia suppresses this signaling. The physiological roles of MIP-3α-CCR6 signaling in the CNS have yet to be fully determined because various roles for chemokines in the brain have recently been proposed, e.g., neurotransmitters and neuromodulators (de Haas et al., 2007; Rostène et al., 2007). The interactions between MIP-3α-CCR6 signaling and other pathways involved in ischemic pathology, e.g., excitotoxicity, acidotoxicity, oxidative stress, and apoptosis, should also be examined.

3.3 Apoptotic proteins

Apoptosis is another important factor for ischemic damage because of its contribution to the cell death subsequent to ischemia/reperfusion injury (Broughton et al., 2009). To date, mitochondrial dysfunction, oxidative stress, and impaired cerebral energy metabolism have been observed during the neuronal cell death that is responsible for much of the poor neurologic outcome from these events. Recent studies using *in vitro* and *in vivo* neuronal cell death models point toward several molecular mechanisms that are either induced or promoted by the oxidative modification of macromolecules, including the consumption of cytosolic and mitochondrial nicotinamide adenine dinucleotide (NAD+) by poly-ADP ribose polymerase (PARP), opening of the mitochondrial inner membrane permeability transition pore, and the inactivation of key, rate-limiting metabolic enzymes, such as the pyruvate dehydrogenase complex. In addition, the relative abundance of proapoptotic proteins in immature brains and neurons, and particularly within their mitochondria, predisposes these cells to the intrinsic, mitochondrial pathway of apoptosis, which is mediated by the Bax- or Bak-triggered release of proteins into the cytosol through the mitochondrial outer membrane. On the basis of these cell dysfunction and death pathways, several approaches toward neuroprotection are being investigated that show promise for their future clinical application. These strategies include minimizing oxidative stress to avoid unnecessary hypoxia, promoting aerobic energy metabolism by the repletion of NAD+, and providing alternative oxidative fuels, e.g., ketone bodies, directly interfering with apoptotic pathways in mitochondria, and pharmacologically inducing antioxidant and anti-inflammatory gene expression. Hypothermia is known to reduce oxidative stress, metabolic dysfunction, delayed neuronal death, and short- and long-term neurobehavioral impairment. However, despite successful clinical trials of hypothermia, its neuroprotective mechanisms have not been well investigated.

We investigated the activity of apoptotic proteases, calpains, and caspase-3 in 2 h MCAO rat brains using α-fodrin, a cytoskeletal protein enriched in the synaptosome-rich fraction, as a substrate for each (Fig. 1). α-Fodrin is fragmented by calpains and caspase-3 to 145/150-kDa and 120-kDa cleavage products from its intact 240-kDa protein, respectively (Wang, 2000). Harada et al. (2002) reported that ischemia-reperfusion induced the proteolysis of α-fodrin with the generation of the 150-kDa fragment, while hypothermia inhibited this ischemia-reperfusion-induced proteolysis. As shown in Fig. 1, an anti-α-fodrin antibody detected several bands, including the 240-kDa intact protein and the 145/150-kDa and 120-kDa breakdown products. We also detected the 145/150-kDa bands at 48 h after MCAO in normothermia. Surprisingly, mild hypothermia (34°C) did not change the density of those bands. These finding suggest that the inhibition of protein degradation by hypothermia is not dominant in our model or that α-fodrin is not a suitable substrate to determine apoptotic degradation.

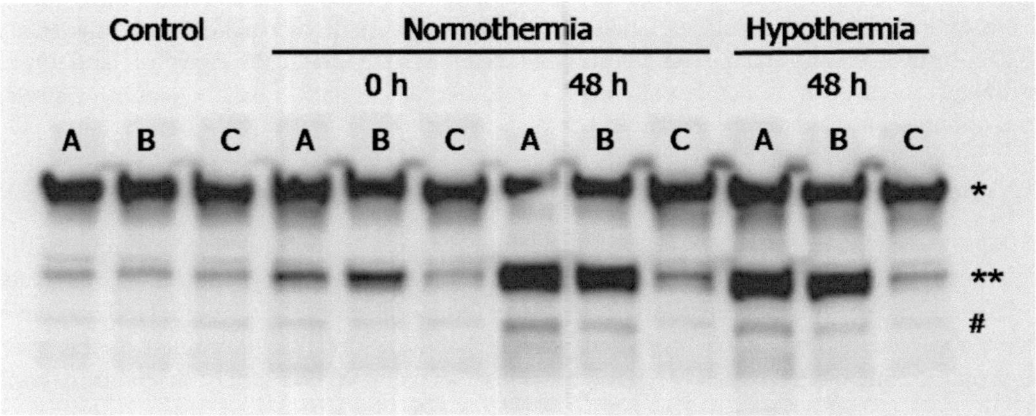

Fig. 1. Lack of effect of hypothermia on α-fodrin degradation in MCAO rat brains

Protease activity in brain tissues of rats maintained at normothermic (37°C) or hypothermic (34°C) conditions at 0 or 48 h after MCAO and in control brain tissues was shown by the degree of α-fodrin degradation. In controls, α- fodrin was detected mainly as a 240-kDa (*) band, although α-fodrin was severely degraded into 145/150- (**) and 120-kDa (#) bands in the core (A) and penumbra (B) regions. Unexpectedly, hypothermia did not affect the degradation of α- fodrin in contralateral (C) regions in MCAO rat brains.

3.4 Cold-inducible RNA-binding proteins

Having some analogy to mild hypothermia, mammalian hibernation serves as a natural model of tolerance to extreme reductions of blood flow, energy consumption, and body temperature, and to the capacity to deliver oxygen to tissues at otherwise lethal levels (Frerichs et al, 1994; Frerichs and Hallenbeck, 1998). As hibernating animals suffer no CNS damage or cellular loss because of these special adaptive changes (Carey et al, 2003; Storey, 2003), the molecular mechanisms that regulate these adaptations are potential targets for drug discovery. Generally, low temperatures reduce the rate of enzymatic reactions, diffusion, and membrane transport due to the inhibition of chemical reaction rates. Although certain microorganisms are able to adapt to cold environments, apart from the over-expression of a defined set of cold-shock proteins, by the modification of enzyme kinetics, in mammalian cells, the molecular mechanisms that govern adaptation to mild hypothermia are not well known, although they may involve a series of events that modulate transcription, translation, the cytoskeleton, the cell cycle, and metabolic processes. Furthermore, it has been demonstrated that cells exposed to moderate hypothermia, even for short periods of time, have broad changes in their gene expression patterns. Upon exposure to moderate hypothermia, certain specific cold-shock proteins respond immediately to ensure that the cell rapidly adapts to the novel environmental conditions (Fujita, 1999). For decades, it was assumed that the proteins activated upon exposure to mild cold temperatures are somehow responsible for the general metabolic deceleration that occurs when mammalian cells are exposed to mild cold shock. Nonetheless, cold-shock proteins actually "facilitate" the accurate and enhanced translation of specific mRNAs at temperatures below physiological temperatures (Durandy, 2008).

Several hypothermia-induced genes have been identified mainly by using *in vitro* cellular models. For example, Tanaka et al. (2010) showed that the exposure of cultured cells to 32°C for as long as 24 h suppressed the hypoxia-induced activation of hypoxia-inducible factor 1 (HIF-1) and the subsequent upregulation of HIF-1 target genes, e.g., vascular endothelial growth factor (VEGF) or glucose transporter 1 (GLUT-1), although HIF-1 protein stability was not affected by hypothermic treatment. Yang et al. (2009) reported that hypothermia markedly reduced OGD-induced apoptosis in human umbilical endothelial cells (HUVEC), which was induced by changing the media of HUVEC subjected to OGD for 5 h with oxygenated media. The cells were then placed in an incubator for 0–20 h at normothermic (37°C) or hypothermic (33°C) conditions to mimic reperfusion. Hypothermia reduced the expression of cleaved caspase-3 and poly(ADP-ribose) polymerase (PARP), but, in contrast, it reversed the OGD-induced activation of Fas/caspase-8, the increase of Bax and decrease of Bcl-2, and the inhibition of JNK1/2 activation via MKP-1 induction, suggesting that hypothermia represses OGD-induced endothelial cell apoptosis by inhibiting the extrinsic- and intrinsic-dependent apoptotic pathways and the activation of JNK1/2. Lee et al. (2009) showed that moderate hypothermia (30°C) reduces ischemic damage after permanent distal MCAO in rats. They showed that early and delayed hypothermia blocked δ-protein kinase C (PKC) cleavage, suggesting that the degradation of proteins by caspases might be suppressed by hypothermia.

We attempted to identify genes whose expression is specifically induced by mild hypothermia using cDNA subtraction between normothermia- and hypothermia-treated rat cortex-derived primary neurons. The culture temperature of the cells was lowered to 32°C for 14 h and then increased to 37°C. The cells were incubated for 0 (A) or 7 h (B) before their mRNA was collected. Control neurons were maintained at 37°C for 21 h (C). We synthesized cDNA from each mRNA sample as described in our previous report (Terao et al., 2009) and generated a subtraction cDNA library for each condition. To identify the specific genes associated with mild hypothermia, we screened subtracted cDNAs between (A) and (C) or (B) and (C). We identified candidate genes whose expression were specifically induced in cells incubated at 32°C compared with controls. All of the genes thus obtained were different gene fragments of the same cold-inducible RNA-binding protein (CIRP) (data not shown). CIRP (also known as A18 hnRNP) is an 18-kD protein that consists of an amino-terminal RNA-binding domain and a carboxyl-terminal glycine-rich domain (Fujita, 1999; Lleonart, 2010). It exhibits structural similarity to a class of stress-induced RNA-binding proteins found in plants. Nishiyama et al. (1997) reported that *cirp* cDNA was induced at a low culture temperature of 32°C in a mouse-derived cell line. In the last decade, a considerable number of cold-shock proteins have been identified in human cells; however, only 2 of these proteins, CIRP and RNA-binding motif protein 3 (RBM3), have been well characterized since their initial discovery. Although *cirp* expression was clearly identified in cell-based experiments, a comprehensive approach comparing MCAO rat brains with or without hypothermia did not identify the modulation of the *cirp* gene by hypothermia. We speculate that the experimental design of the *in vivo* ischemia-hypothermia model may be not appropriate or that the changes in *cirp* gene expression were too small to be detected on oligonucleotide chips. Since Liu et al. (2010) reported that CIRP expression was upregulated even by ischemia alone, it might be difficult to see the difference in *cirp* gene expression between ischemia alone and ischemia plus hypothermia rat brains. We identified several transcription factors, including AP-1, Pax-4, lmo2, MyoD, ADR1, cdxA, and Pax-8, that have potential binding sites on the human and mouse *cirp* genome promoter regions (data not shown). Importantly, cold-inducible RNA-binding proteins, e.g., CIRP and RBM3, are also able to regulate their expression at the level of translation by binding to different transcripts, thus allowing the cell to respond rapidly to environmental signals. The binding of certain proteins to the 5′-untranslated region (UTR) or 3′-UTR of their mRNAs can affect the rate of translation initiation and the stability of the transcript. Xue et al. (1999) reported that hypothermia induces CIRP expression in the CNS *in vitro* (PC12 neuroblastoma cells) and *in vivo* (rat brains) and that oxidative stress and/or ischemia counteract this phenomenon. In contrast, Liu et al. (2010) reported no correlation between CIRP expression and brain metabolism; therefore, it is suggested that CIRP has unrevealed neuroprotective activity.

More recently, Chip et al. (2011) reported that a low temperature (32°C) induces the expression of a small subset of proteins, including RBM3. Immunohistochemistry of the developing postnatal murine brain revealed that the spatiotemporal neuronal expression pattern of RBM3 was very similar to that of doublecortin, a marker of neuronal precursor cells. Mild hypothermia profoundly promoted RBM3 expression and rescued neuronal cells from forced apoptosis, as observed in primary neurons, PC12 cells, and cortical organotypic slice cultures. Blocking RBM3 expression in neuronal cells by specific RBM3 small interfering RNAs significantly diminished the neuroprotective effects of hypothermia, while RBM3 overexpression reduced the cleavage of PARP, prevented internucleosomal DNA fragmentation, and reduced lactate dehydrogenase release, suggesting that RBM3 is a novel apoptosis modulator. Taken together, these data indicate that neuronal RBM3 induction in response to hypothermia apparently accounts for a substantial proportion of hypothermia-induced neuroprotection. Therefore, they concluded that RBM3 may be one of the potential neuroprotective factors induced by hypothermia.

3.5 Chaperones

In response to various forms of stress, cells activate a highly conserved heat shock response in accordance with the induction of a set of heat shock proteins (Hsps) that play important roles in cellular repair and protective mechanisms (Moseley, 2000). Evidence suggests that manipulation of the cellular stress response may offer strategies to protect brain cells from the damage that is encountered following cerebral ischemia or during the progression of neurodegenerative diseases. Hsp70 is a chaperone protein that can fold or refold proteins, coordinate protein trafficking, inhibit protein aggregation or degradation, and exhibits anti-apoptotic and anti-inflammatory activities under physiological conditions. Over-expression of Hsp70 reduced ischemic injury in the mammalian brain (Marber et al., 1995; Zheng et al., 2008). Investigation of the domains within Hsp70 that confer ischemic neuroprotection revealed the importance of the carboxyl-terminal domain (Sun et al., 2006). Arimoclomol, a co-inducer of Hsps, delayed the progression of amyotrophic lateral sclerosis in a mouse model in which motor neurons in the spinal cord and motor cortex degenerate (Kieran et al., 2004). Celastrol, a promising candidate as an agent to counter neurodegenerative diseases, induced the expression of a set of Hsps in differentiated neurons grown in tissue culture (Chow and Brown, 2007). Heat shock "preconditioning" protected the nervous system at the functional level of the synapse, and selective over-expression of Hsp70 enhanced the level of synaptic protection (Ge et al., 2008). Following hyperthermia, constitutively expressed Hsc70 increased in synapse-rich areas of the brain where it associates with Hsp40 to form a complex that can refold denatured proteins. Stress tolerance in neurons is not solely dependent on their own Hsps, but can be supplemented by Hsps from adjacent glial cells; hence, the application of exogenous Hsps at neural injury sites is an effective strategy to maintain neuronal viability.

We identified Hsp70 as a hypothermia-induced protein. Interestingly, we found that Hsp70 was induced by hypothermia after ischemia, although hypothermia by itself did not induce Hsp70 in non-ischemic brains (Terao et al., 2009). Hsp27 was also induced by ischemia, but it was not induced by hypothermia. Other Hsps, including Hsp40, Hsp90, Grp78, Grp94, PDI, and ORP150, were detectable under normal conditions and were not affected by hypothermia or ischemia (Shintani et al., 2011). Recently, Hagiwara et al. (2007) reported that mild hypothermia at 34°C increased the expression level of Hsp70 in LPS-stimulated RAW264.7 cells, although IL-1β, IL-6, and TNF-α expression levels were reduced under the same conditions. In contrast, Tirapelli et al. (2010) recently reported the increase of Hsp70 protein and gene expression in 1 h MCAO rat brains and the role of neuroprotection with hypothermia. However, they showed that the number of Hsp70-positive cells in the ischemic areas was reduced by hypothermia. In our model, we observed the suppression of *Hsp70* mRNA expression by hypothermia, in contrast to the protein level of Hsp70, as observed by Tirapelli et al. (2010). Since we currently do not know the mechanism of induction or the neuroprotective roles of Hsp70 in hypothermia, it is worthy of further investigation.

4. Autophagy

Autophagy is a catabolic process whereby cells respond to energy stress by recycling intracellular components, e.g., proteins, ribosomes, lipids, and even entire organelles (Rabinowitz and White, 2010). In the presence of a sufficient nutrient supply, anabolic reactions predominate within cells, and the autophagy system is maintained at the low levels that are critical for normal cellular homeostasis and survival. Basal levels of autophagic flux are required to degrade long-lived proteins, lipid droplets, and dysfunctional organelles, particularly in post-mitotic cells, e.g., cardiomyocytes and neurons, where the capacity for regeneration is limited. However, autophagy is rapidly activated in response to starvation or is induced either by an inadequate nutrient supply or by defects in growth factor signaling pathways; when cells are exposed to stress, such as starvation and hypoxia, autophagic mechanisms are triggered to liberate energy substrates and eliminate defective organelles. Apart from conditions of emergent nutrient scarcity, the acceleration of autophagy is often observed in other clinically important circumstances, including neurodegenerative disorders, cancer, misfolded protein accumulation, microbial invasion, and cardiovascular diseases (Beau et al., 2011). There is now increasing evidence that, under certain pathological conditions, autophagy is able to trigger and mediate programmed cell death (type II death) (Galluzzi et al., 2008). Such cell death might be involved in the neuronal death observed after global and focal cerebral ischemia (Balduini et al., 2009 and Sheng et al., 2010). Recently, it was reported that delayed neuronal death occurring in the CA1 pyramidal layer of the gerbil hippocampus after ischemia is apoptotic in nature and autophagosomes/autolysosomes are abundant in these neurons before DNA fragmentation (Xu and Zhang, 2011), i.e., under ischemic conditions, autophagy follows the activation of the mitochondrial pathway of apoptosis; cytochrome c is released from mitochondria, and caspase-9/caspase-3 are activated. Turkmen et al. (2011) recently showed that autophagy and autophagic flux are reduced in cold ischemic kidneys treated with bafilomycin A1. Reduced autophagy and autophagic flux were associated with a significant reduction in apoptotic cell death. These results suggest that hypothermia is able to suppress the autophagic phenomena induced by ischemia; however, since the relationship between autophagy and hypothermia is not well understood, detail analyses should be conducted in the future.

5. Conclusion

Hypothermia is a strong therapeutic methodology for delaying and suppressing ischemic damage in the brain. However, as we do not clearly know the precise mechanisms of hypothermia, the technique is still a kind of art and its usage is very limited. To resolve this situation, we should clarify the detailed molecular mechanisms of hypothermia, identify the dominant effectors for neuroprotection, and substitute the function of hypothermia into a therapeutic drug. In this review, we summarized potential drug targets to be considered for developing "hypothermia-like" drugs. At the present time, as we speculate that these targets are still a piece of the whole picture, we should improve our analytical technology and identify critical factors that change the pathophysiological condition induced by ischemia into a hypothermic one.

6. Acknowledgements

The authors express their deep gratitude to Drs. S. Ohkawa, P. Chapman, H. Nagaya, M. Mori, and K. Hirai of the Pharmaceutical Research Division, Takeda Pharmaceutical Company, Ltd., for their stimulating interest and continuing encouragement during the course of this work. They also thank Drs. Y. Kiyota, H. Ohta, Y. Nakagaito, and A. Oda of the Pharmaceutical Research Division, Takeda Pharmaceutical Company, Ltd., who contributed to this project with their elegant techniques.

7. References

K. Akaji, S. Suga, T. Fujino, K. Mayanagi, J. Inamasu, T. Horiguchi, S. Sato, T. Kawase (2003) Effect of intra-ischemic hypothermia on the expression of c-Fos and c-Jun, and

DNA binding activity of AP-1 after focal cerebral ischemia in rat brain, *Brain Res.*, 975: 149-157.

E. Ambrosini, S. Columba-Cabezas, B. Serafini, A. Muscella, F. Aloisi (2003) Astrocytes are the major intracerebral source of macrophage inflammatory protein-3alpha/CCL20 in relapsing experimental autoimmune encephalomyelitis and in vitro, *Glia*, 41, 290–300.

M. Aoki, M. Tamatani, M. Yamaguchi, Y. Bando, L. Kasai, Y. Miyoshi, Y. Nakamura, M.P. Vitek, M. Tohyama, H. Tanaka, H. Sugimoto (2001) Hypothermia treatment restores glucose regulated protein 78 (GRP78) expression in ischemic brain, *Brain Res. Mol. Brain Res.*, 95: 117-128.

W. Balduini, S. Carloni, G. Buonocore (2009) Autophagy in hypoxia-ischemia induced brain injury: evidence and speculations, *Autophagy*, 5: 221-223.

M. Balestrino (1995) Pathophysiology of anoxic depolarization: new findings and a working hypothesis, J. Neurosci. Methods, 59: 99-103.

E. Bandera, M. Botteri, C. Sutton, K.R. Abrams, N. Latronico (2006) Cerebral blood flow threshold of ischemic penumbra and infarct core in acute ischemic stroke: a systematic review, *Stroke*, 37: 1334-1339.

E. Baumann, E. Preston, J. Slinn, D. Stanimirovic (2009) Post-ischemic hypothermia attenuates loss of the vascular basement membrane proteins, agrin and SPARC, and the blood-brain barrier disruption after global cerebral ischemia, *Brain Res.*, 1269: 185-197.

I. Beau, M. Mehrpour, P. Codogno (2011) Autophagosome and human diseases, *Int. Biochem. Cell Biol.*, 43: 460-464.

S.A. Bernard, T.W. Gray, M.D. Buist, B.M. Jones, W. Silvester, G. Gutteridge, K. Smith (2002) Treatment of comatose survivors of out-of-hospital cardiac arrest with induced hypothermia, *N. Engl. J. Med.*, 346: 557–563.

B.R. Broughton, D.C. Reutens, C.G. Sobey (2009) Apoptotic mechanisms after cerebral ischemia, *Stroke*, 40: 331-339.

J. Bureš, O. Burešova, J. Krivănek (1974) Anoxic depolarization. In: *The Mechanisms and Applications of Leao's Spreading Depression of Electroencephalographic Activity*, J. Bureš, O. Burešova, J. Krivănek, 72-86, Academia (Publishing House of the Czechoslovak Academy of Sciences) and Academic Press, Prague and New York.

H.V. Carey, M.T. Andrews, S.L. Martin (2003) Mammalian hibernation: cellular and molecular responses to depressed metabolism and low temperature, *Physiol Rev.*, 83: 1153-81.

A-G. Ceulemans, T. Zgavc, R. Kooijman, S. Hachimi-Idrissi, S. Sarre, Y. Michotte (2010) The dual role of the neuroinflammatory response after ischemic stroke: modulatory effects of hypothermia, *J. Neuroinflamm.*, 7: 74-92.

A-G. Ceulemans, T. Zgavc, R. Kooijman, S. Hachimi-idrissi, S. Sarre, Y. Michotte (2011) Mild hypothermia causes differential, time-dependent changes in cytokine expression and gliosis following endothelin-1-induced transient focal cerebral ischemia, *J. Neuroinflamm.*, 31: 60.

A.S. Charbonnier, N. Kohrgruber, E. Kriehuber, G. Stingl, A. Rot, D. Maurer (1999) Macrophage inflammatory protein 3α is involved in the constitutive trafficking of epidermal langerhans cells. *J. Exp. Med.*, 190: 1755–1768.

T. Cherin, M. Catbagan, S. Treiman, R. Mink (2006) The effect of normothemic and hypothermic hypoxia-ischemia on brain hypoxanthine phosphoribosyl transferase activity, *Neurol. Res.*, 28: 831-836.

S. Chip, A. Zelmer, O.O. Ogunshola, U. Felderhoff-Mueser, C. Nitsch, C. Bűhrer, S. Wellmann (2011) The RNA-binding protein RBM3 is involved in hypothermia induced neuroprotection, *Neurobiol. Dis.*, 43: 388-396.

J.S. Choi, J. Park, K. Suk, C. Moon, Y.K. Park, H.S. Han (2011) Mild hypothermia attenulates intercellular adhesion molecule-1 induction via activation of extracellular signal-regulated kinase-1/2 in a focal cerebral ischemia model, *Stroke Res. Treat.*, 2011: 846716. Epub 2011 Mar 16.

A.M. Chow, I.R. Brown, (2007) Induction of heat shock proteins in differentiated human and rodent neurons by celastrol, *Cell Stress Chaperones*, 12: 237-244

F. Colbourne, S.Y. Grooms, R.S. Zukin, A.M. Buchan, M.W. Bennett (2003) Hypothermia rescues hippocampal CA1 neurons ans attenuates down-regulation of the AMPA receptor GluR2 subunit after forebrain ischemia, *Proc. Natl. Acad. Sci. USA*, 100: 2906-2910.

D.N. Cook, D.M. Prosser, R. Forster, J. Zhang, N.A. Kuklin, S.J. Abbondanzo, X.D. Niu, S.C. Chen, D.J. Manfra, M.T. Wiekowski, L.M. Sullivan, S.R. Smith, H.B. Greenberg, S.K.Narula, M. Lipp, S.A. Lira (2000) CCR6 mediates dendritic cell localization, lymphocyte homeostasis, and immune responses in mucosal tissue, *Immunity*, 12: 495–503.

O. B. Dimitrijevic, S. M. Stamatovic, R. F. Keep, and A. V. Andjelkovic (2006) Effects of the chemokine CCL2 on blood-brain barrier permeability during ischemia-reperfusion injury, *J. Cereb. Blood Flow Metab.*, 26: 797–810.

O. B. Dimitrijevic, S. M. Stamatovic, R. F. Keep, A.V. Andjelkovic (2007) Absence of the chemokine receptor CCR2 protects against cerebral ischemia/reperfusion injury in mice, *Stroke*, 38: 1345–1353.

J.P. Dreier (2011) The role of spreading depression, spreading depolarization and spreading ischemia in neurological disease, *Nat. Med.*, 17: 439-447.

Y. Durandy (2008) Pediatric myocardial protection, *Curr. Opin. Cardiol.*, 23: 85-90.

E. Eberspacher, C. Werner, K. Engelhard, M. Pape, L. Laacke, D. Winner, R. Hollweck, P. Hutzler, E. Kochs (2005) Long-term effects of hypothermia on neuronal cell death and the concentration of apoptotic proteins after incomplete cerebral ischemia and reperfusion in rats, *Acta. Anaesthesiol Scand.*, 49: 477-487.

E. Eberspacher, C. Werner, K. Engelhard, M. Pape, L. A. Gelb, P. Hutzler, J. Henke, E. Kochs (2003) The effect of hypothermia on the expression of the apoptosis-regulating protein Bax after incomplete cerebral ischemia and reperfusion in rats, *J. Neurosurg. Anesthesiol.*, 15: 200-208.

K.U. Freichs, C. Kennedy, L. Sokoloff, J.M. Hallenbeck (1994) Local cerebral blood flow during hibernation, a model of natural tolerance to "cerebral ischemia, *J. Cereb. Blood Flow Metab.*, 14: 193-205.

K.U. Freruchs, J.M. Hallenbeck (1998) Hibernation in ground squirrels induces state and species-specific tolerance to hypoxia and aglycemia: an in vitro study in hippocampal slices, *J. Cereb. Blood Flow Metab.*, 18: 168-75.

J. Fujita (1999) Cold shock response in mammalian cells, *J. Mol. Microbiol. Biotechnol.*, 1: 243-255.

H. Fukuda, T. Tomimatsu, N. Watanabe, J.W. Mu, M. Kohzuki, M. Endo, E. Fujii, T. Kanzaki, Y. Murata (2001) Post-ischemic hypothermia blocks caspase-3 activation in the newborn rat brain after hypoxia-ischemia, *Brain Res.*, 910: 187-191.

O. Fukui, Y. Kinugasa, A. Fukuda, H. Fukuda, E. Tskitishvili, S. Hayashi, M. Song, T. Kanagawa, T. Hosono, K. Shimaya,

M.A. Yenari, H.S. Han (2006) Influence of hypothermia on post-ischemic inflammation: role of nuclear factor kappa B (NfkappaB), *Neurochem. Int.*, 49: 164-169.

O. Fukui, Y. Kinugasa, A. Fukuda, H. Fukuda, E. Tskitishvili, S. Hayashi, M. Song, T. Kanagawa, T. Hosono, K. Shimoya, Y. Murata (2006) Post-ischemic hypothermia reduced IL-18 expression and suppressed microglial activation in the immature brain, *Brain Res.*, 1121: 35-45.

Y. Furuichi, T. Noto, J.Y. Li, T. Oku, M. Ishiye, A. Moriguchi, I. Aramori, N. Matsuoka, S. Mutoh, T. Yanagihara (2004) Multiple modes of action of tacrolimus (FK506) for neuroprotective action on ischemic damage after transient focal cerebral ischemia in rats, *Brain Res.*, 1014: 120–130.

L. Galluzzi, E. Morselli, J.M. Vicencio, O. Kepp, N. Joza, N. Tajeddine, G. Kroemer (2008) Life, death and burial: multifaceted impact of autophagy, *Biochem. Soc. Trans.*, 36: 786-790.

P.F. Ge, T.F. Luo, J.Z. Zhang, D.W. Chen, Y.X. Luan, S.L. Fu (2008) Ischemic preconditioning induces chaperone hsp70 expression and inhibits protein aggregation in the CA1 neurons of rats, *Neurosci. Bull.*, 24: 288-296.

A.R. Green, T. Odergren, T. Ashwood (2003) Animal models of stroke: do they have value for discovering neuroprotective agents? *Trend. Pharmacol. Sci.*, 24: 402–408.

M. Gyngell, E. Busch, B. Schmitz, K. Kohno, T. Back, M. Hoehn-Berlage, K-A. Hossmann (1995) Evolution of acute focal cerebral ischemia in rats observed by localized 1H-MRS, diffusion-weighted MRI, and electrophysiological monitoring, *NMR Biomed.*, 8: 206-214.

A.H. de Haas, H.R. van Weering, E.K. de Jong, H.W. Boddeke, K.P. Biber (2007) Neuronal chemokines: versatile messengers in central nervous system cell interaction, *Mol. Neurobiol.*, 36: 137–151.

S. Hagiwara, H. Iwasaka, S. Matsumoto, T. Noguchi (2007) Changes in cell culture temperature alter release of inflammatory mediators in murine macrophagic RAW264.7 cells, *Inflamm. Res.*, 56: 297–303.

H. S. Han, M. Karabiyikoglu, S. Kelly, R. A. Sobel, M. A. Yenari (2003) Mild hypothermia inhibits nuclear factor-κB translocation in experimental stroke, *J. Cereb. Blood Flow Metab.*, 23: 589-598.

H.S. Han, Y. Qiao, M. Karabiyikoglu, R.G. Giffard, M.A. Yenari (2002) Influence of mild hypothermia on inducible nitric oxide synthase expression and reactive nitrogen production in experimental stroke and inflammation, *J. Neurosci.*, 22: 3921-3928.

K. Harada, T. Maekawa, R. Tsuruta, T. Kaneko, D. Sadamitsu, T. Yamashima, K.K. Yoshida (2002) Hypothermia inhibits translocation of CaM kinase II and PKC-α, β, γ isoforms and fodrin proteolysis in rat brain synaptosome during ischemia-reperfusion, *J. Neurosci. Res.*, 67: 664-669.

M. Holzer (2008) Devices for rapid induction of hypothermia, *Eur. J. Anaesthesiol. Suppl.*, 42: 31-38.

The Hypothermia after Cardiac Arrest Study Group (2002) Mild therapeutic hypothermia to improve the neurological outcome after cardiac arrest, *N. Engl. J. Med.*, 346: 549-556.

W.W. Hu, Y. Du, C. Li, Y.J. Song, G.Y. Zhang (2008) Neuroprotection of hypothermia against neuronal death in rat hippocampus through inhibiting the increased assembly of GluR6-PSD95-MLK3 signaling module induced by cerebral ischemia/reperfusion, *Hippocampus*, 18: 386-397.

J. Inamasu, S. Suga, S. Sato, T. Horiguchi, K. Akaji, K. Mayanagi, T. Kawase (2001) Intra-ischemic hypothermia attenuates intercellular adhesion molecule-1 (ICAM-1) and migration of neutrophil, *Neurol. Res.*, 23: 105-111.

X. Ji, Y. Luo, F. Ling, R.A. Stetler, J. Lan, G Cao, J. Chen (2007) Mild hypothermia diminishes oxidative DNA damage and pro-death signaling events after cerebral ischemia: a mechanism for neuroprotection, 12: 1737-1747.

B. Jieyong, W. Zhong, Z. Shiming, Z. Dai, Y. Kato, T. Kanno, H. Sano (2006) Decompressive craniectomy and mild hypothermia reduces infarction size and counterregulates Bax and Bcl-2 expression after permanent focal ischemia in rats, *Neurosurg. Rev.*, 29: 168-172.

N. Kawai, M. Okauchi, K. Morisaki, S. Nagao (2000) Effects of delayed intraischemic and post-ischemic hypothermia on a focal model of transient cerebral ischemia in rats, *Stroke* 31: 1982–1989.

S. Kelly, D. Cheng, G.K. Steinberg, M.A. Yenari (2005) Mild hypothermia decreases GSK3β expression following global cerebral ischemia, *Neurocrit. Care*, 2: 212-217.

D. Kieran, B. Kalmar, J.R. Dick, J. Riddoch-Contreras, G. Burnstock, L. Greensmith (2004) Treatment with arimoclomol, a coinducer of heat shock proteins, delays disease progression in ALS mice, *Nat. Med.*, 10: 402-405.

H. Y. Kil, J. Zhang, C. A. Piantadosi (1996) Brain temperature alters hydroxyl radical production during cerebral ischemia/reperfusion in rats, *J. Cereb. Blood Flow Metab.*, 16: 100–106.

J.Y. Kim, N. kim, M.A. yenari, W. Chang (2011) Mild hypothermia suppresses calcium-sensing receptor (CaSR) induction following forebrain ischemia while increasing GABA-B receptor 1 (GABA-B-R1) expression, *Transl. Stroke Res.*, 2: 195-201.

Y. Koda, R. Tsuruta, M. Fujita, T. Miyauchi, K. Kaneda, M. Todani, T. Aoki, M. Shitara, T. Izumi, S. Kasaoka, M. Yuasa, T. Maekawa (2010) Moderate hypotheramia suppresses jugular venous superoxide anion radical, oxidative stress, early inflammation, and endothelial injury in forebrain ischemia/reperfusion rats, *Brain Res.*, 1311: 197-205.

P. Laloux (2001) Intravenous rtPA thrombolysis in acute ischemic stroke, *Acta. Neurol. Belg*, 101: 88-95.

C.S. Kidwell, D.S. Kiebeskind, S. Starkman, J.L. Saver (2001) Trends in acute ischemic stroke trials through the 20th century, *Stroke*, 32: 1349–1359.

J.W. Lampe, L.B. Becker (2011) State of art in therapeutic hypothermia, *Annu. Rev. Med.*, 62: 79-93.

S.M. Lee, H. Zhao, C.M. Maier, G.K. Steinberg (2009) The protective effect of early hypothermia on PTEN phosphorylation correlates with free radical inhibition in rat stroke, *J. Cereb. Blood Flow Metab.*, 29: 1589-1600.

J.E. Lee, Y.J. Yoon, M.E. Moseley, M.A. Yenari (2005) Reduction in levels of matrix metalloproteinases and increased expression of tissue inhibitor of metalloproteinase-2 in response to mild hypothermia therapy in experimental stroke, *J. Neurosurg.*, 103: 289-297.

J. Li, S. Benashski, L.D. McCullough (2011) Post-stroke hypothermia provides neuroprotection through inhibition of AMP-activated protein kinase, *J. Neurotrauma.*, 28: 1281-1288.

M. Liebetrau, D. Burggraf, H.K. Martens, M. Pichler, G.F. Hamann (2004) Delayed moderate hypothermia reduces calpain activity and beakdown of its substrate in experimental focal cerebral ischemia in rats, *Neurosci. Lett.*, 357: 17-20.

M.E. Lieonart (2010) A new generation of proto-oncogenes: cold-inducible RNA binding proteins, *Biochim. Biophys. Acta.*, 1805: 43-52.

S. Lin, P.G. Rhodes, Z. Cai (2011) Whole body hypothermia broadens the therapeutic window of intranasally administered IGF-1 in a neonatal rat model of cerebral hypoxia-ischemia, *Brain Res.*, 18: 246-256.

A. Liu, Z. Zhang, A. Li, J. Xue (2010) Effects of hypothermia and cerebral ischemia on cold-inducible RNA-binding protein mRNA expression in rat brain, *Brain Res.*, 1347: 104-110.

L. Liu, J.Y. Kim, M.A. Koike, Y.J. Yoon, X.N. Tang, H. Ma, H. Lee, G.K. Steinberg, J.E. Lee, M.A. Yenari (2008) FasL shedding is reduced by hypothermia in experimental stroke, *J. Neurochem.*, 106: 541-550.

X.C. Lu, A.J. Williams, C. Yao, R. Berti, J.A. Hartings, R. Whipple, M.T. Vahey, R.G. Polavarapu, K.L. Woller, F.C. Tortella, J.R. Dave (2004) Microarray analysis of acute and delayed gene expression profile in rats after focal ischemic brain injury and reperfusion, *J. Neurosci. Res.*, 77: 843–857.

C.M. Maier, G.H. Sun, D. Cheng, M.A. Yenari, P.H. Chan, G.K. Steinberg (2002) Effects of mild hypothermia on superoxide anion production, superoxide dismutase expression, and activity following transient focal cerebral ischemia, *Neurobiol. Dis.*, 11: 28-42.

M.S. Marber, R. Mestril, S.H. Chi, M.R. Sayen, D.M. Yellon, W.H. Dillmann (1995) Overexpression of the rat inducible 70-kD heat stress protein in a transgenic mouse increases the resistance of the heart to ischemic injury, *J. Clin. Invest.*, 95: 1446-1456.

Y. Matsuo, H. Onodera, Y. Shiga, H. Shozuhara, M. Ninomiya, T. Kihara, T. Tamatani, M. Miyasaka, K. Kogure (1994) Role of cell adhesion molecules in brain injury after transient middle cerebral artery occlusion in the rat, *Brain Res.*, 656: 344–352.

P. Meybohm, M. Gruenewald, K.D. Zacharowski, M. Albrecht, R. Lucius, N. Fosel, J. Hensler, K. Zitta, B. Bein (2010) Mild hypothermia alone or in combination with anesthetic post-conditioning reduces expression of inflammatory cytokines in the cerebral cortex of pigs after cardiopulmonary resuscitation, *Crit. Care*, 14: R21.

T. Miyazawa, A. Tamura, S. Fukui, K.A. Hossmann (2003) Effect of mild hypothermia on focal cerebral ischemia. Review of experimental studies, *Neurol. Res.*, 25: 457–464.

P. Moseley (2000) Stress proteins and the immune response, *Immunopharmacology*, 48: 299-302.

Y. Murata (2006) Post-ischemic hypothermia reduced IL-18 expression and suppressed microglial activation in the immature brain, *Brain Res.*, 22: 35-45.

S. Nagel, Y. Su, S. Horstmann, S, Heiland, H. Heiland, H. Gardner, J. Koziol, F.J. Martinez-Torres, S. Wagner (2008) Minocycline and hypothermia for reperfusion injury after focal cerebral ischemia in the rats: effects on BBB breakdown and MMP expression in acute and subacute phase, *Brain Res.*, 1188: 198-206.

H. Nishiyama, H. Higashitsuji, H. Yokoi, K. Itoh, S. Danno, T. Matsuda, J. Fujita (1997) Cloning and characterization of human CIRP (cold-inducible RNA-binding protein) cDNA and chromosomal assignment of the gene, *Gene*, 204: 115-120.

H. Ohta, Y. Terao, Y. Shintani, Y. Kiyota (2007) Therapeutic time window of post-ischemicmild hypothermia and the gene expression associated with the neuroprotection in rat focal cerebral ischemia, *Neuroscience Res.*, 57: 424–433.

N.G. Pabello, S.J. Tracy, A. Snyder-keller, R.W. Keller (2005) Regional expression of constitutive and inducible transcription factors following transient focal ischemia in the neonatal rat: influence of hypothermia, *Brain Res.*, 1038: 11-21.

P.B. Phanithi, Y. Yoshida, A. Santana, M. Su, S. Kawamura, N. Yasui (2000) Mild hypothermia mitigates post-ischemic neuronal death following focal cerebral ischemia in rat brain: immunohistochemical study of Fas, caspase-3 and TUNEL, *Neuropathology*, 20: 105-111.

K. H. Polderman (2004) Application of therapeutic hypothermia in the ICU, *Intensive. Care Med.*, 30: 556-575.

J.D. Rabinowitz, E. White (2010) Autophagy and metabolism, *Science*, 330: 1344-1348.

W. Roste` ne, P. Kitabgi, S.M. Parsadaniantz (2007) Chemokines: a new class of neuromodulator? *Nat. Rev. Neurosci.*, 8: 895–903.

M. Schilling, J-K. Strecker, E.B. Ringelstein, W-R. Schäbitz, R. Kiefer (2009) The role of CC chemokine receptor 2 on microglia activation and blood-borne cell recruitment after transient focal cerebral ischemia in mice, *Brain Res.*, 1289: 79-84.

E. Schutyser, S. Struyf, J. Van Damme (2003) The CC chemokine CCL20 and its receptor CCR6, *Cytokine Growth Factor Rev.*, 14: 409–426.

R. Sheng, L.S. Zhang, R. Han, X.Q. Liu, B. Gao, Z.H. Qin (2010) Autophagy activation is associated with neuroprotection in a rat model of focal cerebral ischemic preconditioning, *Autophagy*, [Epub ahead of print]

T. Shimohata, H. Zhao, J.H. Sung, G. Sun, D. Mochly-Rosen, G.K. Steinberg (2007) Suppression of deltaPKC activation after focal cerebral ischemia contributes to the protective effect of hypothermia, *J. Cereb. Blood Flow Metab.*, 27: 1463-1475.

T. Shimohata, H. Zhao, G.K. Steinberg (2007) Epsilon PKC may contribute to the protective effect of hypothermia in a rat focal cerebral ischemia model, *Stroke*, 38: 375-380.

Y. Shinatni, Y. Terao, H. Ohta (2011) Molecular mechanisms underlying hypothermia-induced neuroprotection, *Stroke Res. Treat.*, 2011: 1-9.

G. G. Somjen (2001) Mechanisms of spreading depression and hypoxic spreading depression-like depolarization, *Physiol. Rev.*, 81: 1065-1096.

K.B. Storey (2003) Mammalian hibernation. Transcriptional and translational controls, *Adv. Exp. Med. Biol.*, 543: 21-38.

Y. Sun, Y.B. Ouyang, L. Xu, A.M. Chow, R. Anderson, J.G. Hecker, R.G. Giffard (2006) The carboxy-terminal domain of inducible Hsp70 protects from ischemic injury in vivo and in vitro, *J. Cereb. Blood Flow Metab.*, 26: 937-950.

T. Tanaka, T. Wakamatsu, H. Daijo, S. Oda, S. Kai, T. Adachi, S. Kizaka-Kondoh, K. Fukuda, K. Hirota (2010) Persisting mild hypothermia suppresses hypoxia-inducible factor-1α protein synthesis and hypoxia-inducible factor-1-mediated gene expression, *Am. J. Physiol. Regul. Integr. Comp. Physiol.*, 298: R661-671.

M. Teilum, M. Krogh, T. Wieloch, G. Mattiasson (2007) Hypothermia affects translocation of numerous cytoplasmic proteins following global cerebral ischemia, *J. proteome Res.*, 6: 2822-2832.

Y. Terao, H. Ohta, A. Oda, Y. Nakagaito, Y. Kiyota, Y. Shintani (2009) Macrophage inflammatory protein-3alpha plays a key role in the inflammatory cascade in rat focal cerebral ischemia, *Neurosci. Res.*, 64: 75-82.

Y. Terao, S. Miyamoto, K. Hirai, H. Kamiguchi, H. Ohta, M. Shimojo, Y. Kiyota, S. Asahi, Y. Sakura, Y. Shintani (2009) Hypothermia enhances heat-shock protein 70 production in ischemic brains, *NeuroReport*, 20: 745–749.

A. Theodorsson, L. Holm, E. Theodorsson (2008) Hypothermia-induced increase in galanin concentrations and ischemic neuroprotection in the rat brain, *Neuropeptides*, 42: 79-87.

D.P. Tirapelli, C.G. Carlotti, J.P. Leite, L.F. Tirapelli, B.O. Colli (2010) Expression of HSP70 in cerebral ischemia and neuroprotective action of hypothermia and ketoprofen, *Arq. Neuropsiquiatr.*, 68: 592-596.

T. Tomimatsu, H. Fukuda, M. Endo, N. Watanabe, J. Mu, M. Kohzuki, E. Fujii, T. Kanzaki, Y. Murata (2001) Effects of hypothermia on neonatal hypoxic-ischemic brain injury in the rat: phosphorylation of Akt, activation of caspase-3-like protease, *Neurosci. Lett.*, 312: 21-24.

K. Turkmen, J. Martin, A. Akcay, Q. Nguyen, K. Ravichandran, S. Faubel, A. Pacic, D. Ljubanović, C.L. Edelstein, A. Jani (2011) Apoptosis and autophagy in cold preservation ischemia, *Transplantation*, 91: 1192-1197.

U. Utans-Schneitz, H. Lorez, W.E. Klinkert, J. Da Silva, W. Lesslauer (1998) A Novel rat CC chemokine, identified by targeted differential display, is upregulated in brain inflammation, *J. Neuroimmunol.*, 92: 179–190.

K.K. Wang (2000) Calpain and caspase: can you tell the difference? *Trends Neurosci.*, 23: 20–26.

G.J. Wang, H.Y. Deng, C.M. Maier, G.H. Sun, M.A. Yenari (2002) Mild hypothermia reduces ICAM-1 expression, neutrophil infiltration and microglia/monocyte accumulation following experimental stroke, *Neuroscience,* 114: 1081–1090.

C.M. Webster, S. Kelly, M.A. Koike, V.Y. Chock, R.G. Giffard, M.A. Yenari (2008) Inflammation and NFκB activation is decreased by hypothermia following global cerebral ischemia, *Neurobiol. Dis.*, 33: 301-312.

P.L. Wood (1995) Microglia as a unique cellular target in the treatment of stroke: potential neurotoxic mediators produced by activated microglia, *Neurol. Res.*, 17: 242–248.

Z. Xiang, S. Thomas, G. Pasinetti (2007) Increased neuronal injury in transgenic mice with neuronal overexpression of human cyclooxygenase-2 is reversed by hypothermia and rofecoxib treatment, *Curr. Neurovasc. Res.*, 4: 274-279.

Y.C. Xie, C.Y. Li, T. Li, D.Y. Nie, F. Ye (2007) Effect of mild hypothermia on angiogenesis in rats with focal cerebral ischemia, *Neurosci. Lett.*, 422: 87-90.

M. Xu, H.L. Zhang (2011) Death and survival of neuronal and astrocytic cells in ischemic brain injury: a role of autophagy, *Acta. Pharmacol.*, doi: 10. 1038/aps.2011.50.

J.H. Xue, K. Nonoguchi, M. Fukumoto, T. Sato, H. Nishiyama, H. Higashitsuji, K. Itoh, J. Fujita (1999) Effects of ischemia and H2O2 on the cold stress protein CIRP expression in rat neuronal cells, *Free Radic. Biol. Med.,* 27: 1238-1244.

Y. Yanagawa, M. Kawakami, Y. Okada (2002) Moderate hypothermia alters interleukin-6 and interleukin-1α reactions in ischemic brain in mice, *Resuscitation*, 53: 93-99.

H. Yanamoto, I. Nagata, Y. Niitsu, Z. Zhang, J.H. Xue, N. Sakai, H. Kikuchi (2001) Prolonged mild hypothermia therapy protects the brain against permanent focal ischemia, *Stroke,* 32: 232–239.

D. Yang, S. Guo, T. Zhang, H. Li (2009) Hypothermia attenuates ischemia/reperfusion-induced endothelial cell apoptosis via alterations in apoptotic pathways and JNK signaling, *FEBS Lett.*, 583: 2500-2506.

M.A. Yenari, T.M. Hemmen (2010) Therapeutic hypothermia for brain ischemia: where have we come and where do we go?, *Stroke*, 41: S72-74.

M.A. Yenari, S. Iwayama, D. Cheng, G.H. Sun, M. Fujimura, Y. Morita-Fujimura, P.H. Chan, G.K. Steinberg, (2002) Mild hypothermia attenuates cytochrome c release but does not alter Bcl-2 expression or caspase activation after experimental stroke, *J. Cereb. Blood Flow Metab.*, 22: 29-38.

M.A. Yenari, H.S. Han (2006) Influence of hypothermia on post-ischemic inflammation: role of NFκB, *Neurochem. Int.*, 49: 164-169.

H. Yoshida, H. Yanai, Y. Namiki, K. Fukatsu-Sasaki, N. Furutani, N. Tada (2006) Neuroprotective effects of edaravone: a novel free radical scavenger in cerebrovascular injury, *CNS Drug Reviews*, 12: 9–20.

Z. Zhang, R.A. Sobel, D. Cheng, G.K. Steinberg, M.A. Yenari (2001) Mild hypothermia increases Bcl-2 protein expression following global cerebral ischemia, *Brain Res. Mol. Brain Res.*, 95: 75-85.

H. Zhang, C. Ren, X. Gao, T. Takahashi, R.M. Sapolsky, G.K. Steinberg, H. Zhao (2008) Hypothermia blocks β-catenin degradation after focal ischemia in rats, *Brain Res.*, 1198: 182-187.

H. Zhang, G. Xu, J. Zhang, S. Murong, Y. Mei, E. Tong (2010) Mild hypothermia reduces ischemic neuron death via altering the expression of p53 and bcl-2, *Neurol. Res.*, 32: 384-389.

H. Zhao, T. Shimohata, J.Q. Wang, G. Sun, D.W. Schaal, R.M. Sapolsky, G.K. Steinberg (2005) Akt contributes to neuroprotection by hypothermia against cerebral ischemia in rats, *J. Neurosci.*, 25: 9794-9805.

H. Zhao, M.A. Yenari, D. Cheng, R.M. Sapolsky, G.K. Steinberg (2005) Moderate hypothermia (30 degree C) for surgery of acute type A aortic dissection, *J. Cereb. Blood Flow Metab.*, 25: 1119-1129.

H. Zhao, M.A. Yenari, D. Cheng, R.M. Sapolsky, G.K. Steinberg (2004) Mild postischemic hypothermia prolongs the time window for gene therapy by inhibiting cytochrome C release, *Stroke*, 35: 572-577.

H. Zhao, J.Q. Wang, T. Shimohata, G. Sun, M.A. Yenari, R.M. Sapolsky, G.K. Steinberg (2007) Conditions of protection by hypothermia and effects on apoptotic pathways in a rat model of permanent middle cerebral artery occlusion, *J. Neurosurg.*, 107: 636-641.

Z. Zheng, J.Y. Kim, H. Ma, J.E. Lee, M.A. Yenari (2008) Anti-inflammatory effects of the 70 kDa heat shock protein in experimental stroke, *J. Cereb. Blood Flow Metab.*, 28: 53-63.

Z. Zheng, M. A. Yenari (2004) Post-ischemic inflammation: molecular mechanisms and therapeutic implications, *Neurological Res.*, 26: 884–892.

C. Zhu, X. Wang, F. Xu, L. Qiu, X. Cheng, G. Simbruner, K. Blomgren (2006) Intraischemic mild hypothermia prevents neuronal cell death and tissue loss after neonatal cerebral hypoxia-ischemia, *Eur. J. Neurosci.*, 23: 387-393.

M. Zoli, R. Grimaldi, R. Ferrari, I. Zini, L.F. Agnati (1997) Short- and long-term changes in striatal neurons and astroglia after transient forebrain ischemia in rats, *Stroke*, 28: 1049–1059.

Part 2

Brain Regeneration After Stroke: Spontaneous Events and Stem Cells Therapy

5

Cortical Neurogensis in Adult Brains After Focal Cerebral Ischemia

Weigang Gu[1,2] and Per Wester[1]
[1]*Umeå Stroke Center, Department of Public Health and Clinical Medicine, Medicine, University of Umeå,*
[2]*Department of Clinical Neuroscience and Neurology, University of Umeå, Umeå, Sweden*

1. Introduction

Stroke ranks as the most common reason to disable adult patients and the second most common reason for mortality. Current treatment of ischemic stroke through trombolysis or trombectomy aims to initiate successful early reperfusion into the ischemic penumbral tissue. If it is promptly and properly performed, it may reverse the ischemic cascade and rescue the ischemic penumbra from being further recruited into infarct and thus improve neurological outcome. Once the infarct is formed, no treatment is currently available to enhance the post stroke brain repair. Under physiological conditions, neurons in the cerebral cortex are terminally differentiated shortly after birth. The phenomenon that a cavity usually forms in the post stroke adult brain elicits the histological postulation that adult brains do not have the capacity to generate new neurons in the cerebral cortex after pathological insults i.e., stroke. Nevertheless, neurological recovery of various degrees is commonly seen in stroke patients with clinical improvement that starts from a week after stroke onset and may last up to 18 months. The underlying mechanisms for this recovery are only sparsely understood, although many factors have been suggested such as de-afferentiating, activity-dependent synaptic changes, altered membrane excitability, and outgrowth of axons and dendrites.

We have been interested in the post stroke brain repair through cell regeneration in the adult brains. One animal model being used for this purpose is the photothrombotic ring stroke in adult rats (Gu et al. 1999a; Hu et al. 1999; Hu et al. 2001; Wester et al. 1995) and mice (Jiang et al. 2006). This stroke model features a large anatomically predefined cortical penumbra (spatially confined by a ring-shaped ischemic locus) in the somatosensory cortex that undergoes critical hypoperfusion and subsequently spontaneous reperfusion (Gu et al. 1999a). Quick induction of immediate early genes in the penumbral cortex (Johansson et al. 2000) is followed by neuronal necrosis and apoptosis with progressively altered neuropil and nerve cell morphology that reach their maximum severity at 48 hours after stroke onset (Gu et al. 1999b; Hu et al. 2002). With a spontaneous reperfusion into the penumbral cortex starting at 72 hours, a remarkable morphologic restoration of the nerve cells starts, that evolves into a chiefly unremarkable cytologic appearance at 7 to 28 days after stroke induction (Gu et al. 1999b). To investigate the mechanism behind the dramatic morphological recovery in the reperfused penumbral cortex, we examine the cell proliferation process in the post stroke cerebral cortex by in vivo delivery of a DNA duplication marker 5-bromodeoxy-uridine (BrdU) into the post stroke animals that is detected by single/double immunohistochemistry / immunofluorescence (Gu et al. 2000). The thymidine analogue BrdU is incorporated into cell DNA during the S-phase of a cell cycle, which can be detected immunohistochemically. BrdU cell labelling is currently a standard method to identify cell proliferation. Surprisingly, widespread BrdU-incorporated cells are consistently observed in the penumbral cortex at 48h and 72h after photothrombotic ring stroke. While the majority of these BrdU-immunopositive cells are proliferating astrocytes and macrophages, 3% to 6% of them are double-immunolabelled by BrdU and one of the neuron-specific marker Map-2 or beta-tubulin III at 7 and 100 days after stroke onset. Three dimension confocal analyses show colocalization of the neuron-specific marker NeuN and the BrdU in the same cells, suggesting stroke induced cortical neurogenesis in the adult cerebral cortex (Gu et al. 2000). To examine the generalizability of this novel finding, the rat model of reversible focal cerebral ischemia by middle cerebral artery (MCA) suture occlusion is also studied (Jiang et al. 2001). In this stroke model, reperfusion is induced mechanically through a withdrawal of the suture at 2h after MCA occlusion. Similarly with our previous findings, widespread BrdU single-immunopositive cells appear in the postischemic cerebral cortex, corpus callosum, striatum and dentate gyrus of the hippocampus ipsilateral to the ischemic infarct at 30 and 60 days after MCA occlusion. Approximately 6-10% of the BrdU positive cells are double immunopositive to one of the neuron markers Map-2, beta-tubulin III or NeuN in the penumbral cortex, indicating a long time survival of the newborn neurons in the post stroke cortex (Jiang et al. 2001).

To trace the origin of the post stroke newborn neurons in the photothrombotic ring stroke model in adult rats (Gu et al. 2009), BrdU is repeatedly injected. Brain sections are collected at different time points after stroke induction and examined by BrdU immunohistochemistry so that the initial spatial appearance of the proliferating cells and their possible migration inside the brains can be evaluated. To detect ongoing cell mitosis or cell death, the M-phase specific marker phosphorylated histone H_3 (Phos H_3) and the spindle components α-tubulin/γ-tubulin are examined by double immunofluorescence with the DNA duplication marker BrdU or nuclear apoptosis marker TUNEL. Cell type is ascertained by double immunolabeling with the neuronal markers Map-2ab/β-tubulin III and NeuN/Hu or the astrocyte marker GFAP. From 16h poststroke, BrdU-immunolabeled cells appear initially in the penumbral cortex. From 24h after stroke induction, Phos H_3 starts to be expressed in the penumbral cortex that is colocalized with BrdU in the same nuclei. Meanwhile, mitotic spindles immunolabeled by α-tubulin/γ-tubulin appear inside the cortical cells containing BrdU-immunopositive nuclei. Unexpectedly, the markers of neuronal differentiation, Map-2ab/β-tubulin III/NeuN/Hu, are expressed in the Phos H_3-immunolabeled cells, and NeuN is detected in some cells containing spindles. These date suggest that endogenous cells with neuronal immunolabeling may duplicate their nuclear DNA and commit cell mitosis to generate daughter neurons in the penumbral cortex, which contribute to the very early cortical neurogenesis in this stroke model in adult rats (Gu et al. 2009). This is in contrast with the reports of cortical neurogenesis after MCA occlusion where neural progenitor cells migrate from subventricular zone into post stroke cortex at 7-14 days after MCA occlusion that contributes to a late stage post stroke neurogenesis (Kreuzberg et al. 2010; Tsai et al. 2006). Therefore, stroke induced cortical neurogenesis may originate either from remote progenitor cell migration or in situ cell division depending on different stroke models used and various phases studied.

The functional status of the observed post stroke neurogenesis is one of our major interests. To elucidate a conceivable functional capacity, neurotransmitter synthesis is studied in the post ischemic rat brains subjected to photothrombotic ring stroke and subsequent BrdU delivery (Gu et al. 2010). In order to detect a possible synthesis of neurotransmitters in the newborn cortical neurons, single/double/triple immunofluorescence cell labelling is performed with the neurotransmitter marker acetycholine (Ach) and its substrate enzyme choline acetyltransferase (ChAT), the neurotransmitter marker GABA and the substrate enzyme glutamic acid decarboxylase (GAD) and BrdU. Among the BrdU-immunolabeled newborn cells at 48h, 5 days, 7 days, 30 days, 60 days and 90 days after stroke, some of these are doubly immunopositive to the cholinergic neuron-specific marker ChAT or GABAergic neuron-specific marker GAD. As analyzed by 3-D confocal microscopy, the neurotransmitters Ach and GABA are colocalized with BrdU in the same cortical cells. In order to confirm the neuronal identity of these neurotransmitter synthesizing newborn cells, NeuN, BrdU and GABA triple immunofluorescence is performed. Under 3-D confocal analyses, the BrdU-immunolabeled newborn cell which is synthesizing GABA is further triple-immunolabeled by NeuN. These data suggest that the newborn neurons are capable of synthesizing the neurotransmitters acetylcholine and GABA in the penumbral cortex (Gu et al. 2010), which is one of the fundamental requisites for these neurons to function during the poststroke recovery.

Stroke is the most common reason to handicap humans in the adult life and the second most common reason for clinical mortality. Post stroke patient care is one of the largest economical burdens in modern society. About 85% of clinical stroke is ischemic origin. A sudden occlusion or severe stenosis of a large or small cerebral artery leads to quick decrease of local cerebral blood flow in the brain tissue down to two critical penumbral thresholds, i.e., electrophysiological and membrane thresholds that differs the ischemic penumbra from ischemic core. If not promptly treated, ischemic penumbra is quickly recruited into the ischemic core and paninfarct of the brain tissue ensues. Aphasia, hemiplegia, and hemianopia are among the most common neurological deficits that handicap stroke patients. Intravenous thrombolysis within the 4.5 hours time window and thrombectomy within the 6 hours time window in MCA or vertebrobasilar stroke are the current treatment for acute ischemic stroke. Prompt recanalization may be achieved in some of the stroke patients that is associated with better clinical outcome. Once a cerebral infarct is formed, no specific treatment is currently available to enhance the poststroke brain repair.

Historically neurons were believed not to be regenerated in the adult brains (Ramón y Cajal 1913). In 1960s, Altman and Das reported neurogenesis in the hippocampus in post natal rats (Altman and Das 1965a). Through [^{3}H]-thymidine DNA labelling and histology, they found that some cells that were [^{3}H]-thymidine labelled in their nuclei had neuron morphological appearance in the postnatal hippocampus. Using the same DNA labelling technique, Altman reported further that neurogenesis persists in olfactory bulbs (Altman 1969). Proliferation and migration of neural stem cell from subventricular zone along a rostral migrating stream into olfactory bulbs was traced to be the cell resource for a life time continuous neurogenesis in the olfactory bulbs (Altman and Das 1965b). Proliferation of neural stem cells in the subgranular layer and their migration into granule layer contributed to the continuous neurogenesis in the dentate gyrus of the hippocampus (Altman and Bayer 1990). In recent years neurogenesis in the dentate gyrus of hippocampus and olfactory bulbs has been extensively studied in different species including humans. Neurogenesis in these brain regions becomes enhanced or inhibited under different physiological and pathological conditions such as physical exercise, learning, stress, depression, drugs and stroke (Brown et al. 2003; Eriksson et al. 1998; Kempermann et al. 1997; Liu et al. 1998; Samuels and Hen. 2011; van Praag et al. 1999).

Ischemic stroke within the middle cerebral artery territory (MCA) is the most common form of stroke in humans. With a MCA stroke, tissue infarct occurs in part of cerebral cortex, striatum, or both. Aphasia, hemiplegia, apraxia, and hemineglect are among the most common neurological deficits to handicap the patients. Under physiological conditions, neurons in the cerebral cortex are terminally differentiated in the adult brains (Rakic 1985). Several weeks after ischemic stroke, a cavity usually appears in the infarct brain region, suggesting an inadequate capacity of the adult brain to compensate the cell loss through regeneration of neurons and glial cells. Nevertheless, spontaneous neurological improvement of various degrees occurs in the majority of poststroke patients. It starts from days after stroke, maximizes in the first 3 months, and may last up to 18 months (Hankey et al. 2007; Twitchell 1951). The pathophysiological mechanisms responsible for the poststroke spontaneous neurological improvement are sparsely understood. Emission of ischemic edema, spontaneous reperfusion of the ischemic penumbra, reversal of ischemic neuronal damage, sprouting of collateral axons and local dendrites, unmasking of potential neuronal pathways, induction of activity dependent synaptic activity, altered membrane excitability have been proposed to be the potential pathways.

In order to meet various purposes of experimental stroke studies, different animal models of focal cerebral ischemia have been established. One of the experimental stroke models being used is photothrombotic stroke in adult rodents. Hence, reproducible thrombosis can be induced photochemically in the cortex of adult rodents, wherein the ischemic lesion may be placed in any desirable location (Watson et al. 1985). In the photothrombotic "ring" version of this model, a large anatomically predefined cortical penumbra is induced in the somatosensory cortex of adult rats (Wester et al. 1995) within the ischemic annulus of the ring. The interior of the ring annulus is concentrically encroached by the radially expanding ring annulus, thus simulating penumbral stroke-in-evolution in an inherently reproducible fashion (Fig. 1). Microvascular platelet thrombi appear in the cortical lesion (Gu et al. 1999a; Wester et al. 1995), reminiscent of clinical thromboembolic stroke. Delayed, but consistent, spontaneous reperfusion of the cortical penumbra can be presaged in the ring model by manipulation of the irradiating laser beam intensity (Fig. 1) (Gu et al. 1999a), by which local cerebral blood flow in the penumbra first decreases to 59, 34, 26, and 33% of baseline values at 1, 2, 24, and 48h after ischemia and then gradually recovers to 56 and 87% of baseline values at 72 and 96h (Gu et al. 1999c).

Dramatic changes in tissue morphology take place in the cortical penumbra (Fig. 2). Neuronal necrosis and apoptosis with a progressively altered neuropil prevail at 24–48h post ischemia (Gu et al. 1999b; Hu et al. 2002), reaching their maximum severity at 48h after stroke with most of the neurons exhibiting eosinophilia and pyknosis (Gu et al. 1999b). Meanwhile, VEGF-mediated angiogenesis is initiated in the same penumbral cortex to facilitate a late spontaneous reperfusion (Gu et al. 2001). At 72h after ischemia, a remarkable morphologic restoration of the nerve cells in the penumbra cortex starts, which evolves into a chiefly unremarkable cytologic appearance at 7 to 28 days after stroke induction (Gu et al. 1999b).

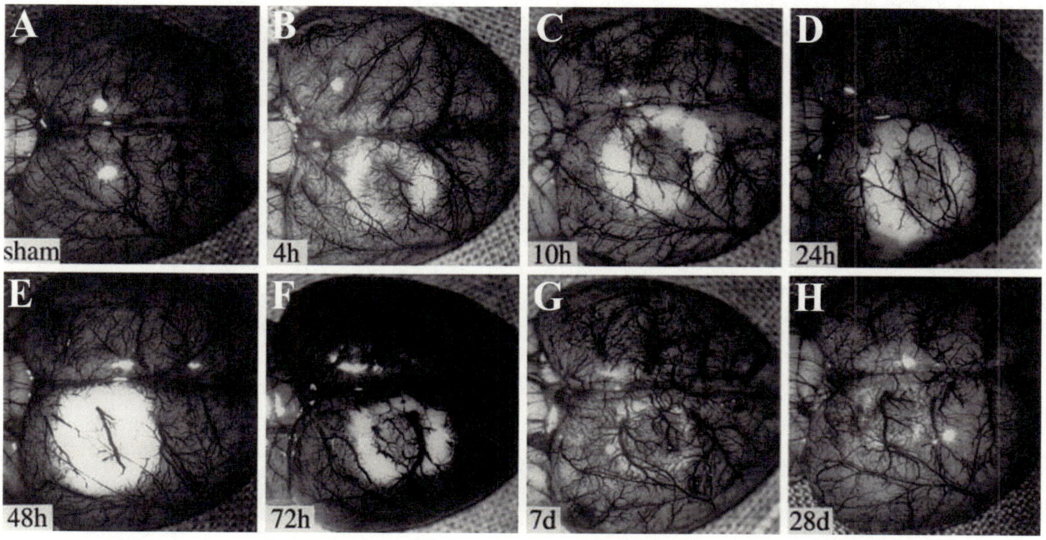

Fig. 1. Photograph of brain samples from post stroke (B-H) and sham-operated (A) rats that are transcardially perfused with carbon-black. A ring-shaped cortical-perfusion deficit is consistently observed on the surface of somatosensory cortex at 4h post ischemia (B). It progressively increases at 10h (C), 24h (D), and reaches its maximum at 48h (E). Thus, the centrally located penumbral cortex enclosed by the ring-shaped ischemic locus looks pale, with a branch of the distal middle cerebral artery being narrowed, but not completely occluded. At 72h postischemia (F), the distal MCA with its small branches in the penumbral cortex becomes patent, representing reperfusion. It becomes more pronounced at 7 days (G) and 28 days (H) after stroke induction. (From Gu W, Jiang W, Wester P., Exp Brain Res 125:163-170, 1999. With permission from Springer-Verlag Heidelberg)

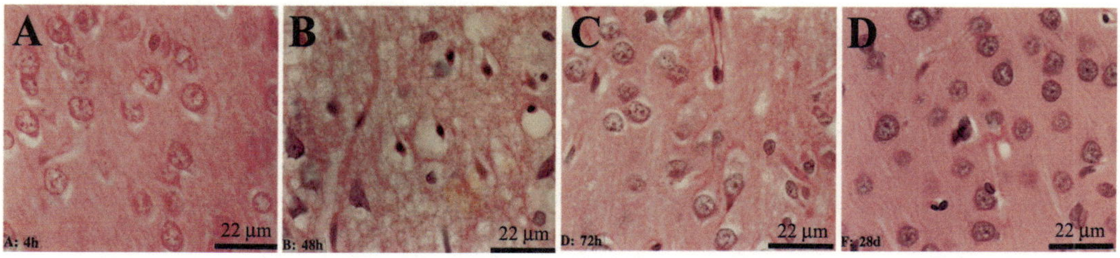

Fig. 2. Hematoxylin and Eosin (HE) staining of the coronal brain sections through the epicenter of the ring lesion showing tissue morphology in the central penumbral cortex. (A) At 4h post stroke induction, the neuropil is slightly pale and the nerve cells are mildly swollen. (B) At 48h after stroke, the majority of the nerve cells are eosinophilic in their cytoplasm and pyknotic and hyperchromatic in their nuclei. The neuropil is pale and edematous. (C) At 72h after stroke, dramatic recovery of tissue morphology is seen in the penumbral cortex, where only mild cytological changes, e.g., somal and nuclear swelling are observed. (D) At 28 days after ischemia, tissue morphology becomes unremarkable. (From Gu el al., Exp Brain Res 125:171-183, 1999. With permission from Springer-Verlag Heidelberg)

To explain this remarkable morphologic restoration, we assumed that cell proliferation might have occurred in the reperfused cortical penumbra in this stroke model (Gu et al. 2000). To test this hypothesis, the cell proliferation-specific marker 5-bromodeoxyuridine is delivered into the poststroke rats through repeated intraperitoneal injections (Gu et al. 2000). As a thymidine analogue, BrdU is incorporated into the cell nuclei when proliferating cells duplicate their DNA during the S-phase, which can be detected immunohistochemically (Gratzner 1982; Miller and Nowakowski 1988). After each injection, BrdU lasts for about two hours for cell uptake (Nowakowski et al. 1989). Repeated BrdU injection maximizes the chance for brain cells that proliferate at different times post stroke to incorporate BrdU into their nuclei. To minimize the possibility of any potential neurotoxic effect of BrdU to the ischemic penumbral tissue, a relative low dose of BrdU injections is used, i.e., 10 mg/kg at each delivery as compared with 75 to 120 mg/kg used elsewhere (Craig et al. 1999). To follow up a possible long-term survival of the newborn cells in the post stroke cortex, the animals are sacrificed either on the same day after the last BrdU delivery or on day 14 or on day 20 after the last BrdU injection.

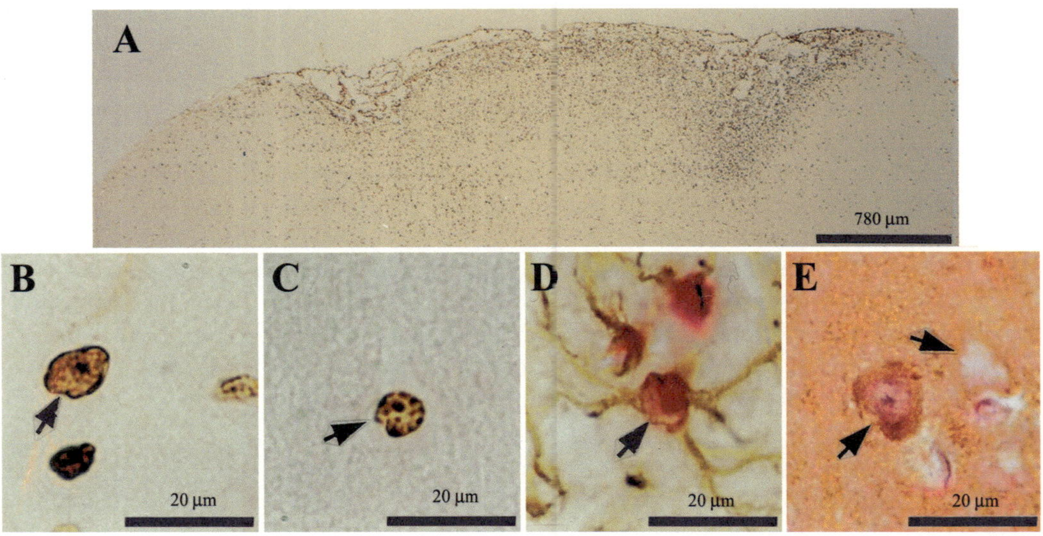

Fig. 3. BrdU immunohistochemistry in rat brains after photothrombotic ring stroke (A) BrdU single immunohistochemistry at 7 days after photothrombotic ring stroke. Widespread BrdU-immunolabeled cells (brown) are observed in the cerebral cortex near the two wedge-shaped infarct cores and the penumbral cerebral cortex that are lying between. (B) A BrdU-immunolabeled cell (arrow, brown) in the penumbral cortex exhibits a large round cell nucleus with a single nucleolus in the center at 7 days after stroke. (C) A BrdU-immunolabeled cell (arrow, brown) in the penumbral cortex exhibits a large round cell nucleus with a single nucleolus in the center at 100 days after stroke. (D) A newborn astrocyte in the penumbral cortex is BrdU-immunolabeled in the nucleus (arrow, red) and GFAP- immunopositive in the cytoplasm (brown) at 7 days post ischemia. (E) A newborn cell in cortical layer II in the penumbral cortex evinces a BrdU-immunolabeled cell nucleus (arrow left, red) and Map-2 immunopositive cytoplasm (brown) at 100 days after stroke induction. (From Gu W, Brannstrom T, Wester P., J Cereb Blood Flow Metab 20:1166-1173, 2000. With permission from Nature Publishing Group)

Detected through BrdU immunohistochemistry, widespread BrdU-incorporated cells are consistently observed in the penumbral cortex and the ischemic core at 7 days after stroke induction (Fig. 3A). At 100 days after ischemia, these cells are still seen, though the BrdU nuclear labeling is generally faded as compared with that observed at 7 days post ischemia. In the penumbral cortex, the majority of the BrdU incorporated cells at 7 days post ischemia represent glial cells (62 ± 9.5% of the total 113 ± 34 BrdU single labeled cells counted per brain), macrophages (20 ± 9.7%), and endothelial cells (12 ± 3.7%). The corresponding proportions at 100 days after stroke are 84 ± 2.1% for glial cells (out of a total 82 ± 30 BrdU single-labeled cells counted per brain), 3.2 ± 3.9% for macrophages, and 8.8 ± 5.1% for endothelial cells. However, some of the BrdU-immunolabeled cell nuclei exhibit neuronal morphologic characteristics, that is, a large round nucleus with a single nucleolus in the center (Fig. 3B-C). To identify the cell lineage of these newborn cortical cells, the neuron-specific marker Map-2ab and the astrocyte specific marker GFAP are employed for double immunohistochemistry with BrdU (Fig. 3D-E). Among the dominant newborn astrocytes (Fig. 3D) in the post stroke penumbral cortex , we are surprised to see that some cells that are BrdU-immunolabeled in their nuclei are further Map-2-immunopositive in their cytoplasm (Fig. 3E), which suggests that these newborn cells are neurons (Gu et al. 2000). These cells are scattered randomly in cortical layers II-VI in the penumbral cortex and count for 3.3 ± 0.3% of the total 1405 ± 108 BrdU single-labeled cells per brain counted at 7 days, and 5.8 ± 1.4% of the total 745 ± 95 BrdU-positive cells counted per brain at 100 days after stroke induction.

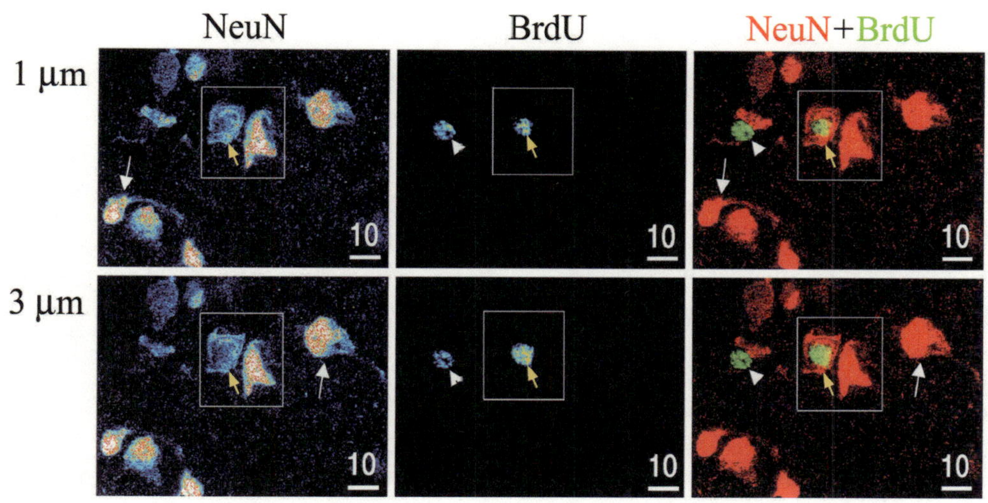

Fig. 4. 3D confocal analyses of NeuN and BrdU double immunofluorescence in the penumbral cortex at two adjacent Z-series planes at 30 days after photothrombotic ring stroke. The left column shows the signal intensity for NeuN, the middle column displays the signal intensity for BrdU, and the right column exhibits a merged image of the NeuN (red) and BrdU (green). The cell designated by the yellow arrows in square is both NeuN-immunopositive (left) and BrdU-immunopositive (middle), resulting in a yellow NeuN and BrdU double-immunopositive nucleus in the merged image (right). White arrowheads indicate a neighbouring NeuN-negative but BrdU-positive nonneuronal newborn cell. White arrows outside the square are pointed at two NeuN-immunopositive but BrdU-negative mature neurons. (From Gu W, Brannstrom T, Wester P., J Cereb Blood Flow Metab 20:1166-1173, 2000. With permission from Nature Publishing Group)

To confirm the co-localization of BrdU with a neuron-specific marker within the same cortical cells after stroke, three-dimensional confocal analyses of BrdU and NeuN double immunofluorescence is conducted (Fig. 4). At 3 days and 30 days after stroke induction, the neuron-specific marker NeuN is colocalized with BrdU in the same cortical cells under 3-D confocal analyses. These data suggest an occurrence of neurogenesis in the penumbral cerebral cortex in adult rats after photothrombotic ring stroke (Gu et al. 2000).

An obvious question raised about this finding is whether or not this phenomenon is stroke model specific. In other word, can cortical neurogenesis be found in any other stroke models? To explore the generality of poststroke cortical neurogenesis, cell regeneration is further explored in adult rats subjected to unilateral MCA suture occlusion (Jiang et al. 2001)- an animal model being widely used in stroke research (Longa et al. 1989; Memezawa et al. 1992). To induce the reperfusion, the intraluminal filament is withdrawn at 2 hours after MCA occlusion. BrdU is injected with a similar schedule as the previous study. Brain samples are collected at 30 days and 60 days after stroke induction.

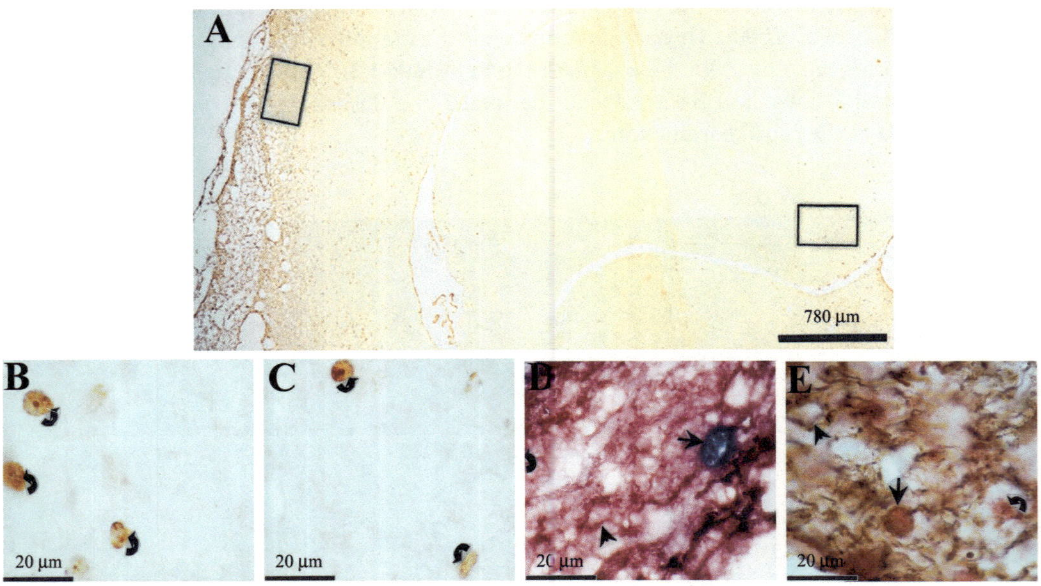

Fig. 5. BrdU immunohistochemistry in rat brains after MCA occlusion (A) BrdU single immunohistochemistry of coronal brain section at 30 days after stroke. Widespread BrdU-immunolabeled cells (brown) are observed in the ipsilateral cortex, with densest distribution in the boundary zone close to the pannecrotic region. BrdU-positive cells are also seen in the corpus callosum and the dentate gyrus of the hippocampus. (B) High magnification photograph of the periinfarct cortex from the left square in A, showing that some cells in the cortical layer II have a large round BrdU-immunolabeled cell nucleus with a single nucleolus in the center (arrow, brown). (C) High magnification photography of the dentate gyrus of the hippocampus from the right square in A, showing that some cells are BrdU-immunopositive in their nuclei (brown, arrow). (D) BrdU and Map-2 double immunohistochemistry at 30 days after ischemia. In a cell in the cortical layer IV, the BrdU-immunolabeled nucleus (arrow, blue) is surrounded by a Map-2–immunopositive cytoplasm (purple) with dendritic processes. The arrowhead shows a Map-2–immunopositive mature neuron. (E) BrdU and β-tubulin III double

immunohistochemistry at 30 days after stroke. A cell in the cortical layer V has its BrdU-immunolabeled nucleus (arrow, red) surrounded by β-tubulin III-immunopositive cytoplasm (brown) with an extending neurite. (From Jiang et al., Stroke 32:1201-1207, 2001. With permission from Wolters Kluwer Health)

Widespread BrdU single-immunopositive cells are observed in the post ischemic cerebral cortex, striatum, corpus callosum, and dentate gyrus of the hippocampus ipsilateral to the ischemic infarct at 30 and 60 days after stroke induction (Fig. 5A). Some of them exhibit neuronal morphologic characteristics, i.e., a large round nucleus with a single nucleolus in the center (Fig. 5B). Three different neuron-specific markers, Map-2, β-tubulin III, and NeuN, are used in conjunction with BrdU to perform double-labeling immunohistochemistry. The cells doubly labelled by BrdU and Map-2 (Fig. 5D), β-tubulin III (Fig. 5E), or NeuN are randomly distributed through cortical layers II through VI, at higher density in the peri-infarct regions than in remote cortical regions. Some BrdU-immunopositive cells, doubly labelled by Map-2, β-tubulin III, or NeuN, are also found in the striatum close to the ischemic lesion. These cells varied in shapes and sizes and often have one or more recognizable Map-2– or β-tubulin III-immunopositive dendrites extending from their cell bodies. They count for approximately 6% to 10% of the total BrdU-immunopositive cells in the penumbral cortex. In 3-D confocal analysis, co-localization of BrdU and NeuN immunofluorescence is detected in the cortical cells at 30 days (Fig. 6) and 60 days after stroke induction. In these cells, the intense BrdU immunofluorescent signal in the cell nuclei is completely merged with the nuclear NeuN-immunopoitive signal, from which the NeuN -immunopositive proximal dendrites are extended (Fig. 6). Therefore, neurogenesis occurs also in the penumbral cortex and striatum in adult rats after MCA occlusion (Jiang et al. 2001). In agreement with this study, cortical neurogenesis has been reported by several studies in MCA stroke in adult rats and mice (Chen et al. 2004; Jiang et al. 2001; Jin et al. 2003; Kreuzberg et al. 2010; Leker et al. 2007; Ohab et al. 2006; Tsai et al. 2006).

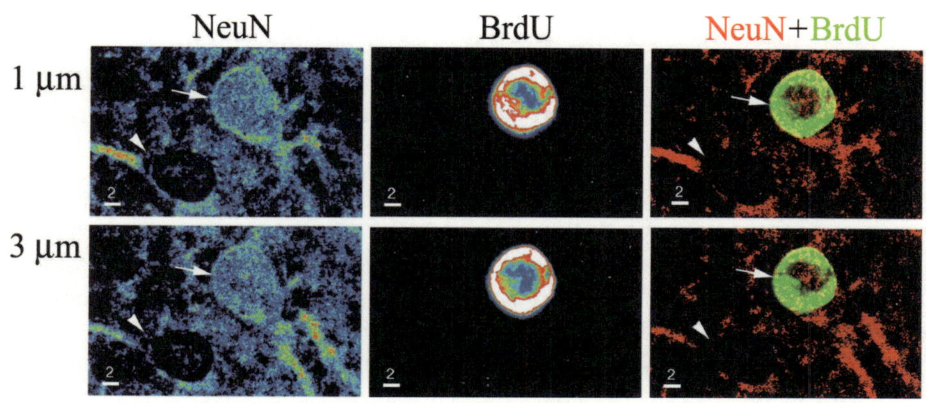

Fig. 6. 3D confocal analyses of NeuN and BrdU double immunofluorescence of cortical cells at two adjacent Z-series planes at 30 days after MCA occlusion. The left column shows the signal intensity for NeuN, the middle column displays the signal intensity for BrdU, and the right column exhibits a merged image of the NeuN and BrdU immunolabeling. In the right column, NeuN (red) and BrdU (green) are co-localized in the same nucleus (arrow) that results in a yellow appearance in the nucleus, which is surrounded by the NeuN cytoplamic labelling with a neurite-like extension. (From Jiang et al., Stroke 32:1201-1207, 2001. With permission from Wolters Kluwer Health)

The reliability of using BrdU nuclear incorporation as the final judgement of adult neurogenesis, in contrast to DNA repair, is questioned especially in the poststroke brain tissue (Kuhn and Cooper-Kuhn 2007; Nowakowski and Hayes 2000). To address this debate, the cellular origin of the progenitor cells that divide and give birth to poststroke newborn cortical neurons must be clarified. In adult mice after MCA stroke, neural stem cells are traced to proliferate and migrate from subventricular zone into the post stroke penumbral cortex. It takes 7-14 days for the stem cells to arrive at the peri-infarct cortex where they differentiate into cortical neurons expressing the neuron-specific marker NeuN (Kreuzberg et al. 2010; Ohab et al. 2006). In contrast, newborn neurons are already identified in the reperfused penumbral cortex at 72 hour after photothrombotic ring stroke in rats (Gu et al. 2000). This time frame is far beyond the migrating speed that SVZ stem cells could make. Therefore we postulate that the newborn cortical neurons may have a cortical origin in this cortical stroke model (Gu et al. 2000).

The crucial point to prove this hypothesis is to determine whether or not cell division has occurred in the poststroke brain tissue, and if it does, when, where and who start to divide. To achieve this purpose, the S-phase marker BrdU is injected intraperitoneally every 4h after stroke induction up to 72h after stroke, then two times daily, and is ended at poststroke day 7. The brain samples are collected at 4, 10, 16, 24, 48, and 72h and 7 and 14 days after stroke induction for analysis (Gu et al. 2009). To explore a possible cell mitosis which is theoretically anticipated during the poststroke cortical neurogenesis, cell mitosis specific marker phosphorylated histone H_3 (Phos H_3) (Hans and Dimitrov 2001; Hendzel et al. 1997) is investigated through immunohistochemistry/immunofluorescence in the brain sections. During a cell cycle, massive phosphorylation of the nuclear protein histone H_3 takes place immediately after the proliferating cells complete their DNA duplication in order to initiate the cell mitosis (Hans and Dimitrov 2001; Hendzel et al. 1997; Van Hooser et al. 1998), and the phosphorylated histone H3 becomes quickly dephosphorylated after cell division (Hendzel et al. 1998). Therefore the Phos H_3 detects specifically mitotic cell nuclei. To visualize mitotic spindles, spindle components α-tubulin (Wittmann et al. 2001) and γ-tubulin (Lajoie-Mazenc et al. 1994) are detected by immunocytochemistry and immunofluorescence. TUNEL labelling is used to detect DNA damage.

In 4h and 10h poststroke rats, a few BrdU-immunolabeled cells are randomly scattered in the brain sections. At 16h poststroke, BrdU-immunolabeled cells are consistently observed in the penumbral cortex. When consecutive sagittal brain sections are examined at this time, a few BrdU-immunolabeled cells are observed on the tangential migrating pathway from the SVZ toward the olfactory bulbs, but not toward the penumbral cortex. At 24h, 48h, 72h, 7 days, and 14 days poststroke, the number of BrdU-immunolabeled cells increases gradually in the penumbral cortex. From at 24h poststroke, Phos H_3 starts to be detected in the same cell nuclei doubly immunolabeled by BrdU (Fig. 7B). Meanwhile, α-tubulin and γ-tubulin immunolabeled mitotic spindles are observed inside the cortical cells containing BrdU- and Phos H3-immunolabeled cell nuclei (Fig. 7C-D). Some of these cells exhibit anaphase or telophase morphology (Fig. 8B-C, Fig. 8F), indicating the completion of cell division. The cell density of Phos H_3-immunopositive cells reaches its maximum at 48h and 72h poststroke and then declines at 7 days. These mitotic cells are dispersed among but distinctly separated from TUNEL-labelled cells (Fig. 7E). To identify a possible cell lineage of the mitotic cells, the neuronal marker Map-2ab, β-tubulin III, NeuN, and Hu and the astrocyte marker GFAP

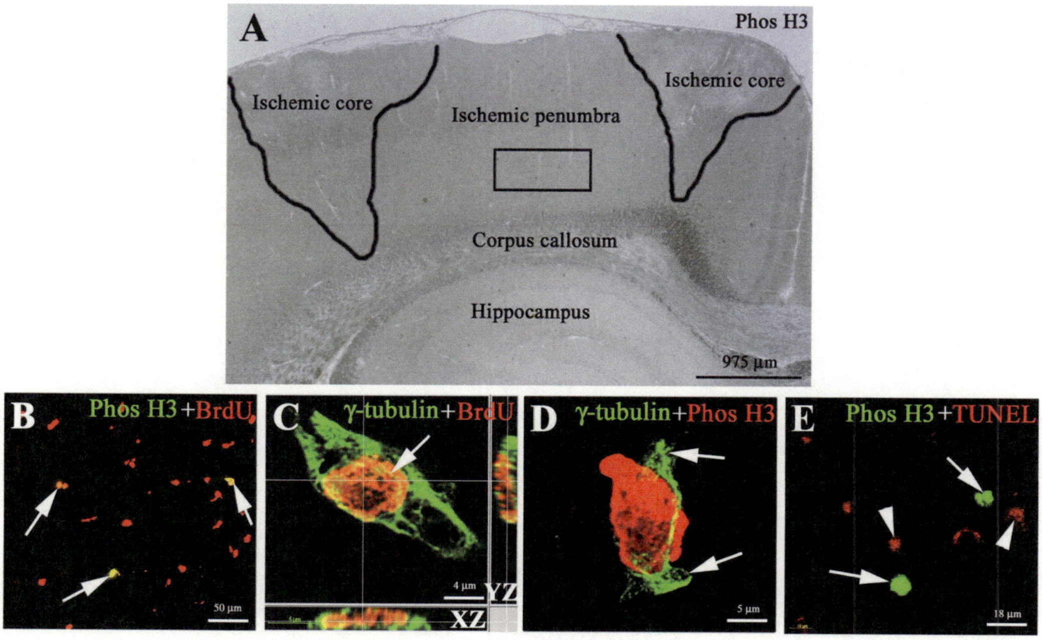

Fig. 7. (A) Low-magnification photograph of Phos H_3 single immunohistochemistry in a coronal brain section through the ischemic lesion at 48h after photothrombotic ring stroke. Two wedge-shaped cortical lesions corresponding to the annular ischemic core are demarcated in the cortex, between which lies the ischemic penumbra. Phos H_3-immunopositive cells (brown) are observed in the penumbral cortex, ipsilateral corpus callosum, and subgranule layer of the hippocampus. (B) Confocal analyses of Phos H_3 and BrdU double immunofluorescence in the penumbral cortex at 48h post stroke. Numerous cells are BrdU-immunolabeled (red) in the cortical penumbra. Among which, three cells are doubly immunolabeled by Phos H_3 (green) in this field, which yields a yellow appearance in the BrdU and Phos H3 double-immunopositive nuclei (arrows). (C) 3D confocal analyses of BrdU and γ-tubulin double immunofluorescence in a cortical cell at 24h poststroke. The γ-tubulin (green) appears as a spindle that is co-localized with the BrdU-immunolabeled nucleus (red) in the same cell. The γ-tubulin-immunolabeled microtubules (green) are penetrating through the BrdU-immunolabeled nucleus (red), which in combination produces a yellow color in the nucleus (arrow). (D) Maximal projection confocal image of Phos H_3 and γ-tubulin double immunofluorescence in the penumbral cortex at 48h post stroke. In a cell from cortical layer II, the Phos H_3-immunopositive nucleus (red) is superimposed on a γ-tubulin-immunolabeled spindle (green) with the pole bodies at the opposite ends (arrows) and the linking microtubules stretching in between. (E) Confocal microscopy of TUNEL and Phos H_3 double immunofluorescence in the penumbral cortex at 48h after stroke. The Phos H_3-immunolabeled cells (arrows, green) are separate from the TUNEL-positive cells (arrowheads, red). (From Gu el al., Stem Cell Res 2:68-77, 2009. With permission from Elsevier)

are used for double immunohistochemistry and immunofluorescence. Some Phos H_3-immunolabeled cell nuclei are enveloped by the cytoplasm immunopositive to the neuron-specific markers β-tubulin III and Map-2 (Fig. 8A-C), or the astrocyte marker GFAP. Under 3D confocal analysis, NeuN or Hu is co-localized with Phos H_3 (Fig. 8D), α-tubulin- or γ-tubulin (Fig. 8E-F) in the same cells.

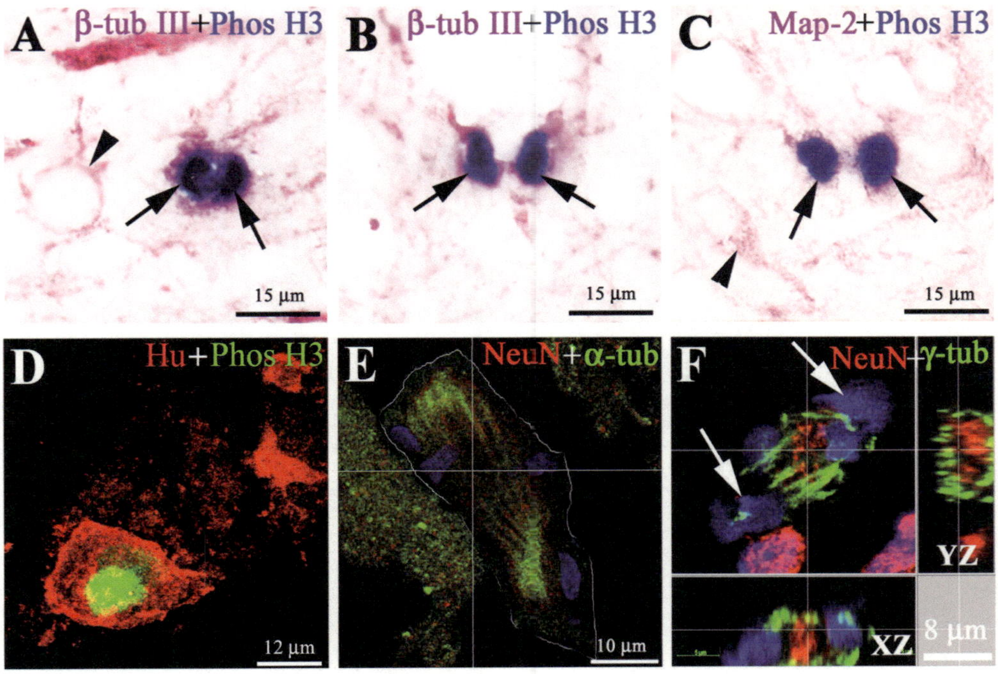

Fig. 8. (A) Phos H_3 and β-tubulin III double immunohistochemistry in the penumbral cortex at 48h after stroke. In a cortical cell, the Phos H_3-immunolabeled nucleus appears in a partially duplicated shape (arrows, blue), which is surrounded by β-tubulin III-immunolabeled cytoplasm (purple). The arrowhead points to a cortical neuron that is singly immunolabeled by β-tubulin III. (B) Phos H_3 and β-tubulin III double immunohistochemistry in the penumbral cortex at 48h after stroke. A pair of Phos H_3 (arrows, blue) and β-tubulin III (purple) double-immunolabeled cells is nearly separated but still slightly connected through β-tubulin III-immunolabeled cytoplasm. (C) Phos H_3 and Map-2 double immunohistochemistry in the penumbral cortex at 48h after stroke. A pair of Phos H_3 (blue) and Map-2 (purple) double-positive cells (arrows) is splitting while slightly connected through Map-2-immunolabeled cytoplasm (purple). The arrowhead points to a cortical neuron singly immunolabeled by Map-2. (D) Projection confocal image of Phos H_3 and Hu double immunofluorescence in the penumbral cortex at 24h after stroke. A large cortical cell that is Hu-immunolabeled in the cytoplasm (arrow, red) contains a Phos H_3-immunopositive nucleus in a duplicated shape (green). (E) Confocal microscopy of α-tubulin, NeuN and DAPI triple immunofluorescence in the penumbral cortex at 24h post stroke. In a large cortical cell in layer III (delineated), α-tubulin (green) appears as a bipolar spindle while the nuclear DNA is transformed into chromosomes (blue) randomly scattered inside the NeuN-immunopositive cytosol (red). (F) 3D confocal image of NeuN, γ-tubulin and DAPI triple immunofluorescence in layer III of the cortical penumbra at 48 h poststroke. In a cortical cell containing a γ-tubulin-immunolabeled spindle (green), the cell DNA (arrows, blue) is pulled toward its opposite ends while NeuN immunoreactivity (red) is detected at the central part of the cytosol. (From Gu el al., Stem Cell Res 2:68-77, 2009. With permission from Elsevier)

The initial appearance of the BrdU-immunolabeled cells within the penumbral cortex rather than in other parts of the brains suggests that the cells starting to proliferate belong to endogenous cortical cells. Their initial entrance into S-phase occurs at 16h poststroke. Their transition from S-phase cells into M-phase, as hallmarked by the nuclear Phos H_3 expression (Fig. 7B) and cytoplasmic spindle formation (Fig. 7C), occurs at 24h poststroke. The length of the S-phase is therefore estimated about 8h and the length of the whole cell cycle is about 8–10h. This is in agreement with the cell cycle calculation performed in normal mouse brains (Nowakowski et al. 1989). The concurrence whereas distinct separation of these mitotic cells from TUNEL labelling suggests that cell regeneration concurs with but differs distinctly from DNA damage and cell death in the same penumbral cerebral cortex after the photothrombotic ring stroke (Fig. 7E). These data provide morphological evidence to support our claim that the sustained BrdU-incorporated neurons in the poststroke penumbral cortex represent neurogenesis (Gu et al., 2000; Gu et al., 2009) rather than DNA repair (Kuhn and Cooper-Kuhn 2007; Nowakowski and Hayes 2000).

It is not clear why differentiated neuronal markers are expressed in cells in metaphase during the poststroke cortical neurogenesis (Fig. 8E-F). In cell culture, differentiated neuron markers are expressed in newborn neurons immediate after their birth from the mother cell, i.e., after cytokinesis (Svendsen et al. 1995). A possible explanation of this phenomenon is that in response to ischemic stroke the endogenous progenitor cells may undergo cell division with a quickened cell differentiation. In agreement with this explanation, neural progenitor cells isolated from the poststroke penumbral cerebral cortex are cultured to generate neurons expressing differentiated neuronal markers (Shimada et al. 2010). An alternative possibility is that after ischemic insult somatic cortical neurons may reprogram themselves and thus function as pluripotent stem cells so that they start to divide and give birth to newborn neurons while they keep their neuronal identity throughout cell division. Future studies are needed to address on this issue.

Whether or not the adult cortical neurogenesis may contribute to the poststroke neurological improvement is one of our major interests. For the newborn neurons to function in the poststroke cerebral cortex, they must fulfil many basic criteria. For example, they must be able to establish synaptic connections with their surrounding neurons, to synthesize various neurotransmitters and receptors, and in response to afferent stimuli to generate action potentials and to release the corresponding neurotransmitters in order to activate their target receptors. Proper neurotransmitter deactivation system is also needed in order to terminate the action. To explore the function potential of poststroke cortical neurogenesis, we start with an examination on the possible biosynthesis of the excitatory neurotransmitter acetycholine (ACh) and its substrate enzyme choline acetyltransferase (ChAT) and the inhibitory neurotransmitter γ-aminobutyric acid (GABA) and the corresponding substrate enzyme glutamic acid decarboxylase (GAD) in adult rats after photothrombotic ring stroke. To detect the newborn cells, BrdU is repeatedly delivered as mentioned previously. To detect neurotransmitter synthesis, brain sections are examined through double immunohistochemistry or immunofluorescence with BrdU and the neurotransmitter markers. To ascertain the cell identify, neuron-specific marker NeuN is used for triple immunofluorescence. With ACh, ChAT, GABA and GAD single immunohistochemistry, ACh, ChAT, GABA and GAD-immunolabeled neurons are frequently observed in the cerebral cortex outside the penumbral region. In contrast, all of the cells in the pannecrotic ischemic core and the majority of the cells inside the cortical penumbra are not immunolabeled by the neurotransmitters ACh and GABA. However, some BrdU-immunolabeled cortical cells in the penumbral cortex are ChAT, ACh, GAD, or GABA- immunolabeled in their cytoplasm. These cells are randomly distributed in cortical layers II–VI in the penumbral cortex at 48h, 5 days, 7 days, 30 days, 60 days, and 90 days after stroke. Under 3-D confocal analyses, BrdU is colocalized with ACh (Fig. 9A) or GABA (Fig. 9B) in the same cells in the corresponding double immunofluorescence, in which the BrdU-immunolabelled cell nucleus is surrounded by ACh or GABA-immunolabelled cytoplasm. In the GABA, NeuN, and BrdU triple immunofluorescence (Fig. 9D-F), the neuron-specific marker NeuN is colocalized with the BrdU in the same nucleus (Fig. 9E), around which the GABA-immunoreactive cytoplasm is detected (Fig. 9F). This observation verifies the neuronal identity of these newborn cortical cells synthesizing corresponding neurotransmitters. In addition, it also helps to address on a theoretical concern that the presence of the neuronal marker NeuN inside the dividing cells might be a consequence of phagocytosis, i.e., the cells that look like dividing neurons are dividing macrophages that have taken neuronal markers. In the setting of focal cerebral ischemia the CBF threshold for neurotransmitter release (corresponding to 40–60% of the baseline CBF) is higher than that of the membrane threshold (corresponding to 20–40% of the baseline CBF) beyond which morphological damage of the ischemic tissue occurs (Hossmann 1994). Timewise, postischemic neurotransmitter release occurs earlier than cell apoptosis and necrosis through which ischemic cell death takes place. In the photothrombotic ring stroke the local CBF in the cortical penumbra drops to 34% of the baseline level at 2h after ischemia (Gu et al. 1999a). This CBF level is already well below the neurotransmitter threshold at which a nonspecific neurotransmitter release is triggered, while the tissue morphology remains intact (Gu et al. 1999b). Consequently, all of the neurons in the ischemic core and the majority of neurons inside the penumbral cortex are not immunolabeled by Ach or GABA at 24-48h (Gu et al. 2010). While some of the cortical cells in the penumbral cortex are DNA damaged in their nuclei, i.e., TUNEL-positive, they are distinctly separated from the dividing cells expressing the mitotic marker Phos H_3 (Fig. 7E). Thus, neurons lose their neurotransmitters at 2-16h after ischemia, and they die through apoptosis/necrosis at 24-48h after the photothrombotic ring stroke. Meanwhile, local cells in the penumbral cortex are induced to proliferate from at 16h after ischemia incorporating BrdU into their nuclei, and then they start to divide at 24-48h poststroke that give birth to daughter neurons expressing various neuronal markers and synthesizing neurotransmitters (Gu et al. 2009; Gu et al. 2000; Gu et al. 2010). It is not until from at 72h postischemia and later ED-1-immunopositive microglias start to appear in the postischemic cortex to clean up the dead cells through phagocytosis (Gu et al. 1999b). Therefore, cell division starts earlier than phagocytosis in the setting of photothrombotic ring stroke. Being a functional marker for living GABAergic neurons, the neurotransmitter GABA should never appear in the debris of dead neurons. The homogeneous colocalization of GABA and NeuN in the same cytoplasm including the proximal dendrites of the newborn neurons (Fig. 9F) speaks directly against the hypothesis that the NeuN immunoreactivity in the dividing cells comes from exogenous fragments of dead neurons. It is conceivable that one or several pieces of neuronal fragments may be engulfed by an activated microglia into its cytoplasm (Takahashi et al. 2005). In that case, it should be debris of dead neurons that becomes engulfed, not a whole living neuron exhibiting intact cell morphology including neurite-like extensions. In the newborn neurons, the neuronal marker NeuN labels both the BrdU-immunolabeled cell nucleus and the perinuclear cytoplasm including the proximal dendrites in the newborn cortical neurons (Fig. 9E) (Gu et al. 2009; Gu et al. 2000; Gu et al. 2010). This NeuN cell labelling pattern agrees with what has been reported in normal adult neurons (Wolf et al. 1996). Therefore, the NeuN nuclear labelling itself speaks also against the exogenous NeuN hypothesis. It is because if a neuronal fragment is engulfed by a dividing microglia, it should barely be taken into the cytoplasm of the dividing microglia. It has no chance to integrate its immunoreactivity into the BrdU/Phos H_3-immunopositive dividing nucleus of the phagocyte. Thus, the neuronal markers and neurotransmitters observed belong indeed to the same cells exhibiting dividing cell nuclei.

Therefore, the capability of neurotransmitter biosynthesis by these newborn neurons qualifies themselves for one of the fundamental prerequisites for their further function during the poststroke neurological and neuropsychological improvement.

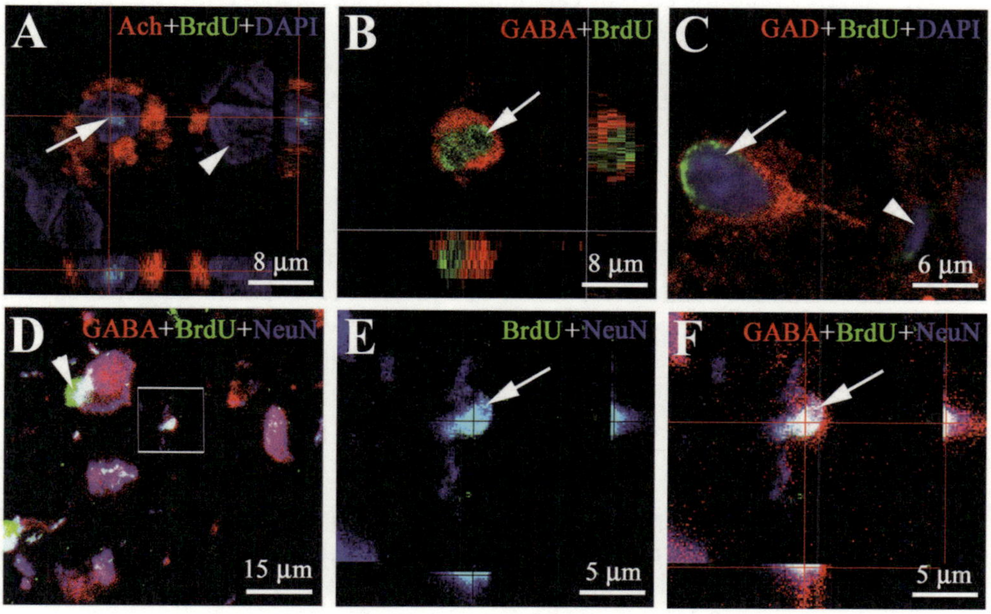

Fig. 9. (A) 3-D confocal analysis of Ach, BrdU, and DAPI triple immunofluorescence from layer IV of the cortical penumbra at 30 days after photothrombotic ring stroke. In a large cortical cell, the cell nucleus is BrdU-immunolabeled (green) and DAPI counterstained (arrow, blue), which is surrounded by Ach-immunopositive cytoplasm (red). Many non-cholinergic cells are seen in the same field (arrowhead, blue). (B) 3-D confocal analysis of GABA and BrdU double immunofluorescence in layer III of cortical penumbra at 48h after stroke. A cortical cell that is BrdU-immunolabeled in the nucleus (arrow, green) is GABA-immunopositive in the cytoplasm (red). (C) A section confocal image of GAD, BrdU, and DAPI triple immunofluorescence in layer III of penumbral cortex at 7 days after stroke. A cortical cell that contains a BrdU- immunolabeled (arrow, green) DAPI counterlabeled nucleus (arrow, blue) is GAD-immunopositive (red) in its cytoplasm. (D-F) GABA, BrdU and NeuN triple immunofluorescence (D) Projection image of GABA (red), BrdU (green), and NeuN (blue) triple immunofluorescence at low magnification. A cortical cell (framed) is triply immunolabeled by GABA (red), BrdU (green), and NeuN (blue). The arrowhead

points at a newborn nonneuronal cell singly immunolabeled by BrdU (green) superimposing on a GABAergic (red) NeuN-immunopositive (blue) cortical neuron. (E) BrdU and NeuN double-channel image of the cortical cell framed in panel D. The BrdU immunolabeling (arrow, green) is colocalized with NeuN (blue) in the same cell nucleus. (F) 3-D analysis of GABA, BrdU, and NeuN triple-channel image from the same scanning section as panel E. The GABA (red) is colocalized with NeuN (blue) and BrdU (green) in the same cell (arrow). (From Gu el al., Stem Cell Res 4:148-154, 2010. With permission from Elsevier)

In summary, neurogenesis occurs in the penumbral cerebral cortex in adult rodents after photothrombotic ring stroke and unilateral middle cerebral artery occlusion. Quick cell division within the penumbral cerebral cortex contributes to an early in situ cortical neurogenesis in the photothrombotic ring stroke model in rats. In contrast, neural stem cell migration from SVZ into the cortical penumbra provides a remote cell resource for a late stage cortical neurogensis in the periinfarct cerebral cortex after middle cerebral artery occlusion. As early as at 48h after the photothrombotic ring stroke, the newborn cortical neurons start to synthesize the neurotransmitters GABA and acetycholine exhibiting neurite-like extensions, which optimizes their further function in the post stroke recovery.

2. Acknowledgements

This study was supported by the Swedish Research Council (11624); King Gustaf V's and Queen Victoria's foundation, the Swedish Heart and Lung Foundation; the Swedish Stroke Foundation; the "Spjutspetsprojekt" of the County of Västerbotten; the Northern Sweden Stroke Foundation; the Medical Faculty of Umeå University, Umeå University Hospital, Umeå, Sweden; and the Swedish Society for Medicine.

3. References

Altman J. (1969) Autoradiographic and histological studies of postnatal neurogenesis. IV. Cell proliferation and migration in the anterior forebrain, with special reference to persisting neurogenesis in the olfactory bulb. *J Comp Neurol* 137:433-457

Altman J, Bayer SA. (1990) Migration and distribution of two populations of hippocampal granule cell precursors during the perinatal and postnatal periods. *J Comp Neurol* 301:365-381

Altman J, Das GD. (1965a) Autoradiographic and histological evidence of postnatal hippocampal neurogenesis in rats. *J Comp Neurol* 124:319-335

Altman J, Das GD. (1965b) Post-natal origin of microneurones in the rat brain. *Nature* 207:953-956

Brown J, Cooper-Kuhn CM, Kempermann G, Van Praag H, Winkler J, Gage FH, Kuhn HG. (2003) Enriched environment and physical activity stimulate hippocampal but not olfactory bulb neurogenesis. *Eur J Neurosci* 17:2042-2046

Chen J, Magavi SS, Macklis JD. (2004) Neurogenesis of corticospinal motor neurons extending spinal projections in adult mice. *Proc Natl Acad Sci U S A* 101:16357-16362

Craig CG, D'sa R, Morshead CM, Roach A, van der Kooy D. (1999) Migrational analysis of the constitutively proliferating subependyma population in adult mouse forebrain. *Neuroscience* 93:1197-1206

Eriksson PS, Perfilieva E, Bjork-Eriksson T, Alborn AM, Nordborg C, Peterson DA, Gage FH. (1998) Neurogenesis in the adult human hippocampus [see comments]. *Nat Med* 4:1313-1317

Gratzner HG. (1982) Monoclonal antibody to 5-bromo- and 5-iododeoxyuridine: a new reagent for detection of DNA replication. *Science* 218:474-475

Gu W, Brannstrom T, Jiang W, Bergh A, Wester P. (2001) Vascular endothelial growth factor-A and -C protein up-regulation and early angiogenesis in a rat photothrombotic ring stroke model with spontaneous reperfusion. *Acta Neuropathol (Berl)* 102:216-226.

Gu W, Brannstrom T, Rosqvist R, Wester P. (2009) Cell division in the cerebral cortex of adult rats after photothrombotic ring stroke. *Stem Cell Res* 2:68-77

Gu W, Brannstrom T, Wester P. (2000) Cortical neurogenesis in adult rats after reversible photothrombotic stroke. *J Cereb Blood Flow Metab* 20:1166-1173

Gu W, Gu C, Jiang W, Wester P. (2010) Neurotransmitter synthesis in poststroke cortical neurogenesis in adult rats. *Stem Cell Res* 4:148-154

Gu W, Jiang W, Wester P. (1999a) A photothrombotic ring stroke model in rats with sustained hypoperfusion followed by late spontaneous reperfusion in the region at risk. *Exp Brain Res* 125:163-170

Gu WG, Brannstrom T, Jiang W, Wester P. (1999b) A photothrombotic ring stroke model in rats with remarkable morphological tissue recovery in the region at risk. *Exp Brain Res* 125:171-183

Gu WG, Jiang W, Brannstrom T, Wester P. (1999c) Long-term cortical CBF recording by laser-Doppler flowmetry in awake freely moving rats subjected to reversible photothrombotic stroke. *J Neurosci Methods* 90:23-32

Hankey GJ, Spiesser J, Hakimi Z, Bego G, Carita P, Gabriel S. (2007) Rate, degree, and predictors of recovery from disability following ischemic stroke. *Neurology* 68:1583-1587

Hans F, Dimitrov S. (2001) Histone H3 phosphorylation and cell division. *Oncogene* 20:3021-3027.

Hendzel MJ, Nishioka WK, Raymond Y, Allis CD, Bazett-Jones DP, Th'ng JP. (1998) Chromatin condensation is not associated with apoptosis. *J Biol Chem* 273:24470-24478

Hendzel MJ, Wei Y, Mancini MA, Van Hooser A, Ranalli T, Brinkley BR, Bazett-Jones DP, Allis CD. (1997) Mitosis-specific phosphorylation of histone H3 initiates primarily within pericentromeric heterochromatin during G2 and spreads in an ordered fashion coincident with mitotic chromosome condensation. *Chromosoma* 106:348-360

Hossmann KA. (1994) Viability thresholds and the penumbra of focal ischemia. *Ann Neurol* 36:557-565

Hu X, Brannstrom T, Gu W, Wester P. (1999) A photothrombotic ring stroke model in rats with or without late spontaneous reperfusion in the region at risk. *Brain Res* 849:175-186

Hu X, Johansson IM, Brannstrom T, Olsson T, Wester P. (2002) Long-lasting neuronal apoptotic cell death in regions with severe ischemia after photothrombotic ring stroke in rats. *Acta Neuropathol* 104:462-470

Hu X, Wester P, Brannstrom T, Watson BD, Gu W. (2001) Progressive and reproducible focal cortical ischemia with or without late spontaneous reperfusion generated by a ring-shaped, laser-driven photothrombotic lesion in rats. *Brain Res Brain Res Protoc* 7:76-85

Jiang W, Gu W, Brannstrom T, Rosqvist R, Wester P. (2001) Cortical neurogenesis in adult rats after transient middle cerebral artery occlusion. *Stroke* 32:1201-1207.

Jiang W, Gu W, Hossmann KA, Mies G, Wester P. (2006) Establishing a photothrombotic 'ring' stroke model in adult mice with late spontaneous reperfusion: quantitative measurements of cerebral blood flow and cerebral protein synthesis. *J Cereb Blood Flow Metab* 26:927-936

Jin K, Sun Y, Xie L, Peel A, Mao XO, Batteur S, Greenberg DA. (2003) Directed migration of neuronal precursors into the ischemic cerebral cortex and striatum. *Mol Cell Neurosci* 24:171-189

Johansson IM, Wester P, Hakova M, Gu W, Seckl JR, Olsson T. (2000) Early and delayed induction of immediate early gene expression in a novel focal cerebral ischemia model in the rat. *Eur J Neurosci* 12:3615-3625

Kempermann G, Kuhn HG, Gage FH. (1997) More hippocampal neurons in adult mice living in an enriched environment. *Nature* 386:493-495.

Kreuzberg M, Kanov E, Timofeev O, Schwaninger M, Monyer H, Khodosevich K. (2010) Increased subventricular zone-derived cortical neurogenesis after ischemic lesion. *Exp Neurol* 226:90-99

Kuhn HG, Cooper-Kuhn CM. (2007) Bromodeoxyuridine and the detection of neurogenesis. *Curr Pharm Biotechnol* 8:127-131

Lajoie-Mazenc I, Tollon Y, Detraves C, Julian M, Moisand A, Gueth-Hallonet C, Debec A, Salles-Passador I, Puget A, Mazarguil H, et al. (1994) Recruitment of antigenic gamma-tubulin during mitosis in animal cells: presence of gamma-tubulin in the mitotic spindle. *J Cell Sci* 107:2825-2837.

Leker RR, Soldner F, Velasco I, Gavin DK, Androutsellis-Theotokis A, McKay RD. (2007) Long-lasting regeneration after ischemia in the cerebral cortex. *Stroke* 38:153-161

Liu J, Solway K, Messing RO, Sharp FR. (1998) Increased neurogenesis in the dentate gyrus after transient global ischemia in gerbils. *J Neurosci* 18:7768-7778

Longa EZ, Weinstein PR, Carlson S, Cummins R. (1989) Reversible middle cerebral artery occlusion without craniectomy in rats. *Stroke* 20:84-91

Memezawa H, Minamisawa H, Smith ML, Siesjo BK. (1992) Ischemic penumbra in a model of reversible middle cerebral artery occlusion in the rat. *Exp Brain Res* 89:67-78

Miller MW, Nowakowski RS. (1988) Use of bromodeoxyuridine immunohistochemistry to examine the proliferation, migration and time of origin of cells in the central nervous system. *Brain Res.* 457:44-52

Nowakowski RS, Hayes NL. (2000) New neurons: extraordinary evidence or extraordinary conclusion? *Science* 288:771.

Nowakowski RS, Lewin SB, Miller MW. (1989) Bromodeoxyuridine immunohistochemical determination of the lengths of the cell cycle and the DNA-synthetic phase for an anatomically defined population. *J Neurocytol* 18:311-318

Ohab JJ, Fleming S, Blesch A, Carmichael ST. (2006) A neurovascular niche for neurogenesis after stroke. *J Neurosci* 26:13007-13016

Rakic P. (1985) DNA synthesis and cell division in the adult primate brain. *Ann N Y Acad Sci* 457:193-211

Ramón y Cajal S. (1913) *Degeneration and Regeneration of the Nervous System*. London: Oxford Univ. Press

Samuels BA, Hen R. (2011) Neurogenesis and affective disorders. *Eur J Neurosci* 33:1152-1159

Shimada IS, Peterson BM, Spees JL. (2010) Isolation of locally derived stem/progenitor cells from the peri-infarct area that do not migrate from the lateral ventricle after cortical stroke. *Stroke* 41:e552-560

Svendsen CN, Fawcett JW, Bentlage C, Dunnett SB. (1995) Increased survival of rat EGF-generated CNS precursor cells using B27 supplemented medium. *Exp Brain Res* 102:407-414

Takahashi K, Rochford CD, Neumann H. (2005) Clearance of apoptotic neurons without inflammation by microglial triggering receptor expressed on myeloid cells-2. *J Exp Med* 201:647-657

Tsai PT, Ohab JJ, Kertesz N, Groszer M, Matter C, Gao J, Liu X, Wu H, Carmichael ST. (2006) A critical role of erythropoietin receptor in neurogenesis and post-stroke recovery. *J Neurosci* 26:1269-1274

Twitchell TE. (1951) The restoration of motor function following hemiplegia in man. *Brain* 74:443-480

Van Hooser A, Goodrich DW, Allis CD, Brinkley BR, Mancini MA. (1998) Histone H3 phosphorylation is required for the initiation, but not maintenance, of mammalian chromosome condensation. *J Cell Sci* 111:3497-3506

van Praag H, Kempermann G, Gage FH. (1999) Running increases cell proliferation and neurogenesis in the adult mouse dentate gyrus. *Nat Neurosci* 2:266-270

Watson BD, Dietrich WD, Busto R, Wachtel MS, Ginsberg MD. (1985) Induction of reproducible brain infarction by photochemically initiated thrombosis. *Ann Neurol* 17:497-504

Wester P, Watson BD, Prado R, Dietrich WD. (1995) A photothrombotic 'ring' model of rat stroke-in-evolution displaying putative penumbral inversion. *Stroke* 26:444-450.

Wittmann T, Hyman A, Desai A. (2001) The Spindle: a dynamic assembly of microtubules and motors. *Nature Cell Biology* 3:E28-E34

Wolf HK, Buslei R, Schmidt-Kastner R, Schmidt-Kastner PK, Pietsch T, Wiestler OD, Bluhmke I. (1996) NeuN: a useful neuronal marker for diagnostic histopathology. *J Histochem Cytochem* 44:1167-1171

6

Brain Plasticity Following Ischemia: Effect of Estrogen and Other Cerebroprotective Drugs

Edina A. Wappler[1,2], Klára Felszeghy[3], Mukesh Varshney[4], Raj D. Mehra[4], Csaba Nyakas[3] and Zoltán Nagy[5]

[1]*Department of Pharmacology, Tulane University, New Orleans*
[2]*Department of Anaesthesiology and Intensive Therapy, Semmelweis University, Budapest*
[3]*Neuropsychopharmacology Research Unit of Semmelweis University and Hungarian Academy of Sciences, Budapest*
[4]*Department of Anatomy, All India Institute of Medical Sciences, New Delhi*
[5]*Cardiovascular Center, Department Section of Vascular Neurology, Semmelweis University, Budapest*
[1]*USA*
[4]*India*
[2,3,5]*Hungary*

1. Introduction

Cytoprotection and brain regeneration are both potential future therapies in the treatment of cerebral ischemia, both based on animal research data. Since Kaas and colleagues first described the reorganization of the sensory cortex after peripherial nerve injury it has become clear that "we can no longer consider the injured brain as a normally wired brain with a missing puzzle piece" (Kaas et al., 1983 as cited in Nudo, 2006) and thus there has been intense research subsequently focused on understanding the regeneration processes following different brain injuries.

In rodent models, permanent and transient middle cerebral artery occlusions (MCAO) are the most relevant for representing human ischemic stroke. The lesioned brain areas in these experimental cases are well comparable to those found in humans. Global cerebral ischemia has been investigated using several models including (1) hypotension with bilateral carotid occlusion, or (2) four vessel (2 carotid arteries + 2 vertebral arteries) occlusion with normotension, (3) hypoxia, and (4) cardiac arrest models in rats or mice, or (5) the transient carotid occlusion in gerbil. These global brain ischemia models mimic severe hypotension, cardiac arrest, hypoxia, cardiac surgery, among other conditions in the human, however, while in rodent models the most vulnerable region is the hippocampus, in human both cortical and subcortical lesions are common. This is one reason why these models are difficult to translate into human studies in global cerebral ischemia, however, longer ischemic periods in rodents also result in patchy cortical and subcortical damage together with the hippocampal damage (Back et al., 2004; Erdő and Hossmann, 2007; Wappler et al.,

2009 and 2010). Though, these models are useful to investigate the pathophyshiological mechanisms of brain ischemia, but it is important to note that in human stroke patients there are usually several other factors to consider, such as obesity, hypertension, diabetes, age and medications a patient may be using (Li and Carmichael, 2006; Popa-Wagner et al., 2007; Wappler et al., 2010) which may also determine the size of the lesion together with the regenerative potential of the tissue. This emphasizes the importance of experiments that utilize disease model and not just young, healthy male subjects to investigate brain injury, which will hopefully bring us closer to an understanding of the complex clinical conditions that result in stroke (Nudo, 2007; Popa-Wagner et al., 2007; Wappler et al., 2010). In addition to understanding the process of ischemic brain injury, examining the potential effects of the drugs that afford protection against ischemia is just as crucial under different conditions.

In this chapter we present an overview of the studies describing the regenerative potential of the brain due to ischemic damage. Furthermore, we discuss drugs that increase brain plasticity after ischemic insult in animal models, focusing on estrogen. In addition, we describe a brief study examining acute 17β-estradiol treatment on synaptic plasticity in the brain with a short (4 days) and a long term (25 days) follow up in young (4 months old) and old (18 months old) gerbils after global cerebral ischemia. The aim of this study was to investigate whether changes in synaptic density can be maintained after a longer period of single dose estrogen treatment, as we have demonstrated an early induction of neuronal plasticity using this model (Wappler et al., 2011b). Maintenance of synaptic density may be an important factor underlying the previously described better functional outcome using this model that was investigated up to the 2nd week after brain ischemia (Wappler et al., 2010). In our current study, we examined two different age groups because synaptic reorganization is known to decrease with advancing age (Kim and Jones, 2010). Our results can help elucidate how synaptic reorganization progresses in the brain after global cerebral ischemia due to a single treatment of a cerebroprotective drug and how brain plasticity is influenced by age.

2. Brain regeneration after cerebral ischemia

Cerebral plasticity is the ability of the brain to change its structure and function during maturation, learning, environmental challenges or pathology (Di Filippo et al., 2008; Lledo et al., 2006). The exact mechanism of spontaneous brain regeneration after brain ischemia is not fully understood; however, there is a remarkable number of publications that describe several mechanisms participating in this process.

Here we discuss the following features in brain regeneration: 1. neural plasticity 2. vascular plasticity and 3. glial plasticity.

2.1 Neuronal plasticity

For almost one hundred years neuroscientists have believed that the adult primate brain, and therefore the human brain, is structurally stable and does not form new synapses or grow new cells (Gould et al., 1999). It is clear by now that certain brain regions generate new cells, and that the continuous "rewiring" of the brain is an important physiological function.

Cortical interneurons but not pyramidal cells have been described to have intense arborization as axonal sprouting, dendritic growth and branching under physiological conditions (Lee et al., 2006) throughout adulthood. The "baseline" cerebral plasticity; however, is much more limited in the mature brain, compared to the developing brain, where high activity takes place. This phenomenon is generated by those structural and functional "brakes", such as myelin, and several neuromodulators, that actively suppress plasticity in the adulthood (Bavelier et al., 2010).

Following cerebral ischemia, or other types of brain insult; however, neuronal plasticity is reactivated in the surviving cells in order to compensate for cell death and to preserve functionality of the damaged but not dead areas (Blizzard et al. 2011; Carmichael, 2006). The possible mechanisms of neural plasticity include dendritic reorganization, axonal sprouting, and activation of endogenous pluripotent cells that can differentiate into neurons. While it has been shown that interneurons undergo structural remodeling in post-traumatic cortical lesions, signs of neural plasticity have not been detected in pyramidal neurons (Blizzard et al., 2011).

2.1.1 Dendritic, axonal and synaptic plasticity

Early studies demonstrated that following entorhinal lesion, as the result of neuronal projections from the contralateral hippocampus, new synapses are formed in the damaged cortex (Lynch et al., 1973). In addition, several subsequent studies have shown dendritic and axonal reorganization after experimental brain ischemia with dynamic changes of synaptic density in the injured brain region (Benowitz and Carmichael, 2010; Brown et al., 2010; Lu et al., 2004; Mostany et al., 2010; Scheff et al., 2005; Sulkowski et al., 2006; Takatsuru et al., 2009; etc.). The most active neuronal regeneration occurs up to 2-3 weeks after brain injury (Blizzard et al. 2011; Jones and Schallert, 1992), which provides a wide therapeutic window in cerebral ischemia.

Cerebral ischemia induces axonal sprouting within the peri -infarct zone and contralateral side. This post-ischemic axonal sprouting establishes new neuronal connection pattern for the damaged brain areas (Carmichael, 2003). Axonal sprouting after different central nervous system injuries can be detected by using growth-associated protein-43 (GAP-43) as it is a marker of regeneration in the adulthood (Benowitz and Routtenberg, 1987; Benowitz and Perrone-Bizzozero, 1991; Simon et al., 2001; Wappler et al., 2011b). During brain development GAP-43 is highly associated with the elongating axons (Benowitz and Routtenberg, 1987; Benowitz and Perrone-Bizzozero, 1991), which becomes less concentrated proximal to the cell soma while the axon is growing, but stays in the growth cone (Benowitz and Perrone-Bizzozero, 1991). GAP-43 is also thought to be involved in neurotransmitter release. The highest GAP-43 immunoreactivity was detected on the 4th postnatal day, when the most active synaptogenesis takes place. This is followed by a rapid decrease in the immunoreactivity with only a few brain regions expressing GAP-43 later in the adulthood. These regions are the cerebral cortex, the hippocampus, the hypothalamus, the amygdala, the striatum, the medial substantia nigra, raphe nuclei, locus coeruleus, olfactory bulb, olfactory tubercule, preoptic area, and stria terminalis (Benowitz and Perrone-Bizzozero, 1991; Yao et al., 1993). Each of these regions has a different level of constant reorganization. In the hippocampus, this reorganization is related to synaptic remodeling during memory formation (Benowitz and Perrone-Bizzozero, 1991; Holahan et al., 2007). Astrocytes and perivascular extracellular matrix are involved in guiding axonal growth and provide a scaffolding surface for this growth (Nedergaard and Dirnagl, 2005).

Shortly after focal cerebral ischemia many neurons in the penumbra lose their dendritic spines in an attempt to survive (Brown et al., 2008; Benowitz and Carmichael, 2010). Nevertheless up to two weeks after stroke dendritic density and turnover increases in the peri-infarct cortex and also in the contralateral hemisphere, which plays a major role in brain regeneration (Brown et al., 2010; Mostany et al., 2010; Takatsuru et al., 2009). In addition, following cardiac arrest decreased microtubule-associated protein 2 (MAP-2) expression was detected in the rat reflecting lower dendritic density (Sulkowski et al., 2006). Recent studies have shown dendritic arbor shortening in the ischemic penumbra in the first weeks following stroke with further loss of dendritic branches after the first month in cortical pyramidal cells (Mostany and Portera-Cailliau, 2011). However, dendritic changes in the basilar tree of these cells or in other neuronal cell types in the cortex could not be ruled out. In contrast, enhanced dendritic arborization has been described in cortical pyramidal neurons following photothrombotic brain ischemia (Brown et al., 2010). Thus, the changes in dendrite vary among different cerebral ischemia models.

In different central nervous system injuries synaptogenesis (formation of new synapses) is critical because new connections restore the cell communication and signaling. Following injury, surviving neurons have been described to form new synapses to compensate for the lost contact surfaces even if pre-traumatic synapse number is not achieved in the traumatized area (Lu et al., 2004; Scheff et al., 2005). Therefore examining synaptic density is a widely used technique to track neuronal plasticity following brain lesions. Both synaptophysin (SYP) and synapsin-I are widely used synaptic markers to assess synaptic density. The vesicular transmembrane protein, SYP, is expressed in the presynaptic terminal and its expression seems to be dispensable in neurotransmission (Becher et al., 1996; Eshkind and Leube, 1995; McMahon et al., 1996) but may be involved in fine-tuning of synaptic activity and in vesicle biogenesis (Becher et al., 1996; Janz and Sudhof, 1998). Synapsin-I is a phosphoprotein located on the small synaptic vesicles in the presynaptic terminal (De Camilli et al., 1983; Schiebler et al., 1986), and participates in regulating plasticity (Roshal et al., 1993; Wei et al., 2011). Following axonal sprouting and dendritic reorganization, sometimes just 21 days after the ischemia, synaptic density increases suggesting the development of mature synapses (Stroemer et al., 1995 as cited in Carmichael, 2003). In addition it has been shown in a rodent model that synaptic density steadily decreased up to one week following global cerebral ischemia (Sulkowski et al., 2006).

2.1.2 Endogenous neurogenesis

While it was not believed that the adult human brain was able to form new neurons, Altman and Das provided the first evidence that neurogenesis occurred in the mature rodent brain (Altman and Das 1965 and 1967). These data opened a new chapter in neuroscience research. Besides the physiological functions, such as memory formation and learning, endogenous neurogenesis has become a major focus of research in different brain lesions (Gao et al., 2009; Kokaia and Lindvall, 2003; Shen et al., 2010; etc.) and their potential therapies (Kim et al., 2009; Leker et al., 2009; Xiong et al., 2011). The majority of the endogenous cerebral stem cells are located in the subventricular zone and hippocampal subgranular zone and can generate either neuronal or glial cells (Zhao et al., 2008) in the lesion site using signals from the damaged cells for their activation and migration released.

These pluripotent cells express certain proteins that are typically present during brain development, and therefore are useful markers to track neurogenesis in adulthood following brain injury, such as ischemia. One of the markers is nestin, an intermediate filament protein, which is expressed in the astrocytes and radial glia cells in the developing brain and disappears after the 11th postnatal day in the rat (Kalman and Ajtai, 2001). Although previous data suggest that nestin immunopositivity in the adult bran is associated with immature cells that are involved in neuogenesis (von Bohlen und Halbach, 2007; Yue et al. 2006), it also a marker of reactive gliosis following various brain lesions (see e.g. Duggal et al. 1997). It also have been reported that following focal ischemia in the rat nestin positive cells from the ipsilateral subventricular zone can differentiate into glial cells (Holmin et al., 1997; Nakagomi et al., 2009; Shen et al., 2010). Therefore, in the adult brain, nestin expression recurs in both proliferating cells and in reactive astrocytes.

2.2 Vascular plasticity

Vascular plasticity includes processes such as angiogenesis and arteriogenesis. Angiogenesis is related to hypoxia and results in new capillaries from the pre-existing vessels, whereas arteriogenesis is induced most importantly by increased shear stress that results in newly formed blood vessels (Heil and Sharper, 2004; Heil et al., 2006; Schierling et al., 2009; Xiong et al., 2010); however, the differences in the cause and the result is usually are not this clear cut. Angiogenesis has a major role in brain regeneration after ischemia as increased blood supply directly enhances cell survival and regenerative processes (Font et al., 2010). Blood vessels not only provide metabolic support but also participate in neurogenesis by leading progenitor cells to the site of injury (Jin et al., 2002; Kojima et al., 2010 as cited in Font et al., 2010; Sun et al., 2010; Udo et al., 2008; Xiong et al., 2010; Yang et al., 2010).There is extensive evidence that neovascularization (both angiogenesis and arteriogenesis) is induced following acute (del Zoppo and Mabuchi, 2003; Issa et al., 2005) and chronic (Busch et al., 2003; Wappler et al., 2011a) brain ischemia.

Hypoxia induced hypoxia-inducible factor-1 (HIF-1) and vascular endothelial growth factor (VEGF) are the most common angiogenesis stimulators (Busch et al., 2003; Carmeliet and Collen, 1997; Liu et al., 1995; Levy et al., 1995) and are involved in capillary sprouting rather than in larger collateral vessel remodeling (Busch et al., 2003; Carmeliet and Collen, 1997) because they act on endothelial cells without inducing smooth muscle proliferation (Busch et al., 2003). In addition, several other factors, such as fibroblast growth factor (FGF) (Issa et al., 2005), angiopoetins (Lin et al., 2000), transforming growth factor β (TGFβ) (Haggani et al., 2005; Krupinski et al., 1996), platelet derived growth factor (PDGF) (Issa et al., 2005; Krupinski et al., 1997), tissue-type plasminogen activator (Carmeliet and Collen, 1997), etc. are just as critical in new vessel formation after brain injury. Most of these molecules have separate effects on cerebroprotection and regeneration, such as the TGFβ-s (Vincze et al., 2010). In addition, nitric oxide derived from endothelial nitric oxide synthetase (eNOS) also induces endothelial cell proliferation and migration, smooth muscle cell differentiation, and other angiogenic processes, where ischemia and shear stress are triggering mechanisms (Amano et al., 2003; Cui et al., 2009; Murohara et al., 1998; Papapetropoulos et al., 1997; Prior et al., 2003; Rudic et al., 1998). Angiogenesis inhibitors, such as endostatin, angiostatin, thrombospondin-1 and thrombospondin-2 have also been detected following brain ischemia (Issa et al., 2005), which provide another avenue for therapeutic interventions.

In addition to these growth factors extracellular matrix proteins, such as laminin and dystroglycan complex (DGC) proteins, are also involved in vascular remodeling in the brain (Wappler et al., 2011a). The DGC proteins may have an important function in signal transduction connecting the extracellular signals and laminin itself with a wide variety of intracellular proteins, such as nitric oxide synthase, ion channels, kinases, and actin (Culligan and Ohlenieck, 2002; Wappler et al., 2011a). Thus, certain DGC proteins make good immunohistochemical markers of vascular reorganization (Wappler et al., 2011a) in addition to the more frequently used laminin.

Neuroregenerative agents that increase angiogenesis, such as estrogen (Ardelt et al., 2005), have been described to improve functional outcome in models of cerebral ischemia (Hermann and Zechariah, 2009; Goldstein, 2009). These experimental data correlate with clinical data where higher microvessel density in the brain after ischemia was accompanied by shorter recovery time and longer survival (Christoforidis et al., 2005; Font et al., 2010; Krupinski et al., 1994; Navarro-Sobrino et al., 2011; Wei et al., 2001).

2.3 Glial plasticity

Glial cells in the brain include astrocytes, microcytes, and oligodendrocytes, of which astrocytes are the most numerous. In the last decade glial cells have been recognized not just for structural but for metabolic and for throphic support. By secreting nerve growth factor [NGF], basic fibroblast growth factor [bFGF], transforming growth factor β [TFG-β], platelet-derived growth factor [PDGF], brain-derived neurotrophic factor [BDNF], ciliary neurotrophic factor, Neuropilin-1, vascular endothelial growth factor [VEGF], etc., they modulate the function of neurons and other cell types. Glial cells are active participants of synaptic interactions and higher level of cerebral function; and key elements of cerebral blood flow regulation (Araque and Navarrete, 2010; Attwell et al., 2010; Iadecola and Nedergaard, 2007; Metea and Newman, 2006; Nedergaard and Dirnagl, 2005). Glial cells also play a key role in regulating neuronal survival and regeneration by regulating the extracellular ion homeostasis, supporting other cells` energy metabolism, reducing glutamate toxicity, promoting neurogenesis, synaptogenesis and angiogenesis, activating endothelial cells, disrupting blood brain barrier (BBB), increasing inflammation, etc. (Himeda et al., 2007; Nedergaard and Dirnagl, 2005; Trendelenburg and Dirnagl, 2005). Generating new astrocytes is also an important feature in brain regeneration that has been mentioned previously in this chapter.

The formation of glial scar and its beneficial and non-beneficial properties are also of great interest when investigating astrocytic reaction following focal brain ischemia. Unlike other tissues where injury repair results in a fibrous scar, brain injury is followed by a special scar formed by the activated astorocytes and the extracellular matrix molecules of proteoglycans. These include heparan sulfate, dermatan keratan sulfate, sulfate proteoglycans, and chondroitin sulfate, which are released by reactive astrocytes to compose a barrier of axonal growth. The role glial scar formation following brain ischemia is still unknown and intense research is ongoing to understand if it is harmful or supportive (Rolls et al., 2009; Silver and Miller, 2004). Besides its support on the injured tissue (Rolls et al., 2009), its inhibitory effect on axonal growth is equally important (Cole and McCabe, 1991; Filder et al., 1999; Katoh-Semba et al., 1995; Rudge and Silver, 1990; Smith-Thomas et al., 1994; Snow et al., 1990). Nevertheless, sulphated proteoglycans have also been described as supporting axonal growth (Hikino et al., 2003; Nakanishi et al., 2006), making glial cell based therapeutic strategies more difficult to design. On In contrast, myelin-associated inhibitors from oligodendrocytes and myelin debris, namely myelin-associated glycoprotein (MAG), Nogo-A and oligodendrocyte myelin glycoprotein (OMgp) have clearly inhibitory effect on neurite outgrowth (for review see Yiu and He, 2006) and are under investigation because blocking their function results in enhanced brain regeneration.

3. Inducing cerebral plasticity following brain ischemia

A therapeutic window for drugs that increase neural plasticity is wider than those that target cytoprotection following cerebral ischemia (Zhang and Chopp, 2009) giving hope for improved functional outcomes in more stroke patients. A cerebroprotective drug can increase synaptic density in several ways. Cytoprotection, a process where cells can utilize more energy to maintain features that are not necessary in cell survival, can contribute to increased neuronal survival and therefore plasticity in the injured area. Presumably for the same reason increased oxygen and metabolic support improve cellular plasticity after different ischemic events in the brain as already mentioned above.

Whether anti-apoptotic genes are able to induce neuronal plasticity by themselves other than by improving metabolic status is important to understand the pathophysiology of brain ischemia. In order to investigate this question we used Bcl-Xl or Bcl-2 gene construct transfections in an in vitro hypoxia model and we observed increased expression of synaptophysin-I and nestin mRNAs and proteins under normoxic conditions. Following hypoxia only nestin expression was significantly different from the untreated hypoxic group (Gal et al., 2009). These data indicate increased that anti-apoptotic gene expression itself can contribute to the amelioration of brain plasticity and its effect might be modified under different stress conditions. Several drugs that are known to be cytoprotective against cerebral ischemia, such as (-)-D-Deprenyl (Simon et al, 2001), and 17β-estradiol (Wappler et al., 2011), also participate in brain regeneration. Although both have anti-apoptotic effect, (-)-D-Deprenyl increases GAP-43 expression whereas 17β-estradiol treatment does not, suggesting that similar pathways may mediate enhanced regeneration through different intracellular signaling (Simon et al, 2001; Szilagyi et al., 2009; Wappler et al., 2011) both *in vitro* and *in vivo*. This is supported by btudies on other cerebroprotective drugs, such as erythropoietin (EPO) (Iwai et al., 2010), statins (Céspedes-Rubio et al., 2010), amphetamine (Liu et al., 2011), melatonin (Chen et al., 2009; González-Burgos et al., 2007), and different spices (Kannappan et al., 2011), where different ways of imporoved brain plasticity was described.

3.1 Estrogen

3.1.1 Estrogen in the brain, estrogen receptors

Corpechot and colleges described the first time that the cerebral sex steroid concentration is much higher than the circulating estrogen level both in men and women (Corpechot et al.,

1981). Subsequently, estrogen synthesis (Le Gascone et al., 1987) and the essential enzymes (Hojo et al., 2004) involved were detected in the brain.

The majority of the investigated intracellular effects of estrogen are related to two estrogen receptors (ERs) in the brain, ER-α and ER-β. Both of these receptors are expressed in the central nervous system; however, their distribution show different pattern. While ER-α is highly expressed in the hippocampus, hypothalamus, and preoptic area accompanied by a low expression in the cortex, ER-β is densely expressed in the cortex together with a high receptor density in the hippocampus, amygdala, cerebellum, etc. (Brann et al., 2011; Merchenthaler et al., 2003; Shughrue et al., 1997, Shughrue and Merchenthaler, 2000). Both of these ERs form homo- or hetero dimers after binding an estrogen molecule, such as estrone, estriol, or the most effective 17β-estradiol (E2). These dimers can bind to the estrogen responsive elements of the DNA and regulate the expression of several genes, such as bcl-2, IGF-1 (insulin like growth factor-1), NGF, (BDNF (McKenna and O'Malley, 2002; Merchenthaler et al., 2003; Nilsson et al., 2001; Sharma and Mehra, 2008). However, besides this "classical" signaling pathway, which requires hours to days to take place, estrogen can induce rapid changes via its non-genomic pathways. These non-genomic responses are mediated through extranuclear ERs, and occur within minutes of estrogen exposure through activation of several signaling cascades, such as phosphatidylinositol-3-kinase (PI3K), extracellular signal regulated kinase (ERK), mitogen-activated protein kinase (MAPK) or protein kinase C (PKC) (Bourque et al., 2009; Brann et al., 2011; Koszegi et al., 2011; Lebesgue et al., 2009; Rebas et al., 2005). Extranuclear receptors have been detected in the cytoplasm of the cell body, but also in the dendrites and axons of the neurons, while ER immunoreactivity was also seen in the organelle membranes, and synaptic vesicles (Milner et el., 2005). In addition, other brain cells, such as glial cells, have also been shown to express ERs (Milner et al., 2005; Woolley, 2007).

A third estrogen receptor, the G-protein-coupled ER (GPR30), has also been detected in the brain (Funakoshi et al., 2006); however, limited data is available regarding its role under physiological and pathological conditions. One of its reported functions in the hippocampus is to increase synaptic transmission (Filardo et al., 2002; Lebesgue et al., 2009). This receptor is more likely to be associated with the ERK/CREB intracellular signaling pathway (Lebesgue et al., 2009), and presumably activates other intracellular signaling cascades as well. GPR30 protein expression was described in the neuronal plasma membrane and endoplasmic reticulum in several brain regions, such as the hippocampus (Funakoshi et al., 2006; Matsuda et al., 2008).

There is strong evidence that the three different estrogen receptors can crosstalk, for example regulating gene expression through ERK and Src signaling via transcription factor, and histone phosphorylation (Brann et al., 2011; Madak-Erdogan et al., 2008). GPR30 pathway also can crosstalk with other extranuclear pathways through Akt activation (Lebesgue et al., 2009).

The effect of age on ERs is worth to mention here as the incidence of cerebral ischemia is higher in the elderly and the old, which can modify the effect of estrogen therapy. Both ER-α and ER-β expressing cell number decreased significantly in the hippocampus of the aged rats; however, optical density of immunoreactivity per cell showed a significant increase for both ER-α and ER-β immunoreactivity in the CA1 neurons, whereas in CA3 neurons, it was significantly reduced (Mehra et al., 2005). Increased expression of ERs per cell is supposedly a compensatory phenomenon. ER-β immunoreactivity was, however, found decreased in the CA1 dendritic synapses in old female rats in another study (Waters et al., 2011).

3.1.2 Estrogen: Afforded protection and plasticity following brain ischemia

Estrogens are known to increase synaptic density in the intact brain (McEwen, 2002; Merchenthaler et al., 2003; Rune et al., 2006; Sá et al., 2009; Sharma et al., 2007; Woolley, 2007) even following administration of a single dose of this hormone (Sá et al., 2009; Wappler et al., 2011b). Even during oestrus cycle there is an intense fluctuation in dendritic density in rodents. Furthermore, ovariectomy or menopause itself results in a significant decrease of synaptic and dendritic density (Ojo et al., 2011; Woolley and McEwen, 1992). Data on estrogen effect also suggest that it acts directly at synapses by activating second messenger signaling, resulting in a rapid altering in neuronal excitability, synaptic transmission, and/or synaptic plasticity (Woolley, 2007). There is; however, limited data on neuronal plasticity following brain ischemia (Wappler et al., 2011b).

Several studies have shown that E2 therapy is neuroprotective in cerebral ischemia. Estrogen increases the number of surviving cells following ischemia (Liu et al., 2009; Merchenthaler et al., 2003; Platha et al., 2004; Wappler et al., 2010), reduces excitotoxicity (Connell et el., 2007; Herson et al., 2009; Weaver et al., 1997), inflammation (Herson et al., 2009 ; Stein, 2001; Suzuki et al., 2007), moderates blood–brain barrier dysfunction (Liu et al., 2005), is antioxidant (Connell et el., 2007), increases cerebral blood flow (Herson et al., 2009; Hurn et al., 1995; Pelligrino et al., 1998), reduces spontaneous postischemic hyperthermia (Platha et al., 2004), etc. Cerebral ischemia studies in ER-α and ER-β KO mice models, and pharmacological receptor inhibition have shown that ER-α is the primary mediator of neuroprotection. (Brann et al., 2011; Dubal et al., 2001; Liu et al., 2009; Merchenthaler et al., 2003; Miller et al., 2005). Both genomic and non-genimic effects seem to be involved in estrogen afforded neuroprotection (Merchenthaler et al., 2003). Selective GPR30 agonists have also been found neuroprotective in *in vitro* and *in vivo* models of brain ischemia (Gingerich et al., 2010; Lebesque et al., 2010; Zhang et al., 2010), however, its specific role in the pathophysiology of ischemic attack is still unknown.

Genomic effects of estrogen includes the inhibition of apoptosis (through bcl-2, bax, caspase-3); the diminution of inflammation (e.g. through tumor necrosis factor-α; interleukin 1, and 6); the induction of growth factor, structural protein, and neuropeptide expression; etc. (Merchenthaler et al., 2003). High dose, acute estrogen treatment in global cerebral ischemia also induces cerebral plasticity by increasing synapsin-I and nestin gene expression in gerbils as we described previously (Wappler et al., 2011b). GAP-43 expression was however not elevated further due to the treatment compared to the already increased level after brain ischemia in our model.

Most of the data on estrogen`s effect in brain ischemia were observed following chronic rather than acute treatment in rodents, which is a postmenopausal estrogen supplementation model as opposed to a model of acute therapy. There have also been a small number of studies that used older, or diseased animals, or females, especially investigating long term outcome (Wappler et al., 2010).

4. Estrogen modulates synaptic density age, and subregion dependently in the gerbil hippocampus after global brain ischemia

Our previous reports have demonstrated protective effects of E2 pre-treatment in gerbils following ischemia by increased cell survival, memory function and attention (Wappler et al., 2010). We also showed that increased cerebral plasticity takes place 4 days after the ischemia in the same model (Wappler et al., 2011b). Therefore, we hypothesized that E2 pre-treatment increase the hippocampal synaptic plasticity both in shorter (4 days) and longer (25 days) time points in gerbils at different ages.

4.1 Materials and methods

4.1.1 Animals

Ovariectomized gerbils of 4 (young), and 18 (old) months of age were used in our experiments. The animals were housed in an air-conditioned room at 22±1 ◦C with a 12 h light/dark cycle, and had free access to water and food. All the procedures on animals were approved by the Animal Examination Ethical Council of the Animal Protection Advisory Board at the Semmelweis University, Budapest, Hungary.

4.1.2 Surgery and 17β-estradiol treatment

The gerbils were anaesthetized with halothane (induction: 4%, maintaining: 1.5-2.5%) in a 30% O_2/70% N_2O mixture, breathing spontaneously via a face mask. Bilateral ovariectomy, and 10 min bilateral carotid occlusion or sham neck surgery were performed as previously described (Wappler et al. 2010). Briefly, bilateral ovariectomy was performed through lateral incisions in each animal. Two weeks later transient bilateral carotid artery occlusion (10 min) was established through a midline cervical incision using atraumatic aneurysm clips (Codman, Johnson and Johnson, Le Locle, NE, Switzerland). The neck tissue was reunited in two layers with non-absorbable, 4.0 silk sutures (Ethicon, Johnson and Johnson). Sham surgery consisted of the midline cervical incision and carotid preparation followed by 10 min period, after which the incision was closed. Thirty minutes prior to surgery, estradiol treated group wasgiven 17β-estradiol (Sigma Chemical Co. St Louis, MO, USA) 0.4 ml/100 g (4 mg/kg) body weight. On the other hand, sham-operated and untreated ischemic animals were injected vehicle solution (50% alcohol in normal saline) in a dose of 0.4 ml/100 g body weight intraperitoneally.

4.1.3 Immunohistochemistry

On the post-operative day (POD) 4 or 25 (n=5 in each groups) animals were sacrificed under deep halothane anesthesia, and brains were isolated and immersion fixed first in 10% buffered paraformaldehyde for 2 days, then in 4% buffered paraformaldehyde for another 5 days. The brain tissues were then embedded into paraffin. From each animal five 20μm-thick coronal sections, 100 μm apart from each other were prepared as previously described (Mehra et al., 2005; Mehra et al., 2007). Goat anti-polyvalent IHC Staining Kit (Labvision, Neomaker lab, USA) was used according to manufacturer protocol for the immunohistochemical localization. SYP specific rabbit polyclonal antibody (Santa Cruz Biotechnology, USA) in a 1:200 dilution was used for 48 - 72 hours at 4 °C for the incubation. Sections were then incubated with biotinylated secondary antibody for 24 hours at 4 °C followed by streptavidin-HRP complex for 2 hours at RT. For proper maintenance of the cytoarchitectural integrity including preventing undesirable background staining, the sections were thoroughly rinsed with wash buffer (0.1M PBS, pH 7.4) between each incubation steps. Localization of the antigen-antibody site was done with the substrate-chromogen reaction using DAB. Immunoreactive sites became brownish under the bright field microscope. Adjacent sections were stained with cresyl violet (CV) to facilitate demarcation of various layers and subfields of the hippocampus. Intermittently some IHC stained sections were also counterstained with CV for the same purpose.

To eliminate non-specific staining, negative controls were processed by incubating the sections with species-specific normal serum, whereas human breast cancer tissue served as the positive controls (data not shown). Sections from all the groups and the immunohistochemical controls were processed simultaneously and repeatedly to rule out any procedural variations.

4.1.4 Image analysis

Semi-quantitative estimation of synaptophysin immunoreactivity (SYP-ir) was carried out on every layer (such as the stratum oriens, stratum pyramidale, and stratum radiatum) in the CA1 and CA2-3 subfields of the dorsal hippocampus in all animals. For the semi-quantitative analysis, integrated optical density (IOD) of SYP-ir was measured using image from five hippocampal sections of each animal, 100 μm apart from each other as previously described (Mehra et al., 2005). These images were viewed under the Nikon Microphot -Fx microscope mounted with a Cool Snap Digital camera (Roper Scientific, USA) and attached to the image analysis system driven by Image Pro-Plus software (v 6.2, Media Cybernetics, USA). The size of the sampling field was 5000 μm^2, where 7-9 non-overlapping digital photomicrographs per section were taken. The quantitative analysis was the same as previously described (Mehra et al., 2005). In brief, photomicrographs were first converted to gray scale with proper background correction, and a standard optical density curve was generated for each image prior to analysis (density of corpus callosum devoid of any pre-or postsynaptic protein was measured as background, and substracted from the image). IOD was measured as cumulative sum of the optical density of immunodense areas. Mathematical values of IOD for comparison between the groups were obtained as arbitrary units (mathematical algoritm) by the analysis software. Data from individual animals of each group were pooled together and the results were expressed as mean IOD ± SD.

4.1.5 Statistics

Statistical analysis was performed using GraphPad Prism version 5.02 for Windows (GraphPad Software, San Diego California USA). One-way ANOVA with post-hoc Newman-Keuls Multiple Comparison Test was done to compare mean IOD between groups. The level of significance was set at $p<0.05$.

4.2 Results

4.2.1 Synaptophysin immunoreactivity: Aspect of age

Cell loss in this model of cerebral brain ischemia can be detected both in CA1, and CA3 regions following 10 minutes occlusion, in contrast to the short occlusion times, where only the CA1 region is affected (Wappler et al., 2010), however, there are surviving cells that gives positive staining to synaptic markers. In the present study, the observed changes in SYP-ir were not limited to just one layer (stratum oriens, stratum pyramidale, or stratum radiatum), but the whole hippocampal region in each case, therefore we discuss our data using the CA1 or the CA3 hippocampal regions.

Young animals showed a significantly lower SYP-ir compared to the old animals after ovariectomy, which resulted in a significant difference between the baseline levels ($p<0.01$ young control vs. old control) (Fig.1., panel A). This might have been caused by the more pronounced change in the circulating estrogen level after ovariectomy in the young than in the old, where the estrogen production of the ovaries is low or there is no estrogen production at all. Due to this difference between the baseline values, age-comparisons were more difficult to make between age groups. Changes are therefore presented as percentages of the age-matched controls (Fig.1., panel B). Old gerbils had more severe synaptic loss in the CA1 area than young gerbils where no significant change following the injury was detected (see 4.2.2. for more details and significant differences among each age group). Estrogen treatment had, however, a positive impact on the synaptic density following ischemia in young animals in the CA3 region (see 4.2.2. for more details and significances between treated and non-treated groups), whereas, the same treatment was less effective in the old gerbils, but still helped imporved SYP-ir to a certain extent (Fig.1., panel B).

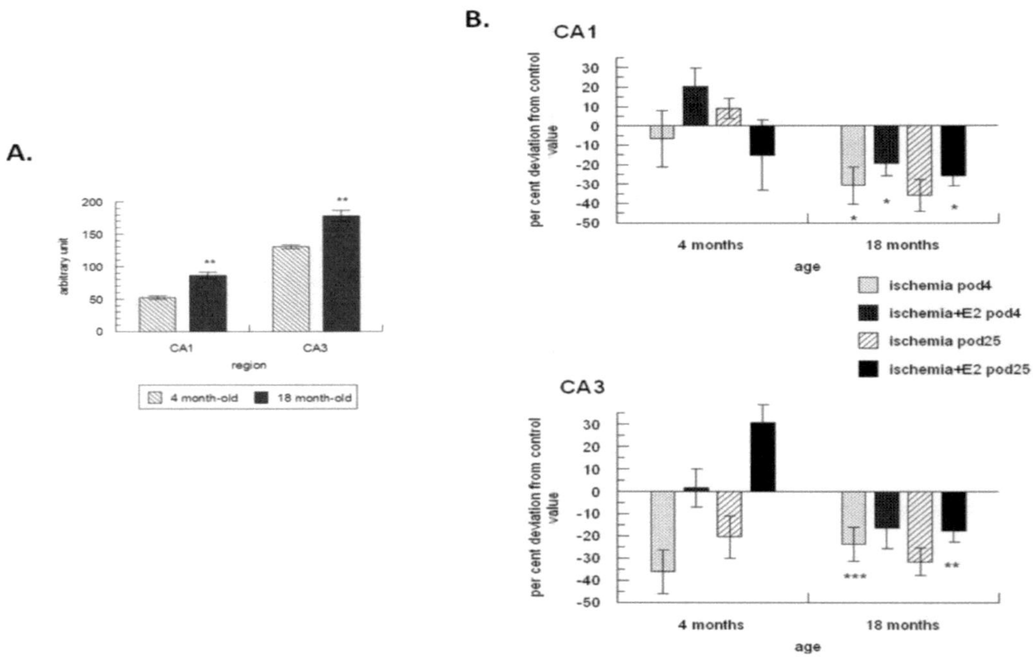

Fig. 1. SYP-ir in the gerbil hippocampus after global cerebral ischemia and estrogen treatment: age comparison. Panel A: control (OVX) groups. Data are expressed as means±SEM. See detailed description in the text. $^{**}p<0.01$ vs. young group same hippocampal region. Panel B: effect of ischemia and ischemia+E2 treatment. Data are expressed as percentages of the age-matched controls±SEM. See detailed description in the text. $^{*}p<0.05$, $^{**}p<0.01$, $^{***}p<0.001$ vs. the same treatment group of the 4 months old animals.

Please note that not every significant change is marked on this figure that you can find in the text.

4.2.2 Synaptophysin immunoreactivity: Aspect of time and synaptic regeneration

Cerebral ischemia itself decreased SYP-ir only in the CA2-3 region in the young, but decreased in both CA1 and CA2-3 regions in the old animals. This can be explained by the fact that ovariectomy itself decreases synaptic density in the young animals (Woolley and McEwen, 1992), and not every area is affected the same way. Areas that are more dependent on E2 to preserve their synapses might show a relatively lower decrease after ischemia, as theire baseline synaptic density is very low, and another stress that is not severe enough, can not cause a significant decrease. In contrast, ovariectomy does not make a significant difference in the circulating E2 levels in the old gerbils, however, there is a slight, but progressive loss of synaptic density and an increased vulnerability to ischemia with age (Ojo et al., 2011; Popa-Wagner et al., 2007), the latter causing a significant decrease even compared to a slightly lower control level. In addition, in the young, at the early time point there was an improvement due to E2 pre-treatment in the CA1 region ($p<0.05$ young ischemia POD4 vs. young ischemia+E2 POD4; significance not shown on figure), but at the later time point there was a decrement in synaptic density in the E2-treated group compared to the ischemic group ($p<0.05$ young ischemia POD25 vs. young ischemia+E2 POD25; significance not shown on figure). This can be explained by a higher estrogen-dependency and vulnerability in the CA1 region compared to the CA2-3 region in gerbils. In the old animals ischemia significantly decreased synaptic density in the CA1 region, however, we did not observe significant improvement with the estrogen pre-treatment in POD4. It only was observed at the late time point, which was close to be significant ($p=0.054$ old ischemia POD25 vs. old ischemia+E2 POD25). This is probably because of the decreasing tendency in SYP-ir following ischemia itself by POD25 that made a more prominent difference between the treated and non-treated group, as E2 treatment seemed to preserve the POD4 synaptic density level. We would like to note that there was no signs of recovery in the old animals following ischemia itself as SYP-ir did not icrease by POD25 compared to the POD4 value in the same group.

Moreover, in this study, estrogen pre-treatment increased synaptic density in the hippocampal CA2-3 region in both age groups, however, the increment was more pronounced in the young. The young group showed further increment in synaptic density following ischemia (Fig.2.). In the CA2-3 region at the early and late time point synaptic density increased following estrogen pre-treatment in the young group ($p<0.01$ young ischemia POD4 vs. young ischemia+E2 POD4; and $p<0.001$ young ischemia POD25 vs. young ischemia+E2 POD25; significances not shown on figure), in addition, following ischemia itself SYP-ir also increased in the young in the CA2-3 area ($p<0.05$ young ischemia POD4 vs. young ischemia POD25; significance not shown on figure). However, no significant changes were detected in the old group with the estrogen pre-treatment, and no regenerative changes were observed after ischemia itself either.

4.2.3 Summarizing our results

Even a single dose of E2 treatment can induce long term changes in synaptic density in the injured hippocampus in our gerbil model of cerebral ischemia. At the same time, differences in age, as well as differences in the investigated brain regions, modulate the degree and the permanence of this E2 effect.

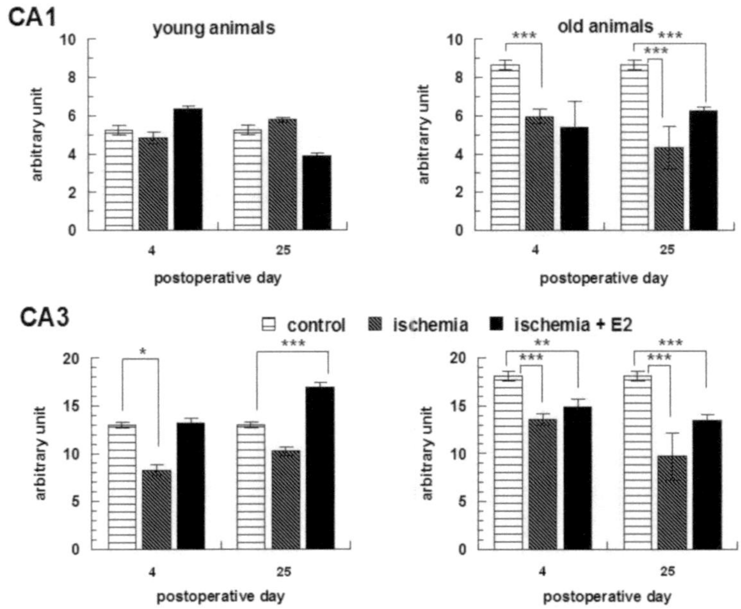

Fig. 2. SYP-ir in the gerbil hippocampus after global cerebral ischemia and estrogen treatment: dynamic changes with time. See detailed description in the text. Data are expressed as means±SEM. *p<0.05, ** p<0.01, ***p<0.001 vs. age-matched control. Please note that only significant changes vs. control groups are shown on the figure, differences between ischemia and ischemia+E2, etc. can be found in the text.

It was an unexpected result for us, that the CA3 region was more vulnerable to ischemia, in terms of synaptic loss, than the CA1 region in the young animals. This result, however, was probably due to the previous ovariectomy that might have had a bigger impact on the CA1 region. This is consistent with our results in the aged animals, where the ischemia decreased synaptic density more in the CA1 region, as expected in this model. It is also possible that aging may predispose to a tendency of diminished synaptogenesis and ability to improve synaptic density, especially in the hippocampal CA3 subfield. Our results emphasize the importance of investigating cerebral regenerative potential in older, female animals as well as at later time points following ischemic injury.

5. Conclusion

Post-ischemic brain regeneration is well documented at the tissue, cellular, and subcellular levels that offer further opportunities for drug development to improve functional outcome. In addition, estrogen, a well known regulator of synaptic density, has a long term impact on regeneration after global cerebral ischemia even as a single, high-dose treatment. Age, however, has influence on its effects, which highlights the importance of using old animals in this field.

6. Acknowledgement

This work was supported by OTKA-NKTH-K-69240. The authors would like to thank Paige Katz, PhD and Nancy Busija, MA for critical reading and editing this paper.

7. References

Altman, J; Das, G.D. (1965) Autoradiographic and histological evidence of postnatal hippocampal neurogenesis in rats. *J Comp Neurol,* Vol. 124, No. 3 (1965 Jun), pp. 319-335.

Altman, J.; Das, G.D. (1967) Postnatal neurogenesis in the guinea-pig. *Nature,* Vol. 214, No. 5093 (1967 Jun), pp. 1098-1101.

Amano, K.; Matsubara, H.; Iba, O.; Okigaki, M.; Fujiyama, S.; Imada, T.; Kojima, H.; Nozawa, Y.; Kawashima, S.; Yokoyama, M; Iwasaka, T. (2003) Enhancement of ischemia-induced angiogenesis by eNOS overexpression. *Hypertension,* Vol. 41, No. 1 (2003 Jan), pp. 156–162.

Araque, A.; Navarrete, M. (2010) Glial cells in neuronal network function. *Philos Trans R Soc Lond B Biol Sci,* Vol. 365, No. 1551 (2010 Aug), pp. 2375-2381.

Ardelt, A.A.; McCullough, L.D.; Korach, K.S.; Wang, M.M.; Munzenmaier, D.H.; Hurn, P.D. (2005) Estradiol regulates angiopoietin-1 mRNA expression through estrogen receptor-alpha in a rodent experimental stroke model. *Stroke,* Vol. 36, No. 2 (2005 Feb), pp. 337-341.

Attwell, D.; Buchan, A.M.; Charpak, S.; Lauritzen, M.; Macvicar, B.A.; Newman, E.A. Glial and neuronal control of brain blood flow. *Nature,* Vol. 468, No. 7321 (2010 Nov), pp. 232-243.

Back, T.; Hemmen, T.; Schüler, O.G. (2004) Lesion evolution in cerebral ischemia. J Neurol, Vol. 251, No. 4 (2004 Apr), pp. 388-397.

Bavelier, D.; Levi, D.M.; Li, R.W.; Dan, Y.; Hensch, T.K. (2010) Removing brakes on adult brain plasticity: from molecular to behavioral interventions. J Neurosci, Vol. 30, No. 45 (2010 Nov), pp. 14964-14971.

Becher, A; Drenckhahn, A; Pahner, I; Margittai, M; Jahn, R; Ahnert-Hilger, G. (1996) The synaptophysin-synaptobrevin complex: a hallmark of synaptic vesicle maturation. *J Neurosci,* Vol. 19, No .6 (1996 Mar), pp. 1922-1931.

Benowitz, L.I., Routtenberg, A. (1987) A membrane phosphoprotein associated with neural development, axonal regeneration, phospholipid metabolism, and synaptic plasticity. *Trends Neurosci,* Vol. 10, No. 12 (1987 Dec), pp. 527-532.

Benowitz, L.I.; Perrone-Bizzozero, N.I. (1991) The relationship of GAP-43 to the development and plasticity of synaptic connections. *Ann N Y Acad Sci,* Vol. 627 (1991), pp. 58-74.

Benowitz, L.I.; Carmichael, S.T. (2010) Promoting axonal rewiring to improve outcome after stroke. *Neurobiol Dis,* Vol. 37, No. 2 (2010 Feb), pp. 259-266.

Blizzard, C.A.; Chuckowree, J.A.; King, A.E.; Hosie, K.A.; McCormack, G.H.; Chapman, J.A.; Vickers, J.C.; Dickson, T.C. (2011) Focal damage to the adult rat neocortex induces wound healing accompanied by axonal sprouting and dendritic structural plasticity. *Cereb Cortex,* Vol. 21, No .2 (2011 Feb), pp.281-291.

Bourque, M.; Dluzen, D.E.; Di Paolo, T. (2009) Neuroprotective actions of sex steroids in Parkinson's disease. *Front Neuroendocrinol,* Vol. 30, No. 2 (2009 Jul), pp. 142-157.

Brann, D.; Raz, L.; Wang, R.; Vadlamudi, R.; Zhang, Q. (2011) Estrogen signaling and neuroprotection in cerebral ischemia. *J Neuroendocrinol*, in press

Brown, C.E.; Wong, C.; Murphy, T.H. (2008) Rapid morphologic plasticity of peri-infarct dendritic spines after focal ischemic stroke. *Stroke*, Vol. 39, No. 4 (2008 Apr), pp. 1286-1291.

Brown, C.E.; Boyd, J.D.; Murphy, T.H. (2010) Longitudinal in vivo imaging reveals balanced and branch-specific remodeling of mature cortical pyramidal dendritic arbors after stroke. *J Cereb Blood Flow Metab*, Vol. 30, No. 4 (2010 Apr), pp. 783-91.

Busch, H.J.; Buschmann, I.R.; Mies, G.; Bode, C.; Hossmann, K.A. (2003) Arteriogenesis in hypoperfused rat brain. *J Cereb Blood Flow Metab, Vol.* 23, No. 5 (2003 May), pp.621–628.

Carmeliet, P.; Collen, D. (1997) Molecular analysis of blood vessel formation and disease. *Am J Physiol*, Vol. 273, No. 5 (1997 Nov), pp. H2091-104.

Carmichael, S.T. (2003) Plasticity of cortical projections after stroke. *Neuroscientist*, Vol. 9, No. 1 (2003 Feb), pp. 64-75.

Carmichael, S.T. (2006) Cellular and molecular mechanisms of neural repair after stroke: making waves. *Ann Neurol*, Vol. 59, No. 5 (2006 May), pp. 735-742.

Céspedes-Rubio, A.; Jurado, F.W.; Cardona-Gómez, G.P. (2010) p120 catenin/αN-catenin are molecular targets in the neuroprotection and neuronal plasticity mediated by atorvastatin after focal cerebral ischemia. *J Neurosci Res*, Vol. 88, No. 16 (2010 Dec), pp. 3621-3634.

Chen, H.Y.; Hung, Y.C.; Chen, T.Y.; Huang, S.Y.; Wang, Y.H.; Lee, W.T.; Wu, T.S.; Lee, E.J. (2009) Melatonin improves presynaptic protein, SNAP-25, expression and dendritic spine density and enhances functional and electrophysiological recovery following transient focal cerebral ischemia in rats. *J Pineal Res*, Vol. 47, no. 3 (2009 Oct), pp. 260-270.

Christoforidis, G.A.; Mohammad, Y.; Kehagias, D.; Avutu, B.; Slivka, A.P. (2005) Angiographic assessment of pial collaterals as a prognostic indicator following intra-arterial thrombolysis for acute ischemic stroke. *AJNR Am J Neuroradiol*, Vol. 26, No. 7 (2005 Aug), pp. 1789–1797.

Cole, G.J.; McCabe, C.F. (1991) Identification of a developmentally regulated keratin sulfate proteoglycan that inhibits cell adhesion and neurite outgrowth. *Neuron*, Vol. 7, No. 6 (1991 Dec), pp. 1007-1018.

Connell, B.J.; Crosby, K.M.; Richard, M.J.; Mayne, M.B.; Saleh, T.M. (2007) Estrogen-mediated neuroprotection in the cortex may require NMDA receptor activation. *Neuroscience*, Vol. 146, No. 1 (2007 Apr), pp. 160–169.

Corpéchot, C.; Robel, P.; Axelson, M.; Sjövall, J.; Baulieu, E.E. (1981) Characterization and measurement of dehydroepiandrosterone sulfate in rat brain. *Proc Natl Acad Sci USA*, Vol. 78, No. 8 (1981 Aug), pp. 4704-4707.

Cui, X.; Chopp, M.; Zacharek, A.; Zhang, C.; Roberts, C.; Chen, J. (2009) Role of endothelial nitric oxide synthetase in arteriogenesis after stroke in mice. *Neuroscience*, Vol. 159, No. 2 (2009 Mar), pp. 744-750.

Culligan, K. and Ohlenieck, K. (2002) Diversity of the brain dystrophin-glycoprotein complex. *J. Biomed. Biotechnol.*, Vol. 2, No. 1 (2002), pp. 31–36.

De Camilli, P.; Harris, S.M. Jr.; Huttner, W.B.; Greengard, P. (1983) Synapsin I (Protein I), a nerve terminal-specific phosphoprotein. II. Its specific association with synaptic vesicles demonstrated by immunocytochemistry in agarose-embedded synaptosomes. *J Cell Biol.* Vol. 96, No. 5 (1983 May), pp. 1355-1373.

del Zoppo, G.J.; Mabuchi, T. (2003) Cerebral microvessel responses to focal ischemia. *J Cereb Blood Flow Metab,* Vol. 23, No. 8 (2003 Aug), pp. 879-894.

Di Filippo, M.; Tozzi, A.; Costa, C.; Belcastro, V.; Tantucci, M.; Picconi, B; Calabresi, P. (2008) Plasticity and repair in the post-ischemic brain. *Neuropharmacology,*Vol. 55, No. 3 (2008 Sep), pp. 353-362.

Dubal, D.B.; Zhu, H.; Yu, J.; Rau, S.W.; Shughrue, P.J.; Merchenthaler, I.; Kindy, M.S.; Wise, P.M. (2001) Estrogen receptor alpha, not beta, is a critical link in estradiol-mediated protection against brain injury. *Proc Natl Acad Sci USA,* Vol. 98, No. 4 (2001 Feb), pp. 1952-1957.

Duggal, N.; Schmidt-Kastner, R.; Hakim, A.M. (1997) Nestin expression in reactive astrocytes following focal cerebral ischemia in rats. *Brain Res,* Vol. 768, No. 1-2 (1997 Sep), pp. 1-9.

Erdő, F.; Hossmann, K.A. (2007) Animal models of cerebral ischemia--testing therapeutic strategies in vivo. Ideggyogy Sz, Vol. 60, No. 9-10 (2007 Sep), pp. 356-369.

Eshkind, L.G.; Leube, R.E. (1995) Mice lacking synaptophysin reproduce and form typical synaptic vesicles. *Cell issue Res,* Vol. 282, No. 3 (1995 Dec), pp. 423– 433.

Fidler, P.S.; Schuette, K.; Asher, R.A.; Dobbertin, A.; Thornton, S.R.; Calle-Patino, Y.; Muir, E.; Levine, J.M.; Geller, H.M.; Rogers, J.H.; Faissner, A.; Fawcett, J.W. (1999) Comparing astrocytic cell lines that are inhibitory or permissive for axon growth: the major axon-inhibitory proteoglycan is NG2. *J Neurosci,* Vol. 19, No. 20 (1999 Oct), pp. 8778-8788.

Filardo, E.J.; Quinn, J.A.; Frackelton, A.R. Jr; Bland, K.I. (2002) Estrogen Action Via the G Protein-Coupled Receptor, GPR30: Stimulation of Adenylyl Cyclase and cAMP-Mediated Attenuation of the Epidermal Growth Factor Receptor-to-MAPK Signaling Axis. *Mol Endocrinol,* Vol. 16, No. 1 (2002 Jan), pp. 70–84.

Font, A.M.; Arboix A.; Krupinski J. (2010) Angiogenesis, neurogenesis and neuroplasticity in ischemic stroke. *Curr Cardiol Rev.* Vol. 6, No 3 (2010 Aug), pp. 238-244.

Funakoshi, T.; Yanai, A.; Shinoda, K.; Kawano, M.M.; Mizukami, Y. (2006) G protein-coupled receptor 30 is an estrogen receptor in the plasma membrane. *Biochem Biophys Res Commun,* Vol. 346, No. 3 (2006 Aug), pp. 904-910.

Gál, A.; Pentelényi, K.; Reményi, V.; Wappler E.A.; Sáfrány G.; Skopál J.; Nagy Z. (2009) Bcl-2 or bcl-XL gene therapy increases neural plasticity proteins nestin and c-fos expression in PC12 cells. *Neurochem Int.* Vol. 55, No. 5 (2009 Sept), pp.349-353.

Gao, X.; Enikolopov, G.; Chen, J. (2009) Moderate traumativ brain injury promotes proliferation of quiescent neural progenitors in the adult hippocampus. *Exp Neurol,* Vol. 219, No. 2 (2009 Oct), pp. 516-523.

Gingerich, S.; Kim, G.L.; Chalmers, J.A.; Koletar, M.M.; Wang, X.; Wang, Y.; Belsham, D.D. (2010) Estrogen receptor α and G-protein coupled receptor 30 mediate the neuroprotective effects of 17β-estradiol in novel murine hippocampal cell models. *Neuroscience,* Vol. 170, No. 1 (2010 Sep), pp. 54-66.

Goldstein, L.B. (2009) Statins and ischemic stroke severity: cytoprotection. *Curr Atheroscler Rep,* Vol. 11, No. 4 (2009 Jul), pp. 296-300.

González-Burgos, I.; Letechipía-Vallejo, G.; López-Loeza, E.; Moralí, G.; Cervantes, M. (2007) Long-term study of dendritic spines from hippocampal CA1 pyramidal cells, after neuroprotective melatonin treatment following global cerebral ischemia in rats. *Neurosci Lett,* Vol. 423, No. 2 (2007 Aug), pp. 162-166.

Gould, E.; Reeves, A.J.; Graziano, M.S.; Gross, C.G. (1999) Neurogenesis in the neocortex of adult primates. *Science,* Vol. 286, No. 5439 (1999 Oct), pp. 548-552.

Haggani, A.S.; Nesic, M.; Preston, E.; Baumann, E.; Kelly, J.; Stanimirovic, D. (2005) Characterization of vascular protein expression patterns in cerebral ischemia/reperfusion using laser capture microdissection and ICAT-nanoLC-MS/MS. *FASEB J,* Vol. 19, No. 13 (2005 Nov), pp. 1809-1821.

Heil, M.; Schaper, W. (2004) Influence of mechanical, and molecular factors on collateral artery growth (arteriogenesis). *Circ. Res.,* 95 (2004), pp. 449–458.

Heil, M.; Eitenmüller, I.; Schmitz-Rixen, T. and Schaper, W. (2006) Arteriogenesis versus angiogenesis: similarities and differences. *J. Cell. Mol. Med.,* Vol. 10, No. 1 (2006 Jan-Mar), pp. 45–55.

Hermann, D.M.; Zechariah, A. (2009) Implications of vascular endothelial growth factor for postischemic neurovascular remodeling. *J Cereb Blood Flow Metab,* Vol. 29, No. 10 (2009 Oct), pp. 1620-1643.

Herson, P.S.; Koerner, I.P.; Hurn, P.D. (2009) Sex, sex steroids and brain injury. *Semin Reprod Med,* Vol. 27, No. 3 (2009 May), pp. 229–239.

Hikino, M.; Mikami, T.; Faissner, A.; Vilela-Silva, A.C.; Pavão, M.S.; Sugahara, K. (2003) Oversulfated dermatan sulfate exhibits neurite outgrowth-promoting activity toward embryonic mouse hippocampal neurons: implications of dermatan sulfate in neuritogenesis in the brain. *J Biol Chem,* Vol. 278, No 44 (2003 Oct), pp. 43744-43754.

Himeda, T.; Tounai, H.; Hayakawa, N.; Araki, T. (2007) Postischemic alterations of BDNF, NGF, HSP 70 and ubiquitin immunoreactivity in the gerbil hippocampus: pharmacological approach. *Cell Mol Neurobiol,* Vol. 27, No. 2 (2007 Mar), pp. 229-250.

Holahan, M.R.; Honegger, K.S.; Tabatadze, N.; Routtenberg, A. (2007) GAP-43 gene expression regulates information storage. *Learn Mem,* Vol. 14, No. 6 (2007 Jun), pp. 407-415.

Hojo, Y.; Hattori, T.A.; Enami, T.; Furukawa, A.; Suzuki, K.; Ishii, H.T.; Mukai, H.; Morrison, J.H.; Janssen, W.G.; Kominami, S.; Harada, N.; Kimoto, T.; Kawato, S. (2004) Adult male rat hippocampus synthesizes estradiol from pregnenolone by cytochromes P45017alpha and P450 aromatase localized in neurons. *Proc Natl Acad Sci USA,* Vol. 101, No. 3 (2004 Jan), pp. 865-870.

Holmin, S.; Almqvist, P.; Lendahl, U.; Mathiesen, T. (1997) Adult nestin-expressing subependymal cells differentiate to astrocytes in response to brain injury. *Eur J Neurosci,* Vol. 9, No. 1 (1997 Jan), pp. 65-75.

Hurn, P.D.; Littleton-Kearney, M.T.; Kirsch, J.R.; Dharmarajan, A.M.; Traystman, R.J. (1995) Postischemic cerebral blood flow recovery in the female: effect of 17 betaestradiol. *J Cereb Blood Flow Metab,* Vol. 15, No. 4 (1995 Jul), pp. 666–672.

Iadecola, C. & Nedergaard, M. (2007) Glial regulation of the cerebral microvasculature. *Nat. Neurosci,* Vol. 10, No. 11 (2007 Nov), pp. 1369–1376.

Issa, R.; AlQteishat, A.; Mitsios, N.; Saka, M.; Krupinski, J.; Tarkowski, E.; Gaffney, J.; Slevin, M.; Kumar, S.; Kumar P. (2005) Expression of basic fibroblast growth factor mRNA and protein in the human brain following ischaemic stroke. *Angiogenesis,* Vol. 8, No. 1 (2005), p. 53-62.

Iwai, M.; Stetler, R.A.; Xing, J.; Hu, X.; Gao, Y.; Zhang, W.; Chen, J.; Cao, G. (2010) Enhanced oligodendrogenesis and recovery of neurological function by erythropoietin after neonatal hypoxic/ischemic brain injury. *Stroke,* Vol. 41, No. 5 (2010 May), pp. 1032-1037.

Janz, R.; Sudhof, T.C. (1998) Cellugyrin, a novel ubiquitous form of synaptogyrin that is phosphorylated by pp60c-src. *J Biol Chem,* Vol. 273, No. 5 (1998 Jan), pp. 2851–2857.

Jin, K.; Zhu, Y.; Sun, Y.; Mao, X.O.; Xie, L.; Greenberg, D.A. (2002) Vascular endothelial growth factor (VEGF) stimulates neurogenesis in vitro and in vivo. *Proc Natl Acad Sci USA,* Vol. 99, No. 18 (2002 Sep), pp. 11946–11950.

Jones, T.A.; Schallert, T. (1992) Overgrowth and pruning of dendrites in adult rats recovering from neocortical damage. *Brain Res,* Vol. 581, No. 1 (1992 May), pp. 156-60.

Kaas, J.H.; Merzenich, M.M.; Killackey, H.P. (1983) The reorganization of somatosensory cortex following peripheral nerve damage in adult and developing mammals. *Annu Rev Neurosci.* Vol. 6 (1983), pp.325–356.

Kálmán, M.; Ajtai, B.M. (2001) A comparison of intermediate filament markers for presumptive astroglia in the developing rat neocortex: immunostaining against nestin reveals more detail, than GFAP or vimentin. *Int J Dev Neurosci,* Vol. 19, No. (2001 Feb), pp. 101-108.

Kannappan, R.; Gupta, S.C.; Kim, J.H.; Reuter, S.; Aggarwal, B.B. (2011) Neuroprotection by spice-derived nutraceuticals: you are what you eat! *Mol Neurobiol,* Vol. 44, No. 2 (2011 Oct), pp. 142-59.

Katoh-Semba, R.; Matsuda, M.; Kato, K.; Oohira, A. (1995) Chondroitin sulphate proteoglycans in the rat brain: candidates for axon barriers of sensory neurons and the possible modification by laminin of their actions. *Eur J Neurosci,* Vol. 7, No. 4 (1995 Apr), pp. 613-621.

Kim, H.J.; Leeds, P.; Chuang, D.M. (2009) The HDAC inhibitor, sodium butyrate, stimulates neurogenesis in the ischemic brain. *J Neurochem,* Vol. 110, No. 4 (2009 Aug), pp. 1226-1240.

Kim, S.Y.; Jones, T.A. (2010) Lesion size-dependent synaptic and astrocytic responses in cortex contralateral to infarcts in middle-aged rats. *Synapse.* Vol. 64, No. 9 (2010 Sep), pp. 659-671.

Kojima, T.; Hirota, Y.; Ema, M.; Takahashi, S.; Miyoshi, I.; Okano, H.; Sawamoto, K. (2010) Subventricular zone-derived neural progenitor cells migrate along a blood vessel scaffold toward the post-stroke striatum. *Stem Cells,* Vol. 28, No 3 (2010 Mar), pp. 545-554.

Kokaia, Z.; Lindvall, O. (2003) Neurogenesis after ischemic brain insults. *Curr Opin Neurobiol,* 2003, Vol. 13, No. 1 (2003 Feb), pp. 127-132.

Koszegi Z, Szego EM, Cheong RY, Tolod-Kemp E, Abrahám IM. (2011) Postlesion Estradiol Treatment Increases Cortical Cholinergic Innervations via Estrogen Receptor-{alpha} Dependent Nonclassical Estrogen Signaling in Vivo. *Endocrinology,* Vol. 152, No. 9 (2011 Sep), pp. 3471-3482.

Krupinski, J.; Kaluza, J.; Kumar, P.; Kumar, S.; Wang, J.M. (1994) Role of angiogenesis in patients with cerebral ischemic stroke. *Stroke,* Vol. 25, No. 9 (1994 Sep), pp. 1794-1798.

Krupinski, J; Kumar, P.; Kumar, S.; Kaluza, J. (1996) Increased expression of TGF-beta 1 in brain tissue after ischemic stroke in humans. *Stroke,* Vol. 27, No. 5 (1996 May), pp. 852-857.

Krupinski, J.; Issa, R.; Bujny, T.; Slevin, M.; Kumar, P.; Kumar, S.; Kaluza, J (1997). A putative role for platelet-derived growth factor in angiogenesis and neuroprotection after ischemic stroke in humans. *Stroke,* Vol. 28, No. 3 (1997 Mar), pp. 564-573.

Lebesgue, D.; Chevaleyre, V.; Zukin, R.S.; Etgen, A.M. (2009) Estradiol rescues neurons from global ischemia-induced cell death: multiple cellular pathways of neuroprotection. *Steroids,* Vol. 74, No. 7 (2009 Jul), pp. 555-561.

Lebesgue, D.; Traub, M.; De Butte-Smith, M.; Chen, C.; Zukin, R.S.; Kelly, M.J.; Etgen, A.M. (2010) Acute administration of non-classical estrogen receptor agonists attenuates ischemia-induced hippocampal neuron loss in middle-aged female rats. *PLoS One,* Vol. 5, No. 1 (2010 Jan), pp. e8642.

Le Goascogne, C.; Robel, P.; Gouézou, M.; Sananès, N.; Baulieu, E.E.; Waterman, M. (1987) Neurosteroids: cytochrome P-450scc in rat brain. *Science,* Vol. 237, No. 4819 (1987 Sep), pp. 1212-1215.

Lee, W.C.; Huang, H.; Feng, G.; Sanes, J.R.; Brown, E.N.; So, P.T.; Nedivi, E. (2006) Dynamic remodeling of dendritic arbors in GABAergic interneurons of adult visual cortex. *PLoS Biol,* Vol. 4, No. 2 (2006 Feb), pp. e29. Erratum in: *PLoS Biol,* Vol. 4, No 5 (2006 May), e126.

Leker, R.R.; Toth, Z.E.; Shahar, T.; Cassani-Ingoni, R.; Szalayova, I.; Key, S.; Bratnicsak, A.; Mezey, E. (2009) Transforming growth factor alpha induces angiogenesis and neurogenesis following stroke. Neuroscience, Vol. 163, No. 1 (2009 Sep), pp. 233-243.

Levy, A.P.; Levy, N.S.; Wegner, S.; Goldberg, M.A. (1995) Transcriptional regulation of the rat vascular endothelial growth factor gene by hypoxia. *J Biol Chem,* Vol. 270, No. 22 (1995 Jun), pp. 13333-13340.

Li, S.; Carmichael, S.T. (2006) Growth-associated gene and protein expression in the region of axonal sprouting in the aged brain after stroke. *Neurobiol Dis,* Vol. 23, No. 2 (2006 Aug), pp. 362-373.

Lin, T.N.; Wang, C.K.; Cheung, W.M.; Hsu, C.Y. (2000) Induction of angiopoietin and Tie receptor mRNA expression after cerebral ischemia-reperfusion. J*Cereb Blood Flow Metab,* Vol. 20, No. 2 (2000 Feb), pp. 387-395.

Liu, H.S.; Shen, H.; Harvey, B.K.; Castillo, P.; Lu, H.; Yang, Y.; Wang, Y. (2011) Post-treatment with amphetamine enhances reinnervation of the ipsilateral side cortex in stroke rats. *Neuroimage,* Vol. 56, No. 1 (2011 May), pp. 280-289.

Liu, M.; Dziennis, S.; Hurn, P.D.; Alkayed, N.J. (2009) Mechanisms of gender-linked ischemic brain injury. *Restor Neurol Neurosci,* Vol. 27, No. 3 (2009), pp. 163-179.

Liu, R.; Wen, Y.; Perez, E.; Wang, X.; Day, A.L.; Simpkins, J.W.; Yang, S.H. (2005) 17β-estradiol attenuates blood–brain barrier disruption induced by cerebral ischemia-reperfusion injury in female rats. *Brain Res,* Vol. 1060, No. 1-2 (2005 Oct), pp. 55–61

Liu, Y.; Cox, S.R.; Morita, T.; Kourembanas, S. (1995) Hypoxia regulates vascular endothelial growth factor gene expression in endothelial cells. Identification of a 5' enhancer. *Circ Res,* Vol. 77, No. 3 (1995 Sep), pp. 638-643.

Lledo, P.M.; Alonso, M.; Grubb, M.S. (2006) Adult neurogenesis and functional plasticity in neuronal circuits. *Nature Reviews Neuroscience,* Vol. 7, No. 3 (2006 Mar), pp. 179-193.

Lu, D.; Goussev, A.; Chen, J.; Pannu, P.; Li, Y.; Mahmood, A.; Chopp, M. (2004) Atorvastatin reduces neurological deficit and increases synaptogenesis, angiogenesis, and neuronal survival in rats subjected to traumatic brain injury. *J Neurotrauma,* Vol. 21, No. 1 (2004 Jan), pp. 21-32.

Lynch, G.; Deadwyler, S.; Cotman, G. (1973) Postlesion axonal growth produces permanent functional connections. *Science,* Vol. 180, No. 4093 (1973 Jun), pp. 1364-1366.

Madak-Erdogan, Z.; Kieser, K.J.; Kim, S.H.; Komm, B.; Katzenellenbogen, J.A.; Katzenellenbogen, B.S. (2008) Nuclear and extranuclear pathway inputs in the regulation of global gene expression by estrogen receptors. *Mol Endocrinol,* Vol. 22, No. 9 (2008 Sep), pp. 2116-2127.

Matsuda, K.; Sakamoto, H.; Mori, H.; Hosokawa, K.; Kawamura, A.; Itose, M.; Nishi, M.; Prossnitz, E.R.; Kawata, M. (2008) Expression and intracellular distribution of the G protein-coupled receptor 30 in rat hippocampal formation. *Neurosci Lett,* Vol. 441, No. 1 (2008 Aug), pp. 94-99.

McEwen, B. (2002) Estrogen actions throughout the brain. *Recent Prog Horm Res,* Vol. 57 (2002), pp. 357-384.

McKenna, N.J.; O'Malley, B.W. (2002) Combinatorial control of gene expression by nuclear receptors andcoregulators. *Cell,* Vol. 108, No. 4 (2002 Feb), pp. 465–74.

McMahon, H.T.; Boshakov, V.Y.; Janz, R.; Hammer, R.E.; Siegelbaum, S.A.; Sudhof, T.C. (1996) Synaptophysin, a major vesicle protein, is not essential for neurotransmitter release. *Proc Natl Acad Sci USA,* Vol. 93, No. 10 (1996 May), pp. 4760–4764.

Mehra, R.D.; Sharma, K.; Nyakas, C.; Vij, U. (2005) Estrogen receptor alpha and beta immunoreactive neurons in normal adult and aged female rat hippocampus: a qualitative and quantitative study. *Brain Res,* Vol. 1056, No. 1 (2005 Sep), pp. 22-35.

Merchenthaler, I.; Dellovade, T.L.; Shughrue, P.J. (2003) Neuroprotection by estrogen in animal models of global and focal ischemia. *Ann N Y Acad Sci,* Vol. 1007 (2003 Dec), pp. 89-100.

Metea, M. R. & Newman, E. A. (2006) Glial cells dilate and constrict blood vessels: a mechanism of neurovascular coupling. *J Neurosci,* Vol. 26, No. 11 (2008 Mar), pp. 2862–2870.

Miller, N.R.; Jover, T.; Cohen, H.W.; Zukin, R.S.; Etgen, A.M. (2005) Estrogen can act via estrogen receptor alpha and beta to protect hippocampal neurons against global ischemia-induced cell death. *Endocrinology,* Vol. 146, No. 7 (2005 Jul), pp. 3070-3079.

Milner, T.A.; Ayoola, K.; Drake, C.T.; Herrick, S.P.; Tabori, N.E.; McEwen, B.S.; Warrier, S.; Alves, S.E. (2005) Ultrastructural localization of estrogen receptor beta immunoreactivity in the rat hippocampal formation. *J Comp Neurol,* Vol. 491, No. 2 (2005 Oct), pp. 81-95.

Mostany, R.; Chowdhury, T.G.; Johnston, D.G.; Portonovo, S.A.; Carmichael, S.T.; Portera-Cailliau C. (2010) Local hemodynamics dictate long-term dendritic plasticity in peri-infarct cortex. *J Neurosci,* Vol. 30, No. 42 (2010 Oct), pp. 14116–14126.

Mostany, R.; Portera-Cailliau, C. (2011) Absence of large-scale dendritic plasticity of layer 5 pyramidal neurons in peri-infarct cortex. *J Neurosci,* Vol. 31, No. 5 (2011 Feb), pp. 1734-1738.

Murohara, T.; Asahara, T.; Silver, M.; Bauters, C.; Masuda, H.; Kalka, C.; Kearney, M.; Chen, D.; Symes, J.F.; Fishman, M.C.; Huang, P.L.; Isner, J.M. (1998) Nitric oxide synthase modulates angiogenesis in response to tissue ischemia. *J Clin Invest,* Vol. 101, No. 11 (1998 Jun), pp. 2567–2578.

Nakagomi, T.; Taguchi, A.; Fujimori, Y.; Saino, O.; Nakano-Doi, A.; Kubo, S.; Gotoh, A.; Soma, T.; Yoshikawa, H.; Nishizaki, T.; Nakagomi, N.; Stern, D.M.; Matsuyama, T. (2009) Isolation and characterization of neural stem/progenitor cells from post-stroke cerebral cortex in mice. *Eur J Neurosci,* Vol. 29, No. 9 (2009 Apr), pp. 1842-1852.

Nakanishi, K.; Aono, S.; Hirano, K.; Kuroda, Y.; Ida, M.; Tokita, Y.; Matsui, F.; Oohira, A. (2006) Identification of neurite outgrowth-promoting domains of neuroglycan C, a brain-specific chondroitin sulfate proteoglycan, and involvement of phosphatidylinositol 3-kinase and protein kinase C signaling pathways in neuritogenesis. *J Biol Chem,* Vol. 281, No. 34 (2006 Aug), pp. 24970-24978.

Navarro-Sobrino, M.; Rosell, A.; Hernández-Guillamon, M.; Penalba, A.; Boada, C.; Domingues-Montanari, S.; Ribó, M.; Alvarez-Sabín, J.; Montaner, J. (2011) A large screening of angiogenesis biomarkers and their association with neurological outcome after ischemic stroke. *Atherosclerosis,* Vol. 216, No. 1 (2011 May), pp. 205-211.

Nedergaard, M.; Dirnagl, U. (2005) Role of glial cells in cerebral ischemia. *Glia*. Vol. 50, No. 4 (2005 Jun), pp. 281-286.

Nilsson, S.; Mäkelä, S.; Treuter, E.; Tujague, M.; Thomsen, J.; Andersson, G.; Enmark, E.; Pettersson, K.; Warner, M.; Gustafsson, J.A. (2001) Mechanisms of Estrogen Action. *Physiol Rev,* Vol. 81, No. 4 (2001 Oct), pp. 1535–1565.

Nudo, R.J. (2006) Mechanisms for recovery of motor function following cortical damage. *Curr Opin Neurobiol,* Vol. 16, No. 6 (2006 Dec), pp. 638–464.

Nudo, R.J. (2007) Postinfarct cortical plasticity and behavioral recovery. *Stroke*. (February 2007); Vol. 38, No. 2 Suppl, pp. 840-845.

Ojo, B.; Rezaie, P.; Gabbott, P.L.; Davies, H.; Colyer, F.; Cowley, T.R.; Lynch, M.; Stewart, M.G. (2011) Age-related changes in the hippocampus (loss of synaptophysin and glial-synaptic interaction) are modified by systemic treatment with an NCAM-derived peptide, FGL. *Brain Behav Immun,* 2011 Oct, in press.

Papapetropoulos, A.; Garcia-Cardena, G.; Madri, J.A.; Sessa, W.C. (1997) Nitric oxide production contributes to the angiogenic properties of vascular endothelial growth factor in human endothelial cells. *J Clin Invest,* Vol. 100, No. 12 (1997 Dec), pp. 3131–3139.

Pelligrino, D.A.; Santizo, R.; Baughman, V.L.; Wang, Q. (1998) Cerebral vasodilating capacity during forebrain ischemia: effects of chronic estrogen depletion and repletion and the role of neuronal nitric oxide synthase. Neuroreport, Vol. 9, No. 14 (1998 Oct), pp. 3285–3291.

Plahta, W.C.; Clark, D. L.; Colbourne, F. (2004) 17beta-estradiol pretreatment reduces CA1 sector cell death and the spontaneous hyperthermia that follows forebrain ischemia in the gerbil. *Neuroscience,* Vol. 129, No. 1 (2004 Sep), pp. 187-193.

Prior, B.M.; Lloyd, P.G.; Ren, J.; Li, Z.; Yang, H.T.; Laughlin, M.H.; Terjung, R.L. (2003) Arteriogenesis: role of nitric oxide. *Endothelium,* Vol. 10, No. 4-5 (2003), pp. 207–216.

Popa-Wagner, A.; Carmichael, S.T.; Kokaia, Z.; Kessler, C.; Walker, L.C. (2007) The response of the aged brain to stroke: too much, too soon? *Curr Neurovasc Res,* Vol. 4, No. 3 (2007 Aug), pp. 216-227.

Rolls, A.; Shechter, R.; Schwartz, M. (2009) The bright side of the glial scar in CNS repair. *Nat Rev Neurosci,* Vol. 10, No. 3 (2009 Mar, pp. 235-241.

Rebas, E.; Lachowicz, L.; Lachowicz, A. (2005) Estradiol modulates the synapsins phosphorylation by various protein kinases in the rat brain under in vitro and in vivo conditions. *J Physiol Pharmacol,* Vol. 56, No. 1 (2005 Mar), pp. 39-48.

Rosahl, T.W.; Geppert, M.; Spillane, D.; Herz, J.; Hammer, R.E.; Malenka, R.C.; Südhof, T.C. (1993) Shortterm synaptic plasticity is altered in mice lacking synapsin I. *Cell,* Vol. 75, No. 4 (1993 Nov), pp. 661–670.

Rudic, R.D.; Shesely, E.G.; Maeda, N.; Smithies, O.; Segal, S.S.; Sessa, W.C. (1998) Direct evidence for the importance of endothelium-derived nitric oxide in vascular remodeling. *J Clin Invest,* Vol. 101, No. 4 (1998 Feb), pp. 731–736.

Rudge, J.S.; Silver, J. (1990) Inhibition of neurite outgrowth on astroglial scars in vitro. *J Neurosci,* Vol. 10, No. 11 (1990 Nov), pp. 3594-3603.

Rune, G.M.; Lohse, C.; Prange-Kiel, J.; Fester, L.; Frotscher, M. (2006) Synaptic plasticity in the hippocampus: effects of estrogen from the gonads or hippocampus? *Neurochem Res,* Vol. 31, No. 2 (2006 Feb), pp.145-155.

Sá, S.I.; Lukoyanova, E.; Madeira, M.D. (2009) Effects of estrogens and progesterone on the synaptic organization of the hypothalamic ventromedial nucleus. *Neuroscience,* Vol. 162, No. 2 (2009 Aug), pp. 307-316.

Sharma, K.; Mehra, R.D.; Dhar, P.; Vij, U. (2007) Chronic exposure to estrogen and tamoxifen regulates synaptophysin and phosphorylated cAMP response element-binding (CREB) protein expression in CA1 of ovariectomized rat hippocampus. *Brain Res,* Vol. 1132, No. 1 (2007 Feb), pp. 10-19.

Sharma, K.; Mehra, R.D. (2008) Long-term administration of estrogen or tamoxifen to ovariectomized rats affords neuroprotection to hippocampal neurons by modulating the expression of Bcl-2 and Bax. *Brain Res,* Vol. 1204 (2008 Apr), pp. 1-15.

Scheff, S.W.; Price, D.A.; Hicks, R.R.; Baldwin, S.A.; Robinson, S.; Brackney, C. (2005) Synaptogenesis in the hippocampal CA1 field following traumatic brain injury. *J Neurotrauma,* Vol. 22, No. 7 (2005 Jul), pp. 719-732.

Schiebler, W.; Jahn, R.; Doucet, J.P.; Rothlein, J.; Greengard, P. (1986) Characterization of synapsin I binding to small synaptic vesicles. *J Biol Chem,* Vol. 261, No. 26 (1986 Sept), pp. 8383–8390.

Schierling, W.; Troidl, K.; Mueller, C.; Troidl, C.; Wustrack, H.; Bachmann, G.; Kasprzak, P.M.; Schaper, W.; Schmitz-Rixen, T. (2009) Increased intravascular flow rate triggers cerebral arteriogenesis. *J Cereb Blood Flow Metab,* Vol. 29, No. 4 (2009 Apr), pp. 726–737.

Shen, C.C.; Yang, Y.C.; Chiao, M.T.; Cheng, W.Y.; Tsuei, Y.S.; Ko, J.L. (2010) Characterization of endogenous neural progenitor cells after experimental ischemic stroke. *Curr Neurovasc Res,* Vol. 7, No. 1 (2010 Feb), pp. 6-14.

Shughrue, P.J.; Lane, M.V.; Merchenthaler, I. (1997) Comparative distribution of estrogen receptor-alpha and -beta mRNA in the rat central nervous system. *J Comp Neurol*, Vol. 388, No. 4 (1997 Dec), pp. 507-525.

Shughrue, P.J.; Merchenthaler, I. (2000) Estrogen is more than just a "sex hormone": novel sites for estrogen action in the hippocampus and cerebral cortex. *Front Neuroendocrinol*, Vol. 21, No. 1 (2000 Jan), pp. 95-101.

Silver, J.; Miller, J.H. (2004) Regeneration beyond the glial scar. Nat Rev Neurosci, Vol. 5, No. 2 (2004 Feb), pp. 146-156.

Simon, L.; Szilágyi, G.; Bori, Z.; Orbay, P.; Nagy, Z. (2001) (-)-D-Deprenyl attenuates apoptosis in experimental brain ischaemia. *Eur J Pharmacol*, Vol. 430, No. 2-3 (2001 Nov), pp. 235-241.

Smith-Thomas, L.C.; Fok-Seang, J.; Stevens, J.; Du, J.S.; Muir, E.; Faissner, A.; Geller, H.M.; Rogers, J.H.; Fawcett, J.W. (1994) An inhibitor of neurite outgrowth produced by astrocytes. *J Cell Sci*, Vol. 107, Pt 6 (1994 Jun), pp. 1687-1695.

Snow, D.M. ; Steindler, D.A.; Silver, J. (1990) Molecular and cellular characterization of the glial roof plate of the spinal cord and optic tectum: a possible role for a proteoglycan in the development of an axon barrier. *Dev Biol*, Vol. 138, No. 2 (1990 Apr), pp. 359-376.

Stein, D.G. (2001) Brain damage, sex hormones and recovery: a new role for progesterone and estrogen? *Trends Neurosci*, Vol. 24, No. 7 (2001 Jul), pp. 386–391.

Stroemer, R.P.; Kent, T.A.; Hulsebosch, C.E. (1995) Neocortical neural sprouting, synaptogenesis, and behavioral recovery after neocortical infarction in rats. *Stroke*, Vol. 26, No. 11 (1995 Nov), pp. 2135–2144.

Sulkowski, G.; Struzyńska, L.; Lenkiewicz, A.; Rafałowska, U. (2006) Changes of cytoskeletal proteins in ischaemic brain under cardiac arrest and reperfusion conditions. *Folia Neuropathol*, Vol. 44, No. 2 (2006), pp. 133-139.

Sun, J.; Zhou, W.; Ma, D.; Yang, Y. (2010) Endothelial cells promote neural stem cell proliferation and differentiation associated with VEGF activated Notch and Pten signaling. *Dev Dyn*, Vol. 239, No. 9 (2010 Sep), pp. 2345-2353.

Suzuki, S.; Brown, C.M.; Wise, P.M. (2009) Neuroprotective effects of estrogens following ischemic stoke. *Front Neuroendocrinol*, Vol. 30, No. 2 (2009 Jul), pp. 201–211.

Szilágyi, G.; Simon, L.; Wappler, E.; Magyar, K.; Nagy, Z. (2009) (-)Deprenyl-N-oxide, a (-)deprenyl metabolite, is cytoprotective after hypoxic injury in PC12 cells, or after transient brain ischemia in gerbils. *J Neurol Sci*, Vol. 283, No. 1-2 (2009 Aug), pp. 182-1866.

Takatsuru, Y.; Fukumoto, D.; Yoshitomo, M.; Nemoto, T.; Tsukada, H.; Nabekura, J. (2009) Neuronal circuit remodeling in the contralateral cortical hemisphere during functional recovery from cerebral infarction. *J Neurosci*, Vol. 29, No. 32 (2009 Aug), pp. 10081-10086.

Trendelenburg, G.; Dirnagl, U. (2005) Neuroprotective role of astrocytes in cerebral ischemia: focus on ischemic preconditioning. *Glia*, Vol.50, No.4 (2005 Jun),pp.307-320.

Udo, H.; Yoshida, Y.; Kino, T.; Ohnuki, K.; Mizunoya, W.; Mukuda, T.; Sugiyama, H. (2008) Enhanced adult neurogenesis and angiogenesis and altered affective behaviors in mice overexpressing vascular endothelial growth factor 120. *J Neurosci*, Vol. 28, No. 53 (2008 Dec), pp. 14522-14536.

Vincze, C.; Pál, G.; Wappler, E.A.; Szabó, E.R.; Nagy, Z.G.; Lovas, G.; Dobolyi, A. (2010) Distribution of mRNAs encoding transforming growth factors-beta1, -2, and -3 in the intact rat brain and after experimentally induced focal ischemia. *J Comp Neurol*, Vol. 518, No. 18 (2010 Sep), pp. 3752-3770.

von Bohlen Und Halbach, O. (2007) Immunohistological markers for staging neurogenesis in adult hippocampus. *Cell Tissue Res*, Vol. 329, No. 3 (2007 Sep), pp. 409-420.

Wappler, E.A.; Szilágyi, G.; Gál, A.; Skopál, J.; Nyakas, C.; Nagy, Z.; Felszeghy, K. (2009) Adopted cognitive tests for gerbils: validation by studying ageing and ischemia. *Physiol Behav*, Vol. 97, No. 1 (2009 Apr), pp. 107-114.

Wappler, E.A.; Felszeghy, K.; Szilágyi, G.; Gál, A.; Skopál, J.; Mehra, R.D.; Nyakas, C.; Nagy, Z. (2010) Neuroprotective effects of estrogen treatment on ischemia-induced behavioural deficits in ovariectomized gerbils at different ages. *Behav Brain Res*, Vol. 209, No. 1 (2010 May), pp. 42-48.

Wappler, E.A.; Adorján, I.; Gál, A.; Galgóczy, P.; Bindics, K.; Nagy, Z. (2011a) Dynamics of dystroglycan complex proteins and laminin changes due to angiogenesis in rat cerebral hypoperfusion. *Microvasc Res*, Vol. 81, No. 2 (2011 Mar), pp. 153-159.

Wappler, E.A.; Gál, A.; Skopál, J.; Nagy, Z. (2011b) Single, high-dose 17β-estradiol therapy has anti-apoptotic effect and induces cerebral plasticity following transient forebrain ischemia in gerbils (Short communication). *Acta Physiol Hung*, Vol. 98, No. 2 (2011 Jun), pp. 189-194.

Waters, E.M.; Yildirim, M.; Janssen, W.G.; Lou, W.Y.; McEwen, B.S.; Morrison, J.H.; Milner, T.A. (2011) Estrogen and aging affect the synaptic distribution of estrogen receptor beta-immunoreactivity in the CA1 region of female rat hippocampus. *Brain Res*, Vol. 1379 (2011 Mar), pp. 86-97.

Weaver, C.E. Jr; Marek, P.; Park-Chung, M.; Tam S.W.; Farb, D.H. (1997) Neuroprotective activity of a new class of steroidal inhibitors of the N-methyl-d-aspartate receptor. *Proc Natl Acad Sci USA, Vol.* 94, No. 19 (1997 Sep), pp. 10450–10454.

Wei, H.; Masterson, S.P.; Petry, H.M.; Bickford, M.E. (2011) Diffuse and specific tectopulvinar terminals in the tree shrew: synapses, synapsins, and synaptic potentials. *PLoS One*, Vol. 6, No 8 (2011 Aug), pp. e23781.

Wei, L.; Erinjeri, J.P.; Rovainen, C.M.; Woolsey, T.A. (2001) Collateral growth and angiogenesis around cortical stroke. *Stroke*, Vol. 32, No. 9 (2001 Sep), pp. 2179-2184.

Woolley, C.S.; McEwen, B.S. (1992) Estradiol mediates fluctuation in hippocampal synapse density during the estrous cycle in the adult rat. *J Neurosci*, Vol. 12, No. 7 (1992 Jul), pp. 2549-54. Erratum in J Neurosci 1992 Oct, Vol. 12, No. 10.

Woolley, C.S. (2007) Acute effects of estrogen on neuronal physiology. *Annu Rev Pharmacol Toxicol*, Vol. 47 (2007), pp. 657-680.

Xiong, M.; Cheng, G.Q.; Ma, S.M.; Yang, Y.; Shao, X.M.; Zhou, W.H. (2011) Post-ischemic hypothermia promotes generation of neural cells and reduces apoptosis by Bcl-2 in the striatum of neonatal rat brain. *Neurochem Int*, Vol. 58, No. 6 (2011 May), pp. 625-33.

Xiong, Y; Mahmood, A.; Chopp, M. (2010) Neurorestorative treatments for traumatic brain injury. *Discov Med*,Vol. 10, No. 54 (2010 Nov), pp. 434-442.

Yang, X.T.; Bi, Y.Y.; Feng, D.F. (2011) From the vascular microenvironment to neurogenesis. *Brain Res Bull*, Vol. 84, No. 1 (2011 Jan), pp. 1-7.

Yao, G.L.; Kiyama, H.; Tohyama, M. (1993) Distribution of GAP-43 (B50/F1) mRNA in the adult rat brain by in situ hybridization using an alkaline phosphatase labeled probe. *Brain Res Mol Brain Res,* Vol. 18, No. 1-2 (1993 Apr), pp. 1-16.

Yiu, G.; He, Z. (2006) Glial inhibition of CNS axon regeneration. *Nat Rev Neurosci*, Vol. 7, No. 8 (2006 Aug), pp. 617–627.

Yue, F.; Chen, B.; Wu, D.; Dong, K.; Zeng, S.E.; Zhang, Y. (2006) Biological properties of neural progenitor cells isolated from the hippocampus of adult cynomolgus monkeys. *Chin Med J,* Vol. 119, No. 2 (2006 Jan), pp. 110-116.

Zhang, B.; Subramanian, S.; Dziennis, S.; Jia, J.; Uchida, M.; Akiyoshi, K.; Migliati, E.; Lewis, A.D.; Vandenbark, A.A.; Offner, H.; Hurn, P.D. (2010) Estradiol and G1 reduce infarct size and improve immunosuppression after experimental stroke. *J Immunol,* Vol. 184, No. 8 (2010 Apr), pp. 4087-4094.

Zhang, Z.G.; Chopp, M. (2009) Neurorestorative therapies for stroke: underlying mechanisms and translation to the clinic. *Lancet Neurol,* Vol. 8, No. 5 (2009 May), pp. 491-500.

Zhao, C.; Deng, W.; Gage, F.H. (2008) Mechanisms and functional implications of adult neurogenesis. *Cell,* Vol. 132, No. 4 (2008 Feb), pp. 645-660.

7

The Promise of Hematopoietic Stem Cell Therapy for Stroke: Are We There Yet?

Aqeela Afzal and J. Mocco
University of Florida, Gainesville, Florida
USA

1. Introduction

Stroke is the leading cause of permanent disability in industrialized nations (Lloyd-Jones et al, 2010). Ischemic stroke occurs secondary to blood flow interruption to the brain, typically secondary to the occlusion of an intra- or extra-cranial artery. This lack of blood supply to the brain results in a paucity of nutrients, glucose, and oxygen, which leads to cerebral ischemia and infarction (Wardlaw et al, 2003). However, intracranial artery occlusion results in varied rates of tissue injury depending on the local anatomy, as well as numerous still-to-be-deciphered physiologic factors. Generally, the ischemic core of the occluded vascular territory rapidly infarcts and becomes unsalvageable tissue. However, the surrounding region of ischemia, known as the penumbra, often receives enough collateral supply that it may be saved, providing adequate perfusion is reestablished in a timely fashion. The concept of restoring normal perfusion to the penumbra, thereby rescuing crucial brain tinssue and hence neurological function, is fundamental.

To date, the significant majority of stem cell stroke research has focused on evaluating the potential of Neural Stem Cells in cerebral ischemic repair (Garzon-muvdi et al, 2009; Miljan et al, 2009; Bersano et al, 2010; Locatelli et al, 2009; Burns et al, 2009). However, the field of Hematopoietic stem cell (HSC) research in stroke is not barren, as a small volume of literature has recently emerged. HSC have recently been shown to mobilize to the peripheral circulation from bone marrow in response to stroke (Hennemann et al, 2008), and increasing circulating HSC levels correlate with improved neurological function following stroke, suggesting a potentially critical role for HSC in limiting stroke injury and/or facilitating stroke recovery (Yip et al, 2008; Taguchi et al, 2009). Moreover, post-ischemic intravascular administration of exogenous HSC has recently been shown to ameliorate ischemic stroke in mice (Schwarting et al, 2008). Additionally, well established therapeutics that are known to mobilize HSC have shown very exciting preliminary results in animal models and are currently undergoing clinical evaluation for other modes of central nervous system injuries (Luo et al, 2009). Increasing levels of circulating HSC have recently been demonstrated to correlate with improved neurological function following stroke, suggesting a potentially critical role for HSC in limiting stroke injury and/or facilitating stroke recovery (Yip et al, 2008; Taguchi et al, 2009).

2. What are hematopoietic stem cells?

A stem cell has the capacity for self-renewal and the ability to differenretiate into multiple cell types (potentcy) (Melton et al, 2004). A progenitor cell has similar characteristics to a stem cell, however, it has limited potential for differentiation (it has limited self renewal capacity and can only differentiate into limited types of cells (Melton et al, 2004). Hematopoietic Stem Cells (HSC) are circulating bone marrow derived mononuclear cells that promote repair in areas of injury (Baum et al, 1998). HSC travel thru peripheral blood from the fetal liver to the bone marrow and seed the bone marrow with immature and maturing cells (Melton et al, 2004); the bone marrow then remains the main site of hematopoiesis in adult life (Melton et al, 2004). During embryogenesis the three embryonic germ layers partition into 3 embryonic layers: ectoderm, mesoderm and endoderm (Hall et al, 2000). The ectoderm gives rise to skin and neural cells and tissues (Hall et al, 2000); the mesoderm gives rise to the blood cells, bone, fat, cartilage and muscle and the endoderm gives rise to the respiratory system and digestive tract (Wells et al 1999). The resulting tissues and organs from these three layers retain their original specification throughout adulthood. The neural crest is the only exception to that rule; it is of ectoderm origin and gives rise to neural, muscle and bone cell lineages. Based on the three germ layers, it would seem that stem cells generate mature cells corresponding to the tissue of that origin only. However, stem cells can transdifferentiate into cells of a completely different lineage. Some tissues in the adult, have been shown to respond very well to re-generation by HSC, for example liver (Varga et al, 2010), and others have been shown to respond poorly, for example, heart (Rumyantsev et al 1987). This may indicate the presence of stem cells within these tissues The brain used to be thought of as a non renewing organ, however, it has now been shown to have a high cell turnover (Kajsutra et al 1999; Altman et al, 1965; Lois et al, 1993).

HSC can self renew themselves at a single cell level and can differentiate to mature progeny of non-renewing and terminally differentiated cells (Seita et al, 2010). In contrast totipotent cells (can give rise to all embyonic and extraembyonic cell types), pluripotent (can give rise to all cell types of the embryo), oligopotent (can give rise to limited cell lineages), or unipotent cells (can give rise to a unique mature cell type), HSC have multipotent developmental potential (can give rise to a subset to cell lineages) (Seita et al, 2010). The HSC in the bone marrow proliferate and differentiate into erythroid, lymphoid and myeloid lineages (Figure 1) (Kondo et al, 2003). Commitment to each lineage is dictated by several growth factors such as VEGF, EGF, IGF, FGF and PDGF. HSC are recruited to the peripheral circulation from bone marrow in response to stress or injury such as stroke (Paczkowska et al, 2005; Machalinski et al, 2006; Henneman et al, 2008). Bone marrow HSC give rise to the hemangiobalst (Urbick 2004; Hristov et al, 2004; Rumpold et al, 2004), which in turn give rise to mature endothelial cells (Hristov et al, 2004).

Bone marrow derived HSC contribute to hematopoietic tissues (tissues which can stimulate the bone marrow) such as: skin (krause et al, 2001), kidney (Kale et al, 2003), central nervous tissue (Brazelton et al, 2000, Weimann et al, 2003) and have also been found to contribute to non hematopoietic tissues such as: myocardium (Orlic et al, 2001) and Skeletal muscle (Ferrari et al, 1998). This phenomenon may be due to circulating HSC lodging in non-hematopoietic tissues or due to lineage conversion of HSC. HSC can either transdifferentiate (alter their lineage specificity by activation of alternate genes), de-differentiate and re-differentiate (HSC can de-differentiate to a more primitive state (multipotent state) and then re-differentiate along a different lineage pathway), come from a homogenous starting population of cells or be a result of cell-cell fusion. Most HSC studies are done by injecting or implanting a large number of cells, many of these cells may be contaminating impurities resuting from harvesting or enriching techniques (described below). The presence of HSC in non-hematopoietic tissues may be due to the impurities present in the starting population (Kanof et al, 2001). Another contributing phenomeon for the presence of HSC in non-hemaoptopoietic tissues may be due to cell-cell fusion (Anderson et al, 2000). Cell-cell fusion is a natural occurence in skeletal myofibres (Anderson et al, 2000) but may also be pathologic, for excample in HIV infection of T-lymphocytes (Mccune et al, 1998). Terada et al (2002) were the first group to show that stem cells may fuse with cells of the central nervous system without commiting to the parent cell lineage. These fusion cells take on the phenotype of the parent cell without complete differentiation to their specific lineage. Alvarez-Dolado et al (2003) also showed that bone marrow derived HSC contribution to non-hematopoietic tissue repair was due to cell-cell fusion rather than transdifferentiation of the HSC.

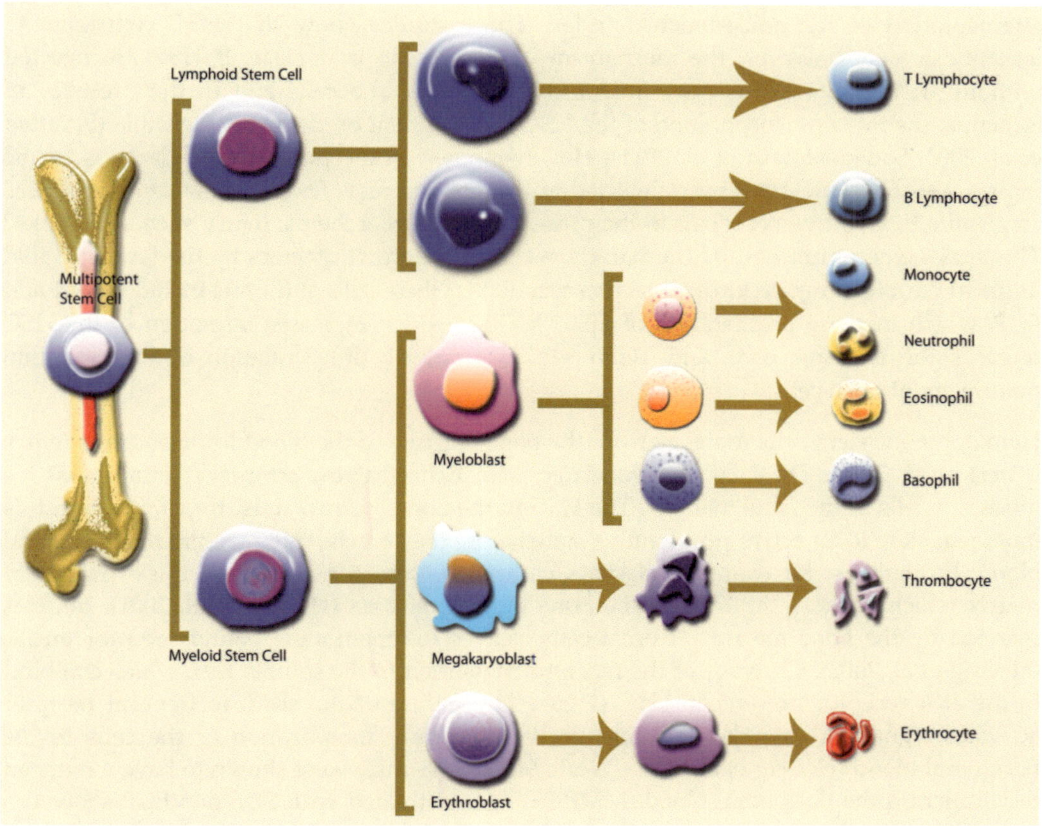

Fig. 1. Multipotent stem cells from the bone marrow can give rise to erythroid, lymphoid and myeloid lineages.

2.1 Mobilization and neovascularization

Mobilization of HSC to the blood occurs via trans-endothelial migration thru the bone marrow. Proteinases, such as elastase, cathepsin G, and matrix metalloproteinases, cleave and release the HSC from the bone marrow stroma and surrounding cells (Heissig et al, 2002). The HSC leave the bone marrow in response to several growth factors, such as SDF1-A, VEGF, EPO and G-CSF. All of these growth factors have been shown to increase levels of circulating HSC in the blood. Neovascularization is the de novo synthesis of blood vessels and it differs from Angiogenesis which refers to sprouting of capillaries from existing blood vessels (Carmeliet et al, 2005). Vasculogenesis/vascularization refers to differentiation of HSC into endothelial cells and was thought to occur only in the embryo (Shi et al, 1998). Bone marrow derived HSC have now been showed to home to a site of neovascularization, proliferate and differentiate into endothelial cells (Masuda et al, 2003). Several groups have shown that neovascularization occurs in response to ischemia in the heart (Hur et al, 2007; Cook et al, 2009; Shintani et al, 2001; Sanganalmath et al 2011). HSC injected into myocardial infarction patients showed an increase in blood flow and an improvement in heart function. In addition, HSC injected into a hind limb ischemia model also results in increased neovascularization in the ischemic limb. Cohorts with ischemic hind limbs were also injected autologous bone marrow which resulted in reduced chest pain and an augmentation of the ankle-brachial index. These studies show that HSC contribute to ischemic rescue, however, the mechanism of this rescue is unclear. If HSC are injected without an injury, there is very little incorporation of the cells and in the presence of ischemia, the rate of incorporation of the HSC is dependent on degree of ischemia (Shintani et al, 2001; Sanganalmath et al, 2011). However, even in the presence of a large ischemic injury, very few HSC have been detected at the site of injury. So then how can a few HSC contribute to blood vessel repair in the presence of a large ischemic injury such as a stroke? This may be accomplished by the paracrine release of growth factors by the few HSC that home to the site of injury. Growth factors secreted by these cells at the site include IGF-1 and FGF, which increase proliferation of HSC, MCP-1, which increases migration of the HSC towards the ischemic core, and TGFb, which promotes differentiation of the cells into mature endothelial cells.

Hematopoietic stem cells mobilize from the bone marrow to the blood in response to injury (Kucia et al, 2004). The HSC are associated with bone marrow stromal cells and exist as quiescent cells in the bone marrow. The HSC in the bone marrow must transform from this quiescent state to an active proliferative state before they can be mobilized to the peripheral blood. Proteinases, for example, elastase, cathpsin G and MMP's cleave the extracellular matrix which anchors the HSC to the bone marrow stroma (Heissig et al, 2002). MMP-9, secreted by the bone marrow stromal cells cleaves the membrane bound receptor mKitL (Heissig et al, 2002). Cleavage of the receptor converts it to the soluble KitL which can bind to the cKit receptor present on HSC (Figure 2). Binding of the sKitL to the cKit receptor activates signalling cascades enabling proliferation and mobilization of the cells to the peripheral blood (Heissig et al, 2002). While MMP-9 -/- mice were shown to have a reduced recruitment o the peripheral blood, MMP-9+/+ mice treated with SDF or VEGF showed a marked increase in mobilization of the HSC from the bone marrow to the peripheral blood (Heissig B, 2002). Presence of HSC in the peripheral blood was first shown in the 1960's and 70's (Korbling et al, 1994). Since then, peripheral blood HSC counts have been used as biomarkers in diseases such as: diabetes (Fadini et al, 2006), Hyperhomocystenemia (Zhu et al, 2006), Aging (Heiss et al, 2005), Hypertension (Pirro et al, 2007), Systemic Scelrosis (Del Papa et al, 2006), Chronic smoking (Kondo et al, 2004) and Coronray Artery Disease (Kunz et al, 2006).

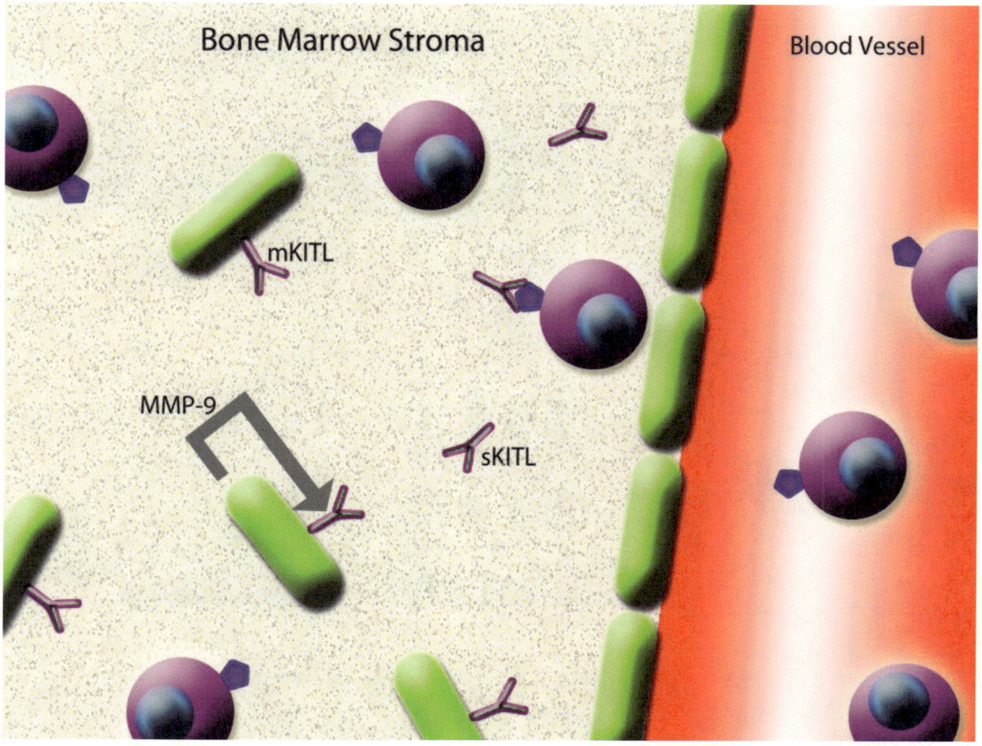

Fig. 2. Bone marrow stromal cells (green) secrete MMP-9, which cleaves the membrane bound cKIT receptor present on HSC (purple). Cleavage of the receptor converts it to the soluble KitL which can bind to the cKIT receptor present on the HSC. Binding of the KitL to the receptor enables proliferation and migration of the HSC to the peripheral blood.

2.2 Growth factors in HSC mobilization

Growth Factors which have remained in the forefront of HSC research include Vascular Endothelial Growth Factor (VEGF) (Leung et al, 1998) and Stromal Derived Growth Factor-1 A (SDF1-A) (Ma Q et al, 1998). VEGF is produced by many cells, not only HSC (Leung et al, 1989) and is formed by alternate splicing of a single gene (Leung et al 1989). Alternate splicing of the parent gene leads to the formation of VEGF (Leung et al 1989), VEGFB (Olofesson et al, 1996), VEGFC (Chilov et al, 1997), VEGFD (Latitinen et al, 1997) and VEGFE (Ogawa et al, 1998), however, VEGF is by far the most studied growth factor (Leung et al, 1998). Carmeliat et al (1996) and Ferrara (1996) both showed that homozygous knockouts of VEGF (VEGF +/-) die in utero due to impaired hematopoiesis (Ferrara et al, 1996) and anigogenesis (Carmeliat et al, 1996). VEGF can bind to 2 receptors: VEGFR1 (flt) (Ortega et al, 1997; Shalaby et al, 1995) and VEGFR2 (kdr) (Terman et al, 1992). Binding of VEGF to VEGFR1 contributes to vascular remodelling (Fang et al, 1996) and binding to VEGFR2 initiates proliferation, migration and differentiation of HSC (Ortega et al, 1997; Matthews et al, 1991) through the PI3K/AKT/NFkb pathway (Byrne M, 2005). Oxygen levels in the bone marrow are typically lower than the peripheral blood (Harrison et al, 2002) thus leading to lower oxidative stress and higher HSC survival and proliferation (Jang et al, 2007).

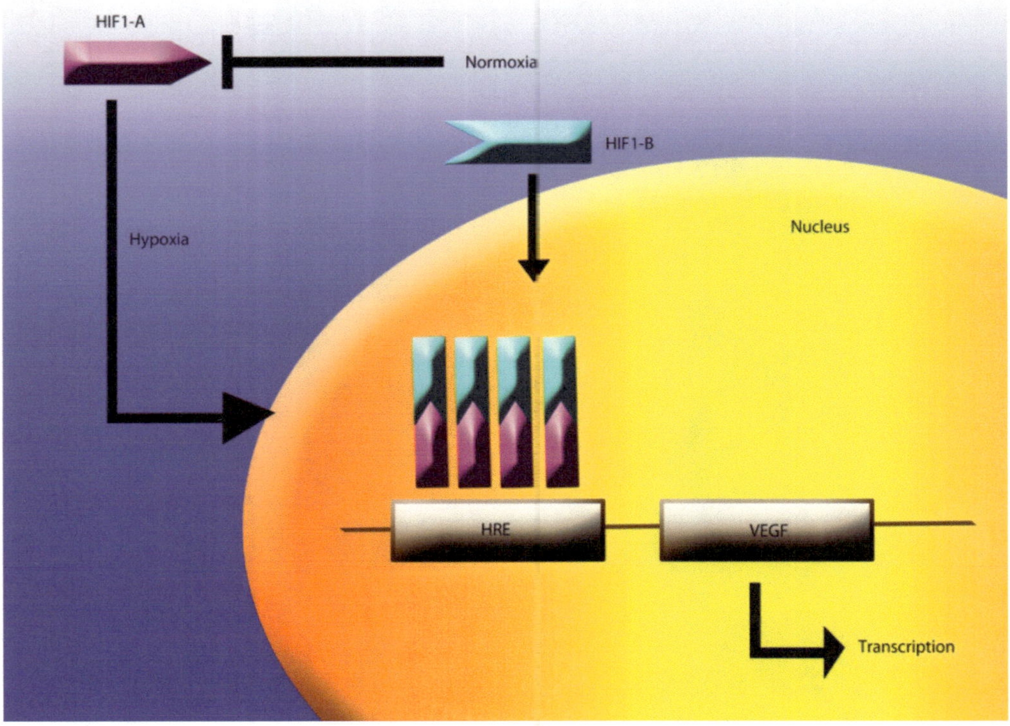

Fig. 3. Hypoxia Inducible Factor (HIF) is composed of two subunits: HIF1-A and HIF1-B. Normoxia leads to degradation of HIF1-A and hypoxia causes it to translocate to the nucleus for activation. In the nucleus, HIF1-A binds with HIF1-B to form heterodimers and bind to hypoxia Response Elements (HRE) to activate transcription of VEGF.

Hypoxic environments activate hypoxia Induced Factor-1 (HIF-1) (Wang et al, 2005). HIF1 is composed of 2 subunits: HIF-1A (the oxygen sensing domain) and HIF-1B (Wang et al, 2005). Under conditons of normoxia, HIF1-A is degraded by proteasomes (Jaakkola et al, 2001) and under hypoxia both units form heterodimers (Figure 3) and bind to Hypoxia Response Genes to activate transcription of VEGF (Wang et al, 1995). Shuweiki et al (1992) showed that hypoxia increases VEGF production which is necessary for hematopoietic activity in the bone marrow (Carmeliat et al, 1992; Gerber et al, 2002). Gerber et al (2002) also showed bone marrow engraftment failure if VEGF negative HSC were injected into lethally irradiated mice. In addition, Hooper et al (2009) showed that impaired VEGFR2 on the HSC also failed to engraft lethally irradiated mice (Hooper et al, 2009). Rehn et al (2011) recently futher confrimed that loss of VEGF expression in VEGF knockout mice increased impaired HSC which were not able to engraft secondary lethally irradiated mice (Rehn et al, 2011).

Stromal Derived Growth Factor-1 Alpha (SDF1-A) is released by cells following stress, injury or hypoxia (Muller et al, 2001). SDF1-A is localized to chromosome 10q11.1 and is highly conserved between species. SDF1-A belongs to the CXC family of chemokines and was originally described as a pre B cell growth stimulating factor. SDF1-A is a ligand for CXCR4, a G protein coupled receptor, and their interaction mediates a chemotactic response followed by cell migration. The receptor for SDF1-A is a 7 transmembrane receptor, CXCR4 (Kucia et al, 2004). Binding of the SDF1-A to its receptor, initiates the PI3K-AKT and NfkB Pathway. This pathway leads to the phosphorylation of MAPK, an intracellular calcium efflux and a subsequent adhesion of the cells to fibronectin (Kucia et al, 2004). SDF1-A and its receptor CXCR4 have been shown to regulate trafficking of HSC in response to injury (Ma et al, 1998; Lapidot O et al, 2002; Pituch-Noworolska et al, 2003). Increasing SDF1-A expression in a hind limb ischemia model has been shown to increase mobilization of HSC and increased angiogensis in the hind limb (Hiasa et al, 2004). An increase in SDF1-A levels in the blood leads to an increase in CXCR4 positive cells to the injured/hypoxic area (Kucia et al, 2004). Following injury, bone marrow levels of SDF1-A decrease (Petit et al, 2002), while those in the peripheral blood increase (Morris et al, 2003). Reduced levels of SDF1-A in the bone marrow mediates secretion of MMP-9 which facilitates mobilization of the cells from the bone marrow to the peripheral blood (Janowska-Wieczoveka et al, 2000). Once in the circulation, the HSC can differentiate into myeloid cells, lymphocytes, erythrocytes, platelets or endothelial progenitor cells (Kondo et al, 2003). Increased levels of SDF1-A help to retain recruited bone marrow HSC in close proximity to angiogenic blood vessel growth (Grunewald et al, 2006).

Eventhough SDF1-A leads to an increase in mobilization of the HSC from the bone marrow, it has also been shown to cultivate a protease rich environment in the bone marrow, which can be both beneficial (Janowska-Wieczoveka et al, 2000; Hiasa et al, 2004) and harmful (McQuibban et al, 2001). An upregulation in SDF1-A levels also leads to an increase in MMP-2 expression which cleaves SDF1-A to a toxic fragment which is incapable of binding to its receptor (CXCR4) and has been shown to be neurotoxic (McQuibban et al, 2001; Zhang et al, 2003). To overcome this unwelcomed cleavage, Segers et al (2011) designed deliverable SDF1-A with mutations which made it resistant to MMP-2 cleavge. This mutated form of SDF1-A sustained local SDF1-A levels, increased SDF1-A levels in the ischemic limb and increased vascular density (Segers et al, 2011). However, the role of SDF1-A in generation of mature vessels is unknown. Therapies using angiogenic growth factors leads to unstable vessel formation which regress following cessation of the therapy (Gounis et al, 2005). In contrast to VEGF, SDF1-A prevents tortuos blood vessel formation (due to extensive proliferation of endothelial cells) which are hyper-permeable (Segers et al, 2010). A study by Moore et al (2001) looked at the synergistic effect of SDF1-A and VEGF; IV delivery of replication incompetent adenovectors expressing the SDF1-A gene incresed plasma SDF1-A levels and increased mobilization of bone marrow derived cells which were positive for VEGFR2 (Moore et al, 2001). Delivery of replication incompetent adenovector expressing VEGF increased plasma VEGF levels and increased moblization of HSC which were positive for VEGFR1. In addition these authors looked at the possibility of combining VEGF with other growth factors such as Angiopoietin-1. The combination treatment prolonged HSC mobilization and increased proliferation of capillary beds (Moore et al, 2001).

3. Sources of hematopoietic stem cells

Very little information is available on human fetal tissue research, which is done mostly in Europe. Gallacher et al (2000) first reported the presence of HSC in aborted human fetuses and subsequent work focused on fetal tissues of vertebral animals (Dzierzak et al, 1999). Embryonic stem cells have also been shown to have high proliferative capacity (Rolletschek et al, 2004) along with the ability of differentiating into several types of blood cells (Hole et al, 1999). Embryonic tissues are a rich source of HSC; however, due to the lack of a source for obtaining these cells, very little data is available on their use in humans. Human cord blood cells (HUCB), produced by the placenta, support a developing fetus and are discarded upon delivery. HUCB are multipotent (able to generate multiple germ layers) and a rich source of HSC which have been used in research studies. The first successful cord blood transplant was done in children with Fanconi's anemia (Laughlin et al, 2001) and more recently have been used in clinical trials for autoimmune disorders, cerebral palsy and Type I diabetes(Laughlin et al, 2001). HUCB offer an ethical source of HSC (McGuckin et al, 2008), have a low risk of host versus graft disease and are readily available from bio banks (Forraz et al, 2011).

Bone marrow stem cells (BMSC) are obtained from a donor from the hip bone (Diefenderfer et al, 2011). The donor is anesthetized and bone marrow drawn up in a syringe (Lee et al, 2004). This procedure is painful and may require a hospital stay (Lee et al, 2004). BMSC contain HSC, stromal and progenitor cells (Diefenderfer et al, 2011). Peripheral blood also has circulating HSC, which can also be induced to mobilize from the bone marrow into the blood in response to GCSF administration (Elfenbein et al, 2004). This procedure is easier on the donor and can be used for autologous and allogeneic administration (Brown et al, 1997; Lickliter et al, 2000). Peripheral blood stem cells have also been used successfully in patients (Verbik et al, 1995; Brown et al, 1997). The biggest difference in obtaining HSC from different sources is the quantity of cells. While HUCB may be a viable source for HSC, very few HSC can be obtained, thereby limiting use to children and not adults. BMSC offers the best viable option, for autologous HSC administration and higher yields.

4. Isolation of hematopoietic stem cells

HSC isolation can be a multistep process requiring the use of Ficoll (Fuss et al, 2009), FACS (Schlenke et al, 1998), Magnetic microbeads (Woywodt et al, 2005), or culturing of Early and late outgrowth cells (Asahara et al, 1997). Depending on the source of sample and the species, protocols may need to be adjusted accordingly. HSC vary in size and density compared to other cells present in the bone marrow or the blood (Fuss et al, 2009). Therefore, they move through specialized density media such as Ficoll (Fuss et al, 2009), at a specific rate when a centrifugal force is applied. The red blood cells and granulucytes are more dense than the HSC and can easily be separated from them following the centrifugation (Kanof et al, 2001; Jaatinen et al, 2007). HSC are isolated from the monouclear cell layer, however, since this layer has impurities, such as lymphocytes (which have overlapping densities with the HSC) (Kanof et al, 2001; Jaatinen et al, 2007), further purification is needed to obtain a pure HSC fraction. In addition to using a density medium, several groups (Schlenke et al, 1998; Ruitenberg et al, 2006) have used a cell preparation tube (CPT) which contains a cell separater solid at the bottom of a vaccutainer tube. The CPT tube requires fewer steps to obtain the mononuclear layer and Ruitenberg et al (2006) reported that there was no significant difference in using either the Ficoll density medium or the CPT tube.

The HSC can be further purified from the bone marrow or peripheral blood and depleted for lineage markers by using either flow cytomtery (Schlenke et al, 1998) or microbeads coated with antibodies (Woywodt et al, 2005). Flow cytometery can be used to select for the presence or absence of markers simultaneously, however, this method cannot process a large number of cells quickly (Schlenke et al, 1998). Column based methods offered a much quicker way of enriching HSC, however, an additional step had to be included to elute cells from the columns (Jaatinen et al 2007). Consequently, magentic bead technology was developed which allowed for either a negative or positive selection of selected markers (Horrocks et al, 1998; Woywodt et al, 2005). A positive selection (an antibody is used to target an antigen which is expressed on the surface of desired cells) yields a higher purity of HSC due to the use of one antibody and a negative selection (use of antibodies to target antigens not expressed on desired cells) requires several antibodies to remove unwanted cells (Woywodt et al, 2005). In contrast to using freshly isolated cells, Asahara et al (1997) first isolated HSC and cultured them in specific media on fibornectin coated dishes. The cells that grew in these cultures were called either early or late endothelial progenitor cells. Both types of cells were positive for acetylated LDL and lectin binding, however, only the Late endothelial progenitor cells were positive for VE-Cadherin, von Willebrand factor and CD31 (all markers of mature endothelial cells) (Asahara et al, 1997).

In addition to a variety of techniques which can be used to isolate the HSC, an array of cell surface markers can be used to enrich for the desired HSC fraction. Asahara et al (1997) were the first to report that mononuclear cells enriched in CD34+ cells can also mature to endothelial cells. They used the peripheral blood monouclear cells to isolate progenitors using magnetic microbeads. The progenitors wre plated onto fibronectin coated plates and confirmed to have endothelal characteristics. Subsequently, cells expressing CD133, CD34 and VEGFR2 were defined as HSC (Asahara et al, 1997; Hristov et al, 2004). CD133 is a membrane glycoprotein which is lost after differentiation and VEGFR2 is present on all endothelial cells. Hemangioblasts in the bone marrow express CD133, CD34 and VEGFR2; once these cells are activated and committed to an endothelial lineage, they gain expression of CD31 and retain CD133, CD34 and VEGFR2. Fully differentiated endothelial cells express CD34, Von Willebrand Factor, VEGF, CD31, VE- Cadherin and E-Selectin (Figure 4). In mice, lineage depletion includes removal of CD5, CD45R, Cd11b, Ly6G and TER119 (Spangrude et al, 1998; Horrocks et al, 1998) and subsequent selection of the SCA and cKIT positive cells (Uchida et al, 1992; Osawa et al, 1996; Lois et al, 2001).

5. Stroke

Chen et al (2001) were the first to report the use of Human cord blood cells (HUCB) following stroke (Chen et al, 2001). The authors used a middle cerebral artery occlusion (MCAO) model in Wistar rats. The HUCB were injected at either day 1 or 7 post stroke. Behavior (rotarod) analysis and neurological score were recorded on days 1, 7, 14, 21, 28 and 35 days post stroke. Behavior was significantly improved in animals which received the HUCB at 1 day post stroke whereas day 7 only showed an improvement in the neurological score and not the behavior testing. The lesion volume was determined by H&E staining on day 35 post stroke and no significant differences were observed in animals which received

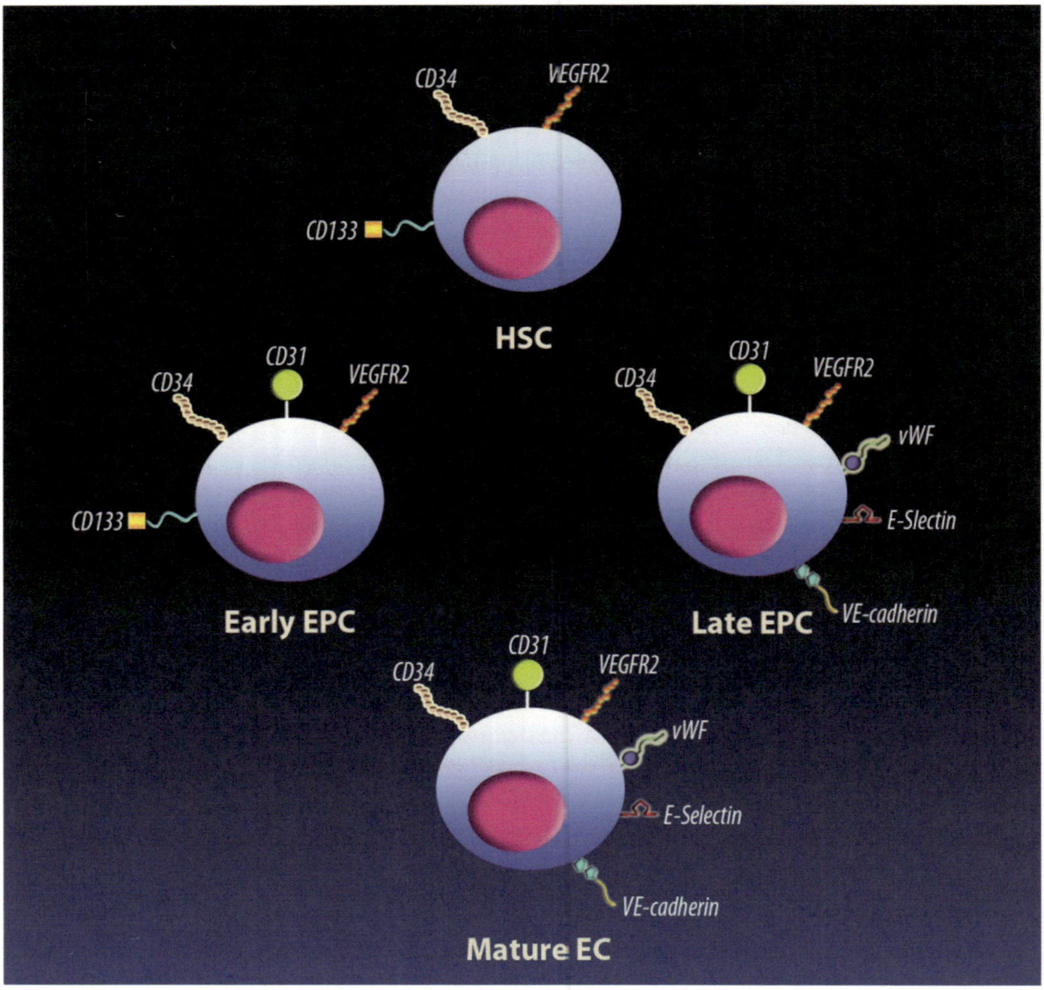

Fig. 4. Hematopoietic stem cells (HSC) express CD133, CD34 and VEGFR2. Mature endothelial cells (EC) express CD34, CD31, VEGFR2, Von Willebrand Factor, E-Selectin and VE-Cadherin. Early endothelial progenitor cells (EPC) express CD31 in addition to the markers carried by the HSC and late EPC express Von Willebrand factor, E-selectin and VE-cadherin in addition to the early EPC markers.

the HUCB and those that did not. HUCB were localized in the ischemic hemisphere by immunostaining for MAB1281 (an anti-human nuclei monoclonal antibody). Additionally co-staining was done for neurons (using micro-tubule-associated protein 2 and NeuN), astrocytes (using glial fibrillar acidic protein) and endothelial cells (using von Willebrand Factor). Animals which did not receive any HUCB showed no positive cells for MAB1281 and of the positive cells in the transplanted animals, 2% of the cells were positive for NeuN markers, 3% for neuronal markers, 6% for astrocytic markers and 8% for endothelial markers. The ability of HUCB to migrate towards an ischemic area was also tested in vitro using a chemotaxis assay. Ischemic brain extracts stimulated migration of HUCB in vitro. This study showed that IV transplantation of HUCB significantly improved behavior when the cells were administered at day 1(Chen et al, 2001) , however, no explanations were offered for the mechanism of action for these cells other than the possibility of release of trophic growth factors from the HUCB.

A possible mechanism by which HUCB may be contributing to a reduction in behavior deficits was suggested by Regelsberger et al (2011). These authors postulated that the HUCB may be modulating delayed apoptosis in the ischemic hemisphere leading to the reduction in behavior deficit. Spontaneously hypertensive rats were used in this study for either a histological evaluation (at 25, 48, 72 and 96 hours) or a molecular biology study (at 6, 24, 36 and 48 hours) post stroke. The HUCB were injected intravenously and histological analysis was done for infarct volumes by assessing the loss of MAP-2 positive neutrophil staining and cleaved caspase-3 positive cells were also counted. The results of this study showed that the number of cleaved caspase-3 cells in the infarct area increased from days 1 to 4 and were not affected by the IV administration of the HUCB. Molecular analysis by real time PCR also showed no differences in the levels of caspase-3 mRNA or survivin (an anti-apoptotic protein) in groups that received or did not receive the HUCB. Unlike other studies using HUCB (Vendrame et al, 2006), the infarct volumes in this study were not affected by the administration of HUCB. The study could not provide any evidence that HUCB contributed to the reduced behavior deficits due to anti-apoptotic effects. Another study (Leonardo CC et al, 2010) suggested that HUCB may lead to a reduced recruitment of pro-inflammatory cells to the ischemic hemisphere and hence faster recovery. Sprague Dawley rats were used to induce an MCAO and HUCB were delivered intravenously. Brain sections were stained for CD11b (inflammatory cell marker) and MMP-9 (causes basement membrane degradation leading to blood brain barrier leakage). Animals which did not get any HUCB had a significantly higher number of CD11b staining cells in the ischemic hemisphere and these cells were also positive for MMP-9 (Leonardo et al, 2010).

In contrast to the previous studies, Makinen et al (2006) have reported that IV administration of HUCB does not contribute to the functional recovery of Wistar rats. These authors suggested that studies evaluating functional behavior outcome following stroke do not utilize a comprehensive battery of tests to evaluate sensorimotor, cognitive and histological tests for the animals. This study used the beam walking test (assesses forelimb and hind limb function), cylinder paw placement (use of impaired contralateral (to lesion) paw to assess forelimb asymmetry, water maze (for length, latency and swimming speed). The HUCB were administered at 24 hours post stroke to avoid an inflammatory response which can complicate recovery if the HUCB are administered immediately after stroke. Previously mentioned studies (Vendrame et al, 2006) had used 10^6 HUCB for IV administration and this study used 1-5x10^7, to ensure that enough HUCB were injected to be detected. None of the behavioral tests showed any significant rescue with IV administration of HUCB. The infarct volumes did not show any difference in either group of animals, however, histological analysis showed the presence of a few HUCB in the ipsilateral hemisphere in close proximity to blood vessels in animals which received the HUCB. In addition to the histological study, Makinen et al injected ^{111}In-Oxine labeled HUCB for Bio-imaging of the live animals. The images revealed that the HUCB were trapped in the liver, lungs, spleen and kidneys. No signal was picked up in the brain. The authors argued that since other studies don't use a comprehensive evaluation method for the animals, they report the presence of HUCB in the brain and a functional recovery following HUCB administration. However, a comprehensive battery of tests performed in this study revealed that HUCB do not contribute to behavior recovery following stroke (Maiken et al, 2006).

Nystedt et al (2006) took a different approach to the previous studies: instead of injecting whole HUCB, they used an enriched CD34+ population from the HUCB. CD34+ cells have been shown to promote angiogenesis (Shyu et al, 2006) and neurogenesis (Taguchi et al, 2004); therefore, the authors reasoned that injection of a pure population of targeted HSC would provide better functional rescue following stroke. Wistar rats were subjected to transient and permanent MCAO, CD34+ cells were injected IV 24 hours post-surgery. The animals were administered CD34+ cells obtained from 2 unrelated HUCB units for better engraftment; one unit acts as a 'fertilizer' and the other unit contributes to hematopoiesis. The animals were pre-trained on the beam walking and cylinder test and evaluated at days 4, 12 and 20; water maze was evaluated at days 22, 23, and 25 post stroke. There was no significant difference in infarct volumes in either model of MCAO and human nuclei specific MAB1281+ cells were not detected in ipsilateral or the contralateral hemisphere at day 25 post stroke. However, both models did show an upward trend in sensorimotor and cognitive performance in animals which received CD34+ cells in both models. Infarct volume did not change in either model following administration of CD34+ cells and these cells were not detected in the ipsilateral hemisphere, yet an improvement in functional recovery was observed. The authors reasoned that recovery may have been due to factors secreted by the CD34+ cells which contributed to the recovery or that the CD34+ cells were indeed present in the brain, but were below the detection limits of the antibodies used for detection. Taguchi et al (2004) also used an enriched CD34+ fraction to show that these cells contribute to new vessel formation following stroke. This study used SCID mice and Brdu labeling to visualize endothelial cell proliferation in the vasculature surrounding the penumbral region. The Brdu label was co-localized with CD31 (expressed by angiogenic endothelial cells) at day 1 and 3 post stroke. The highest density of newly formed vasculature was observed in the ischemic hemisphere of CD34+ transplanted animals. This effect was blocked with the administration of Endostatin (suppresses endothelial proliferation) and erythropoietin (a pro-angiogenic agent) increased the neovascularization. Functional recovery was also seen (at day 90 post transplant) in animals which received the CD34+ cells (Nystedt et al, 2006). These studies indicate that an enriched CD34+ fraction from HUCB may be better suited for regeneration of blood vessels in the brain rather than the whole HUCB fraction.

Use of autologous BMSC stem cells (BMSC) avoids ethical, infectious and immunological concerns. BMSC are abundant and therefore negate a need for cell expansion by culturing prior to administration. Iloshi et al (2004) used BMSC from rats, transfected with a LacZ reporter gene and administered IV following MCAO. The cells were administered at 3 or 72 hours post stroke. Animals which received BMSC had a significant reduction in behavior deficit (Morris water maze and treadmill stress test) at 2 weeks following stroke. Animals which received BMSC within 3 hours of stroke had an almost non-existent lesion and LacZ+ cells were detected in and around the ischemic zone. The authors suggested that the BMSC secrete angiogenic growth factors (e.g. VEGF and bFGF) (Chen et al, 2002; Kurozumi et al, 2005) which cause the functional recovery. Another mechanism suggested by the authors is that the BMSC integrate into the ischemic area and differentiate into mature cells rather than fuse with resident cells. Another study by Keimpema et al (2009) also evaluated the use of

BMSC following MCAO. This study, in contrast to the previous one, injected the BMSC intra-arterially (IA) at 3, 6, 12 and 24 hours post stroke. BMSC were observed in the blood vessels (when administered 3 hours post stroke), at the lesion (when administered at 6 hours post stroke) and phagocytic activity when BMSC were administered 24 hours post stroke. All time points evaluated showed a reduction in infarct volume. The authors defended their choice of arterial and administration of the BMSC at later time points by stating that signals from the lesion are relatively lower immediately following the stroke, thus IV injected BMSC get trapped in the spleen and are not observed in the ischemic regions of the brain. The authors also suggested that BMSC may release angiogenic growth factors which may also contribute to the reduction of infarct size. BMSC have also been engineered to overexpress potentially beneficial genes to the brain. For example, erythropoietin (EPO) is a known antioxidant, anti-apoptotic and anti-inflammatory (Chong et al, 2003). However, EPO cannot cross the blood barrier if injected IV and therefore its beneficial effect on the brain cannot be evaluated (Cho et al, 2010). Cho et al transduced BMSC to produce increased levels of EPO, and sterotaxically implanted $6x10^5$ cells. The transduced cells were shown to secrete higher levels of BDNF, SDF1-A and TGF1-b. Animals which received the transduced BMSC had improved neurological function, lower infarct volumes and higher levels of phosphorylated AKT (a downstream effector of EPO) (Cho et al, 2010).

6. Are HSC a viable avenue for stroke therapy?

It appears likely that a viable stroke therapy may be generated from HSC research, but much investigation is yet to be done. There are discrepancies in evaluation modalities currently being employed for studying the contribution of HSC to recovery following stroke. There is a lack of an overall consensus for which animal strain, stroke model, anesthetic, duration of occlusion, HSC markers, number of HSC to be injected, source of HSC, route of delivery, time of delivery, time for evaluation and kind of end point testing (for behavior), should be used for research. Several studies have either used Wistar rats (Nystedt et al, 2006) or Sprague Dawley (Leonardo et al, 2010) for the research. However, Bardutzky et al (2005) reported that the ischemic lesion evolution was substantially different between the Wister and Spargue Dawley strains. The authors used cortical blood flow and apparent diffusion co-efficient maps and 2,3,5-Triphenyltetrazolium chloride staining to arrive at that conclusion (Bardutzky et al, 2005). Other studies have also confirmed differences in the middle cerebral artery occlusion model in different strains (Fox et al, 1993; Sauter et al, 1995; Oliff et al, 1996). The rat model for middle cerebral artey was first described by Robinson et al in 1975 and has since been modified by several groups (Tamura et al, 1981; Bederson et al, 1986). As this model was refined over the years, researchers have chosen the best adaptation of this model for their research. For example, the anesthetic used for the rats varies from chloral hydrate (Bederson et al, 1986), ketamine hydrochloride (Longa et al, 1989), Halothane (Chen et al, 1992), and isoflurane (Belayev et al, 2009). While these anesthetic agents are used for the surgery, none of the publications report on the effect of the choice of the anesthetic on the injected HSC or the level of neuroprotection imparted by the choice of anesthetic. Culley et al (2011) showed that isoflurane does not kill stem cells, but does affect their proliferative capacity, which is an important feature for stem cell survival and engraftment into the area of injury. The mouse model of middle cerebral artery occlusion has been described by Connolly et al (1996) and has also been been modified and used by several researchers (Olsen et al, 1986; Clark et al, 1997; Belayev et al, 1999). Various anesthetic gases have also been used for the mouse model of middle cerebral artery occlusion (Yanamoto et al, 2003) with no correlation on the effect of the choice of anesthetic to the engraftment potential of the HSC. The duration of occluison also varies from 45 minutes to 120 minutes (Campagne et al, 1999). Campagne et al (1999) showed that the time of occlusion affects the infarct size and hence would affect how quickly tissue can recover following occlusion. Li et al (2005) and Popp et al (2009) also showed that extent of infarct volume increases with occlusion time and leads to varying areas of confinement for neuronal and glial markers.

Studies have also used a variety of sources for the HSC that are injected following stroke: HUCB (Vendrame M et al, 2006) , bone marrow (Iloshi et al, 2004) and Peripheral blood (Lickliter et al, 2000). In addition, there appears to be no consensus as to the number of cells injected for evalaution; Vendrame et al (2006) used $1x10^6$ and Makinen et al (2007) used $1\text{-}5x10^7$. Not only are the cell sources and numbers variable, but methods for enrichment before injection also vary (Iloshi et al, 2004; Taguchi et al, 2004). It seems logical that HSC fractions enriched for markers that are more committed towards an endothelial lineage (for example, CD34+ cells in humans) (Asahara et al, 1997) would offer a faster recovery than those that are whole enriched fractions. Since HSC have an intrinsic potential for proliferation in response to growth factor, such as VEGF (Ortega et al, 1997), it would seem that unnecessarily increasing the number of injected cells may not be needed. If injected cells can contribute to recovery directly and by paracrine secretion of tropic growth factors, then an exorbitant number of HSC may not be needed. Actual HSC numbers needed for adminitration in stroke patients for receovery is critical for those considering an autologus transplantation from their own bone marrow. The route of injection of HSC also varied from IV (Nystedt et al, 2006; Leonardo et al, 2010; Regelsberger et al, 2011), to IA (Keimpema et al, 2009) and direct implantation in the brain (Cho et al, 2010). Maikinen et al (2006) showed that IV injection of HSC lead to them being trapped in the spleen and kidneys and none of the cells were detected in the brain following the stroke. In contrast, Keimpura et al (2004) injected the HSC intra-arterially and showed that if the HSC were injected within 3 hours of the stroke, they would be found in the blood vessels in the brain. People have long focussed on an IV route of delivery of HSC for easier translation to human clincal trials. However, it is logical for the spleen to trap and remove cells which are not part of the normal circulation, hence reducing the number of cells which are available for contribution to recovery in the brain. Intra-arterial administration of the HSC is a viable alternate for the delivery of HSC since it places the cells in close proximity to the injury and avoids clearence of masses of cells by the spleen. Intra-arterial delivery may be challenging in smaller rodents, such as mice, however, it has been successfully used (Torrente et al, 2001).

End point testing after HSC administration has relied on immunohistochemical staining (Kramer et al, 2000) or behavioral outcome (Chen et al, 2001). Infarct volumes are evaluated using a variety of methods: 2,3,5-Tetrazolium Chloride staining (Kramer et al, 2000), Nissl staining (Popp et al, 2009), Cresyl violet staining (Tureyen et al, 2004), or MRI imaging (Lansberg et al, 2001). Immunohsitochemistry of tissues lends itself to artifacts from incomplete fixation, excessive dehydration and overstaining (Werner et al, 2000; Fowler et al, 2008), so the variability in these techniques is well known. However, some studies have used these stains to report no changes in infarct volumes following administration of HSC (Regelsberger et al 2011; Nystedt et al, 2006). This may be due to artifacts inherent in the nature of the staining methods rather than ineffective HSC contribution to reducing infarct volume following stroke. MRI imaging offers the best option for in vivo evaluation of infarct volumes following stroke (Moseley et al, 2001). However, access to MRI imaging for reasearchers may not be as feasible as doing an imunohistochemical staining for infarct volume. Behavior testing has become a hallmark end point testing for stroke in rodents (Chen et al, 2001), however, Makinen et al (2006) were not incorrect in suggesting that the majority of studies do not utilize the full battery of tests to arrive at a conclusion. Even within the battery of tests recommended by Makinen et al (2006), there is plently of room for variability within studies. Behavior tests are not sensitive enough to forgive minor differences in testing protocols which may be due to availability of resources for the research. For example, studies have evaluated end point behavior testing in rodents as early as 24 hours post stroke (Nystedt et al, 2006) and as late as 25 days post stroke (Taguchi et al, 2004). Earlier time points for behavior may not have allowed sufficient times for the HSC to have an affect and the later time points may be too far away from the time of injury and thus allow for spontaneous recovery.

Studies have either reported the presence or absence of of HSC in the ipsilateral hemisphere following stroke (Makinen et al, 2006). However, few studies have addressed issues of cell-cell fusion (Terada et al, 2004) whereby the resident vascular endothelial cells and the injected HSC fuse together and the HSC do not take on the mature phenotype of the resident cell. While it may difficult to differentitate a maturing HSC from a resident endothelial cell, since a maturing cell would carry the same markers as the resident cell, it is important for future stroke HSC therapy to be able to differentiate whether certain treatments lead to integrating HSC or mere cell to cell fusion. HSC also have the ability to transdifferentiate, dedifferentiate and re-differentiate depending on the cues provided by the local environment. It is entirely possible that some HSC integrate into the vasculature and others transdifferentiate into neurons. However, the majority of studies focus either on vascular beds (Taguchi et al, 2009) or neuronal regeneration (Garzon-muvdi et al, 2009) rather than both.

In conclusion, it seems likely that HSC investigation may contribute critical knowledge to our constantly evolving understanding of stroke pathophysiology. However, this enthusiasm must be tempered with realistic expectations that this will only occur with detailed, diligent labortory investigation. It is our hope that with such critical investigation our field will achieve it's goal of providing a viable new therapy for stroke.

7. References

Altman, J., Das, G.D. (1965). Autoradiographic and histological evidence of postnatal hippocampal neurogenesis in rats. *J Comp Neurol.* 124(3):319-35.

Alvarez-Dolado, M., Pardal, R., Garcia-Verdugo, J. M., Fike, J. R., Lee, H. O., Pfeffer, K., Lois, C., Morrison, S. J., Alvarez-Buylla, A. Fusion of bone-marrow-derived cells with Purkinje neurons, cardiomyocytes, and hepatocytes. *Nature.* 2003; 425,968-973.

Anderson, J.M. (2000). Multinucleated giant cells. *Curr. Opin. Hematol. 7(1)*, 40–47.

Asahara T., Murohara T., Sullivan A., Silver M., van der Zee R., Li T., Witzenbichler B., Schatteman G. and Isner J. M. (1997). Isolation of putative progenitor endothelial cells for angiogenesis. *Science.* 275, 964–967.

Bardutzky, J., Shen Q., Henninger, N., Bouley, J., Duong, TQ., Fisher, M. (2005). Differences in ischemic lesion evolution in different strains usinf diffusion and perfusion imaging. *Stroke*. 36; 2000-2005.

Baum, C.M., Weissman, I.L., Tsukamoto, A.S., Buckle, A.M., and Peault, B. (1992). Isolation of a candidate human hematopoietic stem-cell population. *Proc. Natl. Acad. Sci. U. S. A.* 89, 2804–2808.

Bederson, JB., Pitts, LH., Tsuji, M., Nishimura, MC., Davis, RL., Bartkowski, H. (1986). Rat middle cerebral artey occlusion: evaluation of the model and dvelopment of a neurologic examination. *Stroke*. 17(3). 472-476.

Belayev, LR., Busto, et al. (1999). Middle cerebral artery occlusion in the mouse by intraluminal suture coated with poly-L-lysine: neurological and histological validation. *Brain Res.* 833(2): 181-90.

Belayev, L., Khoutoroval, L., Atkins, KD., Bazan, NG. (2009). Robust docosahexaenoic acid-mediated neuroprotection in a rat model of transient, foal cerebral ischemia. *Stroke.* 40 (9):3121-6.

Bersano, A., Ballabio, E., Lanfranconi, S., Boncoraglio, GB., Corti, S., Locatelli, F., Baron, P., Bresolin, N., Parati, E., Candelise, L. (2010). Clinical Studies in Stem Cells Transplantation for Stroke: A Review. *Curr Vasc Pharmacol.* 8(1):29-34

Brazelton, T.R., Rossi, F.M., Keshet, G.I., and Blau, H.M. (2000). From marrow to brain: expression of neuronal phenotypes in adult mice. *Science.* 290, 1775–1779.

Brown, R. A., D. Adkins, et al. (1997). Factors that influence the collection and engraftment of allogeneic peripheral-blood stem cells in patients with hematologic malignancies. *J Clin Oncol* 15(9): 3067-74.

Byrne A. M., Bouchier-Hayes D. J. and Harmey J. H. (2005): Angiogenic and cell survival functions of vascular endothelial growth factor (VEGF). *J. Cell Mol. Med.* 9:777–794.

Campagne, MVL., Thomas, GR., Thibodeaux, H., Palmer, JT., Williams, SP., Lohe, DG., Brugeen, NV. (1999). Time of occlusion affects infarct size and hence would affect how quickly tissue can recover following the occlusion. *J cereb blood flow and metabolism.* 19:1354-1364.

Carmeliet P, Ferreira V, Breier G, et al. (1996) Abnormal blood vessel development and lethality in embryos lacking a single VEGF allele. *Nature.* 380(6573):435-439.

Carmeliat, P. Angiogenesis inlife, disease and medicine. (2005). Nature. 438:932-936.

Chen H, Chopp M, Zhang ZG, Garcia JH. (1992). The effect of hypothermia on transient middle cerebral artery occlusion in the rat. *J Cereb Blood Flow Metab.* 12:621- 628.

Chen, JP., Sanberg, PR., Li, Y., Wang, L., Lu, M., Willing, AE., Sanchez-Ramos, J., Chopp, M.. (2001). Intravenous administration of human umbilical cord blood reduces behavioral deficits after stroke in rats. *Stroke* 32(11): 2682-8.

Chen, X., Y. Li, Y., Wang, L., Katakowski, M., Zhang, L., Chen, J., Xu, Y., Gautam, SC., Chopp, M. (2002). Ischemic rat brain extracts induce human marrow stromal cell growth factor production. *Neuropathology.* 22(4): 275-9.

Chilov D., Kukk, E., Taira, S., Jeltsch, M., Kaukonen, J., Palotie, A., Joukov, V., Alitalo, K. (1997). Genomic organization of human and mouse genes for vascular endothelial growth factor C. *J Biol Chem.* 272: 25176-25183.

Cho, GW., Koh, SH., kim, MH., Yoo, AR., Noh, MY., Oh, S., Kim, SH. (2010). The neuroprotective effect of erythropoietin-transduced human mesenchymal stromal cells in an animal model of ischemic stroke. *Brain Res.* 1353: 1-13.

Chong, ZZ., Kang, JQ., Maiese, K. (2003). Erythropoietin: cytoprotection in vascular and neuronal cells. *Curr Drug Targets Cardiovasc Haematol Disord.* 3(2): 141-54.

Clark, WM., Lessov, NS., Dixon, MP., Eckenstein, F. (1997). Monofilament intraluminal middle cerebral artery occlusion in the mouse. *Neurol Res.* 19(6): 641-8.

Coffin JD, Harrison J, Schwartz S, Heimark R. (1991). Angioblast differentiation and morphogenesis of the vascular endothelium in the mouse embryo. *Dev Biol.* 148: 51-62.

Connolly. ES Jr., Winfree, CJ., Stern, DM., Solomon, RA., Pinsky, DJ. (1996). Procedural and strain-related variables significantly affect outcome in a murine model of focal cerebral ischemia. *Neurosurgery.* 38(3): 523-31; discussion 532.

Cook, MM., K. Kollar, G.P. Brooke, Atkinson, K. (2009). Cellular therapy for repair of cardiac damage after acute myocardial infarction. *Int J Cell Biol.* 2009:906507-906518.

Culley, DJ., Boyd, JD., Palanisamy, A., Xie, X., Kojima, K., Vacanti, CA., Tanzi, RE., Crosby, G. (2011). Isoflurane decreases self-renewal capacity of rat neural stem cell. *Anesthesiology.* Epub ahead of print.

Del Papa N, Quirici, N., Soligo, D., Scavullo, C., Cortiana, M., Borsotti, C., Maglione, W., Comina, DP., Vitali, C., Fraticelli, P., Gabrielli, A., Cortelezzi, A., Lambertenghi-Deliliers, G. (2006). Bone marrow endothelial progenitors are defective in systemic sclerosis. *Arthritis Rheum.* 54(8):2605-15.

Diefenderfer, DL., Osyczka, AM., Garino, JP., Lebov, PS. (2003). Regulation of BMP-induced transcription in cultured human bone marrow stromal cells. *J Bone Joint Surg Am.* 85-A (3): 19-28.

Dzierzak, E. (1999). Embryonic beginnings of definitive hematopoietic stem cells. Ann. N. Y. Acad. Sci. 872, 256–262.

Elfenbein, G. J., R. Sackstein, et al. (2004). Do G-CSF mobilized, peripheral blood-derived stem cells from healthy, HLA-identical donors really engraft more rapidly than do G-CSF primed, bone marrow-derived stem cells? No! *Blood Cells Mol Dis.* 32(1): 106-11.

Fadini, GP., Sartor, S., Albiero, M., Baesso, I., Murphy, E., Menegolo, M., Grego, F., Kreutzenberg, SVD., Tiengo, A., Agostini, C., Avogano, A. (2006). Number of endothelial progenitor cells as a marker of severity for diabetic vasculopathy. *Arterioscle. Thromb. Vasc.* Biol. 26:2140-2146.

Ferrara N, Carver-Moore, K., Chen, H., Dowd, M., Lu, L., O'Shea, KS., Powell-Braxton, L., Hillan, KJ., Moore, MW. (1996). Heterozygous embryonic lethality induced by targeted inactivation of the VEGF gene. *Nature.* 380: 439-442.

Ferrari, G., Cusella-De Angelis, G., Coletta,M., Paolucci, E., Stornaiuolo, A., Cossu, G., Mavillo, F. (1998). Muscle regeneration by bone marrow-derived myogenic progenitors. *Science.* 279(5356):1528-30.

Fong GH, Rossant J, Gertsenstein M, Breitman ML. (1995). Role of the Flt-1 receptor tyrosine kinase in regulating the assembly of vascular endothelium. *Nature.* 376: 66-70.

Forraz, N, McGuckin, CP. (2011). The umbilical cord: a rich and ethical stem cell source to advance regenerative medicine. *Cell Prolif.* 44(1): 60-9.

Fowler CB, O'Leary TJ, Mason JT. (2008). Modeling formalin fixation and histological processing with ribonuclease A: effects of ethanol dehydration on reversal of formaldehyde cross-links. *Lab Invest.* 88(7):785-91.

Fox, G., Gallacher, D., Shevde, S., Loftus, J., Swayne, G. (1993). Anatomic variation of the middle cerebral artery in the Sprague-Dawley rat. *Stroke*. 24: 2087–2092.

Fuss, IJ., Kanof, ME., Smith, PD., Zola, H. (2009). Isolation of whole mononuclear cells from peripheral blood and cord blood. *Curr Protoc Immunol*. 7:Unit 7.1.

Gallacher, L., Murdoch, B., Wu, D., Karanu, F., Fellows, F., and Bhatia, M. (2000). Identification of novel circulating human embryonic blood stem cells. *Blood*. 96, 1740–1747.

Garzon-Muvdi, T. and A. Quinones-Hinojosa. (2009). Neural stem cell niches and homing: recruitment and integration into functional tissues. *ILAR J*. 51(1): p. 3-23.

Gerber, HP., Malik, AK., Solar, GP., Sherman, D., Liang, XH., Meng, G., Hong, K., Marsters, JC., Ferrara, N. (2002). VEGF regulates haematopoietic stem cell survival by an internal autocrine loop mechanism. *Nature*. 417(6892):954-958.

Gill, M., dias, S., Hattori, K., Rivera, ML., Hicklin, D., Witte, L., Girardi, L., Yurt, R., Himel, H., Rafii, S. (2001). Vascular trauma induces rapid but transient mobilization of VEGFR2(+)AC133(+) endothelial precursor cells. *Circ Res*. 88: 167-174.

Gounis, MJ., Spiga, MG., Graham, RM., Wilson, A., Haliko, S., Leiber, BB., Wakhloo, AR., Webster, KA. (2005). Angiogensis is confined to the transient period of VEGF expression that follows adenoviral gene delivery to ischemic muscle. *Gene Ther*. 12:762-771.

Grunewald, M., Avraham, I., Dor, Y., Bachar-Lustig, E., Itin, A., Jung, S., Chimenti, S., Lansman, L., Abramovitvh, R., Kesher, E. (2006). VEGF-induced adult neovascularization: recruitment, retention and role of accessory cells. *Cell*. 124:175-189.

Hall, BK. (2000)The neural crest as a fourth germ layer and vertebrates as quadroblastic not triploblastic. Evol Dev. 2(1):3-5.

Harrison JS, Rameshwar P, Chang V, Bandari P. (2002). Oxygen saturation in the bone marrow of healthy volunteers. *Blood*. 99(1):394.

Heiss, C., Keymel, S., Niesler, U., Ziemann, J., Kelm, M., Kalka, C. (2005). Impaired progenitor cell activity in age-related endothelial dysfunction. *J Am Coll Cardiol*. 3;45(9):1441-8.

Heissig B., Hattori K., Dias S., Friedrich M., Ferris B., E. Miller–Kasprzak et al.: EPCs contributing to vascular repair 257 Hackett N. R., Crystal R. G., Besmer P., Lyden D., Moore M. A., Werb Z. and Rafii S. (2002): Recruitment of stem and progenitor cells from the bone marrow niche requires MMP-9 mediated release of kit-ligand. *Cell*, 109, 625–637.

Hennemann, B., Ickenstein, G., Sauerbruch, S., Luecke, K., Haas, S., Horn, M., Andreesen, R., Bogdahn, U., Winkler, J (2008). Mobilization of CD34+ hematopoietic cells, colony-forming cells and long-term culture-initiating cells into the peripheral blood of patients with an acute cerebral ischemic insult. *Cytotherapy*. 10(3):303-11.

c, K., Ishibashi, M., Ohtani, K., Inoue, S., Zhao, Q., Kitamoto, S., Sata, M., Ichiki, T., Takeshita, A., Egashira, K. (2004). Gene transfer of stromal cell-derived factor-1 alpha enhances ishemic vasculogenesis and angiogenesis via vascular endothelial growth factor/endothelial nitric oxide synthase-related pathway: next generation chemokine therapy for therapeutic neovascularization. *Circulation*. 109: 2454-2461.

Hole, N. (1999). Embryonic stem cell-derived haematopoiesis. *Cells Tissues Organs*. 165, 181–189.

Hooper AT, Butler JM, Nolan DJ, et al. (2009). Engraftment and reconstitution of hematopoiesis is dependent on VEGFR2-mediated regeneration of sinusoidal endothelial cells. *Cell Stem Cell*. 4(3):263-274.

Horrocks, C., Fairhurst, M., Miller, C., Thomas, T. (1998). StemSep: a new method for negative selection of murine hematopoietic progenitors. *Exp Hematol*. 26:733.

Hristov M. and Weber C. (2004): Endothelial progenitor cells: characterization, pathophysiology, and possible clinical relevance. *J. Cell Mol. Med*. 8:498–508.

Hur, J., Yang, HM., Yoon, CH., Lee, CS., Park, KW., Kim, JH., Kim, TY., Kim, JY., Kang, HJ., Chae, IH., Oh, BH., Park, YB., Kim, HS. (2007). Identification of a novel role of T cells in postnatal vasculogenesis: characterization of endothelial progenitor cell colonies. *Circulation*. 116:1671–1682.

Iihoshi, S., Honmou, O., Houkin, K., Hashi, K., Kocsis, JD. (2004). A therapeutic window for intravenous administration of autologous bone marrow after cerebral ischemia in adult rats. *Brain Res*. 1007(1-2): 1-9.

Jaakkola P, Mole DR, Tian YM, Wilson, MI., Gielbert, J., Gaskell, SJ., Kriegsheim, AV., Hebestreit, HF., Mukherji, M., Schofield, CJ., Maxwell, PH., Pugh, CW., Ratcliffe, PJ. (2001). Targeting of HIF-alpha to the von Hippel-Lindau ubiquitylation complex by O2-regulated prolyl hydroxylation. *Science*. 292(5516):468-472.

Jaatinen, T. and Laine, J. 2007. Isolation of Mononuclear Cells from Human Cord Blood by Ficoll-Paque Density Gradient. *Current Protocols in Stem Cell Biology*. 1:2A.1.1–2A.1.4.

Kajstura, J., Leri, A., Finato, N., Di Loreto, C., Beltrami, C.A., and *8*, 607–612. Anversa, P. (1998). Myocyte proliferation in end-stage cardiac failure. *Proc Natl. Acad. Sci. USA*. 95(15):8801-8805.

Jang, YY., Sharkis, SJ. (2007). A low level of reactive oxygen species selects for primitive hematopoietic stem cells that may reside in the low-oxygenic niche. *Blood*. 110(8):3056-3063

Janowska-Wieczorek A, Marquez LA, Dobrowsky A, Ratajczak MZ, Cabuhat ML. (2000). Differential MMP and TIMP production by human marrow and peripheral blood CD34(+) cells in response to chemokines. *Exp Hematol*.28:1274- 1285.

Kale, S., Karihaloo, A., Clark, P.R., Kashgarian, M., Krause, D.S., Cantley, L.G. (2003). Bone marrow stem cells contribute to repair of the ischemically injured renal tubule. *J. Clin. Invest*. 112, 42–49.

Kanof, M E., Smith, PD. and Zola, H.(2001). Isolation of Whole Mononuclear Cells from Peripheral Blood and Cord Blood. Current Protocols in Immunology. Chapter 7:Unit 7.1.

Keimpema, E., Fokkens, MR., Nagy, Z., Agoston, V., Luiten, PG., Nyakas, C., Boddeke, HW., Copray, JC. (2009). Early transient presence of implanted bone marrow stem cells reduces lesion size after cerebral ischemia in adult rats. *Neuropathol Appl Neurobiol*. 35(1): 89-102.

Kondo, M., Wagers, A.J., Manz, M.G., Prohaska, S.S., Scherer, D.C., Beilhack, G.F., Shizuru, J.A., and Weissman, I.L. (2003). Biology of hematopoietic stem cells and progenitors: implications for clinical application. *Annu. Rev. Immunol*. 21, 759–806.

Kondo, T., Hayashi, M., Takeshita, K., Numaguchi, Y., Kobayashi, K., Iino, S., Inden, Y., Murohara, T. (2004). Smoking cessation rapidly increases circulating progenitor cells in peripheral blood in chronic smokers. *Arterioscler Thromb Vasc Biol.* 24(8):1442-7.

Korbling M, Fliender TM. History of blood stem cell transplants. Blood stem cell transplants. In: Gale RP, Juttner CA, Henon P eds. Peripheral blood stem cell autographts. New York: Cambridge University Press; 1994:9

Kramer ,M., Dang, J., Baertling, F., Denecke, B., Clarner, T., Kirsch, C., Beyer, C., Kipp, M. (2010). TTC staining of damaged brain areas after MCA occlusion in the rat does not constrict quantitative gene and protein analyses. *J Neurosci Methods.* 15;187(1):84-9.

Krause, DS., Theise, ND., Collector, MI., Henegariu, O., Hwang, JA., Gardner, R., Neutzel, S., and Sharkis, SJ. (2001). Multi organ, multi-lineage engraftment by a single bone marrow-derived stem muscular dystrophy patient receiving bone marrow transplantation. *Cell. 105*, 369–377.

Kucia M., Jankowski K., Reca R., Wysoczynski M., Bandura L., Allendorf D.J., Zhang J., Ratajczak J. and Ratajczak M. Z. (2004): CXCR4-SDF-1 signaling, locomotion, chemotaxis and adhesion. *J. Mol. Histol.*, 35, 233–245.

Kunz, GA., Liang, G., Cuculi, F., Gregg, D., Vata, KC., Shaw, LK., Goldschmidt-Clermont, PJ., Dong, C., Taylor, DA., Peterson, ED. (2006). Circulating endothelial progenitor cells predict coronary artery disease severity. *Am Heart J.* 152(1):190-5.

Kurozumi, KK., Nakamura, K., Tamiya, T., Kawano, Y., Ishii, K., Kobune, M., Hiraj, S., Uchida, H., Sasaki, K., Ito, Y., Kato, K., Honmou, O., Houkin, K., Date, I., Hamada, H. (2005). Mesenchymal stem cells that produce neurotrophic factors reduce ischemic damage in the rat middle cerebral artery occlusion model. *Mol Ther.* 11(1): 96-104.

Laitinen, M., Ristimaki, A., Honkasalo, M., Narko, K., Paavonen, K., Ritvos, O. (1997). Differential hormonal regulation of vascular endothelial growth factors VEGF, VEGF-B and VEGF-C messenger ribonucleic acid levels in cultured human granulosa-luteal cells. *Endocrinology*. 138: 4748-4756.

Lansberg, MG., O'Brien, MW., Tong, DC., Moseley, ME., Albers, GW. (2001). Evolution of cerebral infarct volume assessed by diffusion-weighted magnetic resonance imaging. *Arch Neurol.* 58:613-617.

Lapidot, T., Kollet, O. (2002). The essential roles of the chemokine SDF-1 and its receptor CXCR4 in human stem cell homing and repopulation of transplanted immune deficient NOD/SCID and NOS/SCID/B2m (null) mice. *Leukemia*. 16:1992-2003.

Laughlin, M.J. (2001). Umbilical cord blood for allogeneic transplantation in children and adults. *Bone Marrow Transplant*. 27, 1–6.

Lee, RH., kim, B., Choi, I., kim, H., Choi, HS., Suh, K., Bae, YC., Jung, JS. (2004). Characterization and expression analysis of mesenchymal stem cells from human bone marrow and adipose tissue. *Cell Physiol Biochem* 14(4-6): 311-24.

Leonardo, CC., hall, AA., Collier, LA., Amjo, CT Jr., Willing, AE., Pennypacker, KR. (2010). Human umbilical cord blood cell therapy blocks the morphological change and recruitment of CD11b-expressing, isolectin-binding proinflammatory cells after middle cerebral artery occlusion. *J Neurosci Res.* 88(6): 1213-22.

Leung, DW., Cachianes, G., Kuang, WJ., Goeddel, DV., Ferrara, N. (1989). Vascular endothelial growth factor is a secreted angiogenic mitogen. *Science*. 246: 1306-1309.

Lickliter, J.D., McGlave, P.B., DeFor, T.E., Miller, J.S., Ramsay, N.K., Verfaillie, C.M., Burns, L.J., Wagner, J.E., Eastlund, T., Dusenbery, K., and Weisdorf, D.J. (2000). Matched-pair analysis of peripheral blood stem cells compared to marrow for allogeneic transplantation. *Bone Marrow Transplant.* 26, 723–728.

Lloyd-Jones, D.M., et al. (2010). Defining and Setting National Goals for Cardiovascular Health Promotion and Disease Reduction: The American Heart Association's Strategic Impact Goal through 2020 and Beyond. *Circulation.* 121(4):586-613.

Locatelli, F., Bersano, A., Ballabio, E., Lanfranconi, S., Papadimitriou, D., Strazzer, S., Bresolin, N., Comi, GP., Corti, S. Stem cell therapy in stroke. *Cell Mol Life Sci.* 66(5): p. 757-72.

Luo, J., Zhang, HT., Jiang, XD., Xue,S., Ke, YQ. (2009). Combination of bone marrow stromal cell transplantation with mobilization by granulocyte-colony stimulating factor promotes functional recovery after spinal cord transaction. *Acta Neurochir (Wien).* 151(11):1483-92.

Burns, T.C., Verfaillie, CM., low, WC. (2009). Stem cells for ischemic brain injury: a critical review. *J Comp Neurol.* 515(1): p. 125-44.

Lois, C., Alvarez-Buylla, A. (1993). Proliferating subventricular zone cells in the adult mammalian forebrain can differentiate into neurons and glia. *Proc. Natl. Acad. Sci.* USA 90, 2074–2077.

Louis, SA., Croome ,A., Eaves, AC., Thomas, TE. (2001). A new and simple nonmagnetic technique to highly and rapidly enrich hematopoietic progenitors from mouse bone marrow. Blood. 98:118b.

Longa, EZ., Weinstein, PR., Carlson, S., Cummins, R. (1989). Reversible middle cerebral artery occlusion without craniectomy in rats. *Stroke.* 20: 84–91.

Ma, Q., Jones, D., Borghesani, PR., Segal, RA., Nagasawa, T., Kishimoto, T., Bronson, RT., Springer, TA. (1998). Impaired B-lymphopoiesis, myelopoiesis and derailed cerebellar neuron migration in CXCR4 and SDF1 deficient mice. *Proc Natl Acad Sci USA.* 95:9448-9453.

Machalinski, B., Paczkowska, E., Koziarska, D., Ratajczak, MZ. (2006). Mobilization of human hematopoietic stem/preogenitor-enriched CD34+ cells into peripheral blood during stress related to ischemic stroke. *Folia Histochem Cytobiol.* 44(2):97-101.

Makinen, S., Kekarainen, T., Nystedt, J., Liimatainen, T., Huhtala, T., Narvanen, A., Laine, J., Jolkkonen, J. (2006). Human umbilical cord blood cells do not improve sensorimotor or cognitive outcome following transient middle cerebral artery occlusion in rats. *Brain Res.* 1123(1): 207-15.

Masuda, H., Asahara, T. (2003). Post natal endothelial progenitor cells for neovascularization is tissue regeneration. *Cardiovasc. Res.* 58: 390-398.

Matthews, W., Jorda, CT., Gavin, M., Jenkins, NA., Copeland, NG., Lemischka, IR. (1991). A receptor tyrosine kinase cDNA isolated from a population of enriched primitive hematopoietic cells and exhibiting close genetic linkage to c-kit. *Proc Natl Acad Sci USA.* 88: 9026-9030.

McCune, JM., Rabin, LB., Feinberg, MB., Lieberman, M., Kosek, JC., Reyes, GR., and Weissman, IL. (1988). Endoproteolytic cleavage of gp160 is required for the activation of human immunodeficiency virus. Cell *53*, 55–67.

McGuckin, C.P., Forraz, N. (2008). Umbilical cord blood stem cells--an ethical source for regenerative medicine. *Med Law* 27(1): 147-65.

McQuibban, GA., Butler, GS., Gong, JH., Bendall, L., Power, C., Clark-Leiws, I, Overall, CM. (2001). Matrix metalloproteinases activity inactivates the CXC chemokine stromal cell-derived factor-1. *J. Biol. Chem.* 276:43503-43508.

Melton DA, Cowan C. (2004). "Stemness": Definitions, Criteria and Standards. In: Lanza R, editor. Handbook of Stem Cells. 1st ed. Burlington, San Diego, USA: Elsevier Academic Press.

Mignatti, P., Tsuboi, R., Robbins, E., Rifkin, DB. (1989) In vitro angiogenesis of the human amniotic membrane: Requirement for basic fibroblast growth factor-induced proteinases. *J. Cell Biol.* 108: 671-682.

Miljan, EA., Sinden, JD. (2009). Stem cell treatment of ischemic brain injury. *Curr Opin Mol Ther.* 11(4): p. 394-403.

Montesano, R., Vassalli, JD., Baird, A., Guillemin, R., Orci, L. (1986) Basic fibroblast growth factor induces angiogenesis in vitro. *Proc. Natl. Acad. Sci.* U.S.A. 83:7297-7301.

Moore, MA., Hattori, K., Heissig, B., Shieh, JH., Dias, S., Crystal, RG., Rafii, S. (2001). Mobilization of endothelial and hematopoietic stem and progenitor cells by adenovector-mediated elevation of serum levels of SDF-1, VEGF and Angiopoietin-1. *Ann NY ACAD Sci.* 938:36-45.

Morris CL, Siegel E, Barlogie B, Cottler-Fox, M., Lin, P., Fassas, A., Zangari, M., Anaissia, E., Tricot, G. (2003). Mobilization of CD34+ cells in elderly patients (>/= 70 years) with multiple myeloma: influence of age, prior therapy, platelet count and mobilization regimen. *Br. J. Haematol.* 120:413-423.

Muller, A., Homey B., Soto H., Ge N., Catron D., Buchanan ME., McClanahan T., Murphy E., Yuan W., Wagner S. N., Barrera J. L., Mohar A., Verastegui E. and Zlotnik A. (2001): Involvement of chemokine receptors in breast cancer metastasis. *Nature*. 410, 50–56.

Nystedt, J., Makinen, S., Laine, J., Jolkkonen, J. (2006). Human cord blood CD34+ cells and behavioral recovery following focal cerebral ischemia in rats. Acta Neurobiol Exp (Wars). 66(4):293-300

Oliff HS, Marek P, Miyazaki B, Weber E. (1996). The neuroprotective efficacy of MK-801 in focal cerebral ischemia varies with rat strain and vendor. *Brain Res.* 731: 208–212.

Olofsson, B., Pajusola, K., Von Euler, G., Chilov, D., Alitalo, K., Eriksson, U. (1996). Genomic organization of the mouse and human genes for vascular endothelial growth factor B (VEGF-B) and characterization of a second splice isoform. *J Biol Chem.* 271: 19310-19317.

Olsen, T. S. (1986). Regional cerebral blood flow after occlusion of the middle cerebral artery. *Acta Neurol Scand.* 73(4): 321-37.

Orlic, D., Kajstura, J., Chimenti, S., Jakoniuk, I., Anderson, S.M., Li, find their niche. Nature 414, 98–104. B., Pickel, J., McKay, R., Nadal-Ginard, B., Bodine, D.M., et al. Stamm, C., Westphal, B., Kleine, H.D., Petzsch, M., Kittner, C., (2001). Bone marrow cells regenerate infarcted myocardium. *Nature*. 410, 701–705.

Ogawa, S., oku, A., Sawamo, A., Yanaguchi, S., Yazaki, Y., Shibuva, M. (1998). A novel type of vascular endothelial growth factor, VEGF-E (NZ-7, VEGF), preferentially utilizes KDR/Flk-1 receptor and carries a potent mitotic activity without heparin-binding domain. *J Biol Chem.* 273: 31273-31282.

Ortega, N., Jonca, F., Vincent, S., Favard, C., Ruchoux, MM., Plouet, J. (1997). Systemic activation of the vascular endothelial growth factor receptor KDR/flk-1 selectively triggers endothelial cells with an angiogenic phenotype. *Am J Pathol.* 151: 1215-1224.

Osawa, M., Hanada, K., Hamada, H., Nakauchi, H. (1996). Long-term lymphohematopoietic reconstitution by a single CD34⁻ low/negative hematopoietic stem cell. *Science.* 273:242–245.

Paczkowska, E., Larysz, B., Rzeuski, R., Karbicka, A., Jalowinski, R., Kornacewicz-Jach, Z., Ratajczak, MZ., Machalinski, B. (2005). Human hematopoietic stem/preogenitor-enriched CD34(+) cells aremobilized into peripheral blood during stress related to ischemic stroke or acute myocardial infarction. Eur. J. Haemtol. 75(6):461-7.

Petit I, Szyper-Kravitz, M., Nagler, A., Lahav, M., Peled, A., Habler, L., Ponomarvov, T., Taichamn, RS., Arezana-Seisdedos, F., Fujii, N., Sandbank, J., Zipori, D., Lapidot, T. (2002). G-CSF induces stem cell mobilization by decreasing bone marrow SDF-1 and upregulating CXCR4. *Nat. Immunol.* 3:687-694.

Pirro, M., Schillaci, G., Menecali, C., Bagaglia, F., Paltricciar, R., Vaudo, G., Mannarino, MR., Mannarino, E. (2007). Reduced numbers of circulating endothelial progeniros and HOXA9 expression in CD34+ cells of hypertensive patients. *J. Hypertens.* 25(10):2093-9.

Pituch-Noworolska A., Majka, M., Janowska-Wieczorek, A., Baj-Krzyworzeka, M., Urbanowicz, B., Malec, E., Ratajczak, MZ. (2003). Circulating CXCR4-positive stem/progenitor cells compete for SDF1 positive niches in bone marrow, muscle and neural tissues: An alternative hypothesis to stem cell plasticity. *Folia Histochemica et Cytobiologica.* 41:13-21.

Popp, A., Jaenisch, N., Witte, OW., Frahm, C. (2009) Identification of Ischemic Regions in a Rat Model of Stroke. *PLoS ONE.* 4(3): e4764.

Rafii, S., Oz, MC., Seldomridge, JA., Ferris, B., Asch, AS., Nachman, RL., Shapiro, F., Rose, EA., Levin, HR. (1995) . Characterization of hematopoietic cells arising on the textured surface of left ventricular assist devices. *Ann Thorac Surg.* 60: 1627-1632.

Ratajska, A., Torry, RJ., Kitten, GT., Kolker, SJ., Tomanek, RJ. (1995). Modulation of cell migration and vessel formation by vascular enodtheial growth factor and basic fibroblast growth factor in cultured embryonic heart. *Dev Dyn.* 203(4):339-407.

Rehn, M., Olsson, A., Reckzeh, K., Diffner, E., Carmeliet, P., Landberg, G., Cammenga, J. (2011). Hypoxic induction of vascular endothelial growth factor regulates murine hematopoietic stem cell function in the low-oxygenic niche. *Blood.* 118(6):1534-43.

Riegelsberger, UM., Deten, A., Posel, C., Zille, M., Kranz, A., Boltze, J., Wagner, DC. (2011). Intravenous human umbilical cord blood transplantation for stroke: impact on infarct volume and caspase-3-dependent cell death in spontaneously hypertensive rats. *Exp Neurol.* 227(1): 218-23.

Robert, B., St John, PL., Hyink, DP., Abrahamson, DR. (1996). Evidence that embryonic kidney cells expressing flk-1 are intrinsic, vasculogenic angioblasts. *Am J Physiol.* 271: F744-F753.

Robinson RG., Shoemaker, WJ., Schlumpf, M. (1975). Effect of experimental cerebral infarction in rat brain on catecholamines and behavior. *Nature.* 255: 332-334.

Rolletschek, AP., Blyszczuk, P., Wobus, AM. (2004). Embryonic stem cell-derived cardiac, neuronal and pancreatic cells as model systems to study toxicological effects. *Toxicol Lett.* 149(1-3): 361-9.

Ruitenberg, JJ., Mulder, CB., Maina, VC., Landay, AL., Ghanekar, SA. (2006). Vaccutainer CPT and ficoll density gradient separation perform equivalently in maintaining the quality and function of PBMC from HIV seropositive blood samples. *BMC Immunol*. 25:7-11.

Rumyantsev, P.P., and Borisov, A. (1987). DNS synthesis in myocytes from different myocardial compartments of young rats in norm, after experimental infarction and in vitro. *Biomed. Biochem. Acta.* 46 (8-9): S510–S615.

Rumpold, H., Wolf, D., Koeck, R. and Gunsilius, E. (2004): Endothelial progenitor cells: a source for therapeutic vasculogenesis? *J. Cell Mol. Med.*, 8, 509–518.

Salcedo, R., Oppenheim, JJ. (2003). Role of chemokines in angiogenesis: CXCL12/SDF-1 and CXCR4 interaction, a key regulator of endothelial cell responses. *Microcirculation*. 10:359-370.

Sanganalmath, SK., Abdel-Latif, A., Bolli, R., Xuan, YT., Dawn, B. (2011). Hematopoietic cytokines foir cardiac repair mobilization of bone marrow cells and beyond. *Basic Res Cardiol*. 106(5):709-33.

Sauter, A., Rudin, M. (1995). Strain-dependent drug effects in rat middle cerebral artery occlusion model of stroke. *J Pharmacol Exp Ther*. 274: 1008–1013.

Schlenke, P., Kluter, H., Muller-Steinhardt, M., Hammers, HJ., Borchert, K., Bein, G. (1998). Evaluation of a Novel Mononuclear Cell Isolation Procedure for Serological HLA Typing. *Clin Diagn Lab Immunol*. 5(6): 808–813.

Schwarting, S., Litwak, S., Hao, W., Bahr, M., Weise, J., Neumann, H. (2008). Hematopoietic stem cells reduce postischemic inflammation and ameliorate ischemic brain injury. *Stroke*. 39(10):2867-75.

Segers, VF., Lee, RT. (2010). Protein therapeutics for cardiac regeneration after myocardial infarction. *J. Cardiovasc. Transl. Res.* 3:469-477.

Segers, VFM., Revin, V., Wu, W., Qiu, H., Lee, RT., Sandrasgara, A. (2011). Protease-resistant stromal cell derived factor-1 for the treatment of experimental peripheral arterial disease. Circulation. 123:1306-1315.

Seita J, Weissman IL. (2010). Hematopoietic stem cell: self-renewal versus differentiation. Wiley Interdiscip Rev Syst Biol Med. 2(6):640-53.

Shalaby, F., Rossant, J., Yamaguchi, TP., Gertsenstein, M., Wu, XF., Breitman, ML., Schuh, AC.. (1995). Failure of blood-island formation and vasculogenesis in Flk-1-deficient mice. *Nature*. 376: 62-66.

Shi, Q., Rafii, S., Wu, MH., Wijelath, ES, Yuc, I. (1998). Evidnece for circulating bone marrow derived endothelial cells. *Blood*. 92:362-367.

Shintani, S., Murohara, T., Ikeda, H., Ueno, T., Honma, T., Katoh, A., Sasaki, K., Shimada, T., Oike, Y., Imaizumi, T. (2001). Mobilization of endothelial progenitor cells in patients with acute myocardial infarction. *Circulation*. 103: 2776–2779.

Shweiki, D., Itin, A., Soffer, D., Keshet, E. (1992). Vascular endothelial growth factor induced by hypoxia may mediate hypoxia-initiated angiogenesis. *Nature*. 359(6398):843-845.

Shyu, W. C., S. Z. Lin, et al. (2006). Intracerebral peripheral blood stem cell (CD34+) implantation induces neuroplasticity by enhancing beta1 integrin-mediated angiogenesis in chronic stroke rats. *J Neurosci*. 26(13): 3444-53.

Spangrude GJ, Heimfeld S, Weissman IL. (1998). Purification and characterization of mouse hematopoietic stem cells. *Science*. 241:58-62.

Taguchi, AT., Soma, T., Tanaka, H., Kanda, T., Nishimura, H., Yoshikawa, H., Tsukamoto, Y., Iso, H., Fujimori, Y., Stern, DM., Naritomi, H., Matsuvama, T. (2004). Administration of CD34+ cells after stroke enhances neurogenesis via angiogenesis in a mouse model. *J Clin Invest.* 114(3): 330-8.

Taguchi, A., Nakagomi, N., Matsuvama, T., Kikuchi-Taura, A., Yoshikawa, H., Kasahara, Y., Hirose, H., Moriwaki, H., Nakagomi, T., Soma, T., Stern, DM., Naritomi, H. (2009). Circulating CD34-positive cells have prognostic value for neurologic function in patients with past cerebral infarction. *J Cereb Blood Flow Metab.* 29(1): 34-8.

Takahashi, T., Kalka, C., Masuda, H., Chen, D., Silver, M., Kearney, M., Mager, M., Isner, JM., Asahara, T. (1999). Ischemia- and cytokine-induced mobilization of bone marrow-derived endothelial progenitor cells for neovascularization. *Nat Med.* 5: 434-438.

Takubo, K., Goda, N., Yamada, W., Iriuchishima, H., Ikeda, E., kubota, Y., Shima, h., Johnson, RS., Hirao, A., Suematsu, M., Suda, T. (2010). Regulation of the HIF-1alpha level is essential for hematopoietic stem cells. *Cell Stem Cell.* 7(3):391-402.

Tamura, A., Graham, DI., McCulloch, J., Tesdale, GM. (1981). Focal cerebral ischemia in the rat. Description of technique and ear;y neuropatholgical consequences following middle cerebral artey occlusion. *J cereb blood flow Metabol.* 1:53-60.

Terada, N., Hamazaki, T., Oka, M., Hoki, M., Mastalerz, D.M., Nakano, Y., Meyer, E.M., Morel, L., Petersen, B.E., and Scott, E.W. (2002). Bone marrow cells adopt the phenotype of other cells by spontaneous cell fusion. *Nature. 416*, 542–545.

Terman BIet al. . Identification of the KDR tyrosine kinase as a receptor for vascular endothelial cell growth factor. *Biochem Biophys Res Commun.* 1992. 187: 1579-1586.

Torrente, Y., Tremblay, JP., Pisati, F., Belicchi, M., Rossi, B., Sironi, M., Fortunato, F., El Fahime, M., D'Angelo, MG., Caron, NJ., Constantin, G., Paulin, D., Scarlato, G., Bresolin, N. (2001). Intraarterial injection of muscle-derived CD34(+)Sca-1(+) stem cells restores dystrophin in mdx mice. *J Cell Biol.* 22;152(2):335-48.

Türeyen, K., Vemuganti, R., Sailor, KA., Dempsey, RJ. (2004). Infarct volume quantification in mouse focal cerebral ischemia: a comparison of triphenyltetrazolium chloride and cresyl violet staining techniques. *J. Neurosci Methods.* 139(2):203-207

Uchida, N., Weissman, IL. (1992). Searching for hematopoietic stem cells: evidence that Thy-1.1lo Lin− Sca-1+ cells are the only stem cells in C57BL/Ka-Thy-1.1 bone marrow. *J Exp Med.* 175: 175–184.

Urbich C., Dimmeler S. (2004): Endothelial progenitor cells: characterization and role in vascular biology. *Circ. Res.* 95: 343–353.

Wang, GL., Jiang ,BH., Rue, EA., Semenza ,GL. (1995). Hypoxia-inducible factor 1 is a basic-helix-loop-helix-PAS heterodimer regulated by cellular O2 tension. *Proc Natl Acad Sci U S A.* 92(12):5510-5514.

Wang, GL., Semenza, GL. (1995). Purification and characterization of hypoxia-inducible factor 1. *J Biol Chem.* 270(3):1230-1237.

Wang, V., Davis, DA., Haque, M., Huang, LE., Yarchoan, R. (2005). Differential gene up-regulation by hypoxia-inducible factor-1A and hypoxia-inducible factor-2A in HEK293T cells. *Cancer Res.* 65:3299-3306.

Wardlaw, J.M., P.A. Sandercock, and E. Berge. (2003). Thrombolytic therapy with recombinant tissue plasminogen activator for acute ischemic stroke: where do we go from here? A cumulative meta-analysis. *Stroke.* 34(6): p. 1437-42.

Weimann, J.M., Charlton, C.A., Brazelton, T.R., Hackman, R.C., and Blau, H.M. (2003). Contribution of transplanted bone marrow cells to Purkinje neurons in human adult brains. *Proc. Natl. Acad. Sci. USA*. 100. 2088–2093.

Wells, J.M., and Melton, D.A. (1999). Vertebrate endoderm development. *Annu. Rev. Cell Dev.* Biol. 15, 393–410.

Werner M, Chott A, Fabiano A, Battifora H. (2000). Effect of formalin tissue fixation and processing on immunohistochemistry. *Am J Surg Pathol*. 24(7):1016-9.

Woywodt, A., Blann, AD., Kirsch, U., Erdbruegger, N., Banzet, N., Haubitz, M., Dignat George, F. (2005). Isolation and enumeration of circulating endothelial cells by immunomagentic isolation: proposal of a definition and a consensus protocol. *Journal of Thrombosis and Haemostasis*. 4: 671–677.

Varga, NL., Bárcena, A., Fomin, ME., Muench, MO. (2010). Detection of human hematopoietic stem cell engraftment in the livers of adult immunodeficient mice by an optimized flow cytometric method. *Stem Cell Stud*. 1(1). pii: e5.

Vendrame, MC. Gemma, C., Pennypacker, KR., Bickford, PC., Davis Sanberg, C., Sanberg, PR., Willing, AE. (2006). Cord blood rescues stroke-induced changes in splenocyte phenotype and function. *Exp Neurol*. 199(1): 191-200

Verbik, DJ, Jackson, JD., Pirruccello, SJ., Patil, KD., Kessinger, A., Joshi, SS. (1995). Functional and phenotypic characterization of human peripheral blood stem cell harvests: a comparative analysis of cells from consecutive collections. *Blood*. 85(7): 1964-70.

Yamahara K, Sone M, Itoh H, Yamashita JK, Yurugi-Kobayashi, T., homma, k., Chao, TH., miyashita, K., Park, k., Ovamada, N., Sawada, N., Taura, D., Fukunaga, Y., Tamura, N., Nakao, k. (2008). Augmentation of Neovascularizaiton in Hindlimb Ischemia by Combined Transplantation of Human Embryonic Stem Cells-Derived Endothelial and Mural Cells. *PLoS ONE*. 3(2): e1666.

Yanamoto, HI., nagata, I., Niitsu, Y., Xue, JH., Zhang, Z., Kikuchi, H. (2003). Evaluation of MCAO stroke models in normotensive rats: standardized neocortical infarction by the 3VO technique. *Exp Neurol*. 182(2): 261-74.

Yip, HK., Chang, LT., Chang, Wn., lu, CH., Liou, CW., Lan, MY., liu, JS., Youssef, AA., Chang, HW. Level and value of circulating endothelial progenitor cells in patients after acute ischemic stroke. *Stroke*. 2008; 39(1): p. 69-74.

Zhang, K., Mcquibban, GA., Silva, C., Butler, GS., Johnson, JB.m, Holden, J., Clark-Lewis, I., Overall, CM, Power, C. (2003). HIV-induced metalloproteinase processing of the chemokine stromal cell derived factor-1 causes neuro-degenration. *Nat Neurosci*. 6:1064-1071.

Zhu, J., Wang, X., Chen, J., Sun, J., Zhang, F. (2006). Reduced number and activity of circulating endothelial progenitor cells from pateints with hyperhomocysteinemia. *Arch Med Res*. 37(4):484-9.

8

Toward a More Effective Intravascular Cell Therapy in Stroke

Bhimashankar Mitkari[1], Erja Kerkelä[2], Johanna Nystedt[2], Matti Korhonen[2], Tuulia Huhtala[3] and Jukka Jolkkonen[1]
[1]*Institute of Clinical Medicine – Neurology, University of Eastern Finland, Kuopio,*
[2]*Finnish Red Cross Blood Service, Advanced Therapies and Product Development, Helsinki,*
[3]*A. I. Virtanen Institute, University of Eastern Finland, Kuopio, Finland*

1. Introduction

Cerebral ischemia remains the main cause of adult disability in Western countries. More than 50% of stroke survivors are left with a motor disability, causing a huge burden for patients, relatives and healthcare systems (Bonita et al., 1997). Cell-based therapies have emerged as some of the most promising experimental approaches to restore brain function after stroke (Bliss et al., 2010; Banerjee et al., 2011; Lindvall & Kokaia, 2011). A wide variety of cell types have been studied, such as neural progenitors from different sources, including bone marrow- and blood-derived stem cells. Preclinical data with cell therapies are promising (Bliss et al., 2007; Hicks & Jolkkonen, 2009; Hicks et al., 2009a; Janowski et al., 2010). The understanding of how transplanted cells exert their therapeutic effect is, however, not clear, but it is believed that the positive outcome is due to paracrine effects with an improved protective cellular environment (e.g., reduced inflammation, neuroprotection, reduced apoptosis, activation of endogenous repair) rather than as a consequence of neuronal differentiation and cell replacement (Zhang & Chopp, 2009).

The robust therapeutic effect shown in the majority of preclinical studies is somewhat surprising given that cell preparations, experimental models and outcome measures have varied greatly (Table 1). More work is definitely needed to establish standard treatment protocols, which in turn should be expected to lead to effective translation of experimental data. The recently published STEPS guidelines are one step forward to guide future cell-based research in stroke (STEPS Participant, 2009; Savitz et al., 2011). In addition to preclinical recommendations, guidelines on designing early-stage clinical trials are included.

The first patient studies have shown the safety and feasibility of systemic cell therapy, but only marginal therapeutic benefit has so far been observed (Bang et al., 2005; Battistella et al., 2011; Honmou et al., 2011). Whether this is related to the type of cells, study design or low engraftment of the delivered cells is not known. This review provides an update of the current progress in intravascular cell therapy in stroke with a particular emphasis on strategies of how to improve the therapeutic effects.

Stroke model	tMCAO, pMCAO, endothelin-1, cortical photothrombosis, hypoxia-ischemia
Species	rats, mice
Cell type	rat/mouse/human cells from BM, UCB or adipose tissue, neural cells, genetically modified cells
Delivery route	intravenous (tail vein, femoral vein), intra-arterial (common carotid artery, internal carotid artery, external carotid artery)
Delivery time	30 min - 1 month after the ischemic event
Outcome measures	histology (e.g., MAB1248), behavioral testing (e.g. sensorimotor, cognitive), imaging (e.g. MRI, SPECT, optical imaging)

BM - bone marrow; MRI - magnetic resonance imaging; UCB - umbilical cord blood; SPECT - single photon emission computed tomography; tMCAO - transient middle cerebral artery occlusion; pMCAO - permanent middle cerebral artery occlusion

Table 1. Variables with cell-based therapy in experimental stroke

2. Special challenges in intravascular cell therapy in stroke

Cell-based therapy after massive ischemic damage in stroke patients can be challenging compared to diabetes or Parkinson's disease, in which a restricted population of cells is lost. Not only neurons, but also glial cells and blood vessels need to be repaired. Severe odema and vascular compression associated with ischemic damage may limit the engraftment of cells, particularly in areas adjacent to infarct. Another distinction is that stroke is an acute injury with little or no degenerative process. Appropriate transplantable cells may not be immediately available for such an emergency. In addition, while early cell transplantation may provide neuroprotection, the hostile environment endangers the long-term survival of transplanted cells. Transplantation at later time points may be more realistic, targeting secondary neurodegeneration and promoting enhancement of the brain's own repair mechanisms (Zhang & Chopp, 2009). Although cell survival may be preferable, scar formation and a lack of functional vasculature may limit the therapeutic benefit. The advantage is, however, that cell transplantation can be combined with other rehabilitative treatments to ensure maximal therapeutic benefit (Hicks et al., 2009b).

Efficient cell delivery and an optimal delivery route are the keys to successful clinical outcomes, especially in all novel forms of cell therapy. Optimal cell delivery will be indication-dependent and local transplantation has until now been considered as the primary choice for regenerative tissue treatments. Systemic introduction should, however, be the ultimate goal for cell therapy, enabling rapid off-the-shelf therapy in any clinic and this would also allow less invasive treatments. Both stereotactic transplantation of cells into the brain and systemic delivery have been applied in experimental stroke (Guzman et al., 2008; Hicks & Jolkkonen, 2009). Given that stroke often produces large ischemic damage, it is not known whether a targeted approach can provide efficient and extensive cell engraftment, even with the aid of anatomical and functional imaging to explore the location of cell transplantation. Another concern is the invasive nature of intracerebral transplantation. In contrast, the systemic introduction is minimally invasive and thus perhaps more easily applied in the clinic.

There are, however, some obstacles in the intravenous delivery route for cellular therapeutics, one of the main ones being massive lung adhesion, which has been observed after intravenous injection (Allers et al., 2004; Barbash et al., 2003; Fischer et al., 2009; Gao et al., 2001; Hakkarainen et al., 2007; Kang et al., 2006; Mäkinen et al., 2006; Meyerrose et al., 2007; Nystedt et al., 2006; Schrepfer et al., 2007; Tolar et al., 2006; Vilalta et al., 2008). In addition to the negative impact this has on the possibility of reaching clinically relevant cell numbers in target organs, lung entrapment of mesenchymal stem cells (MSC) has also been observed causing severe lung damage in mouse models (Anjos-Afonso et al., 2004; Lee et al., 2009a). Importantly, pulmonary toxicity is reported as one of the most common non-hematological complications after autologous bone marrow transplantation in humans, a complication that is also detectable in a mouse model (Bhalla & Folz, 2002). Interestingly, and on the contrary, beneficial effects have been found after MSC lung entrapment, where embolized human MSCs improved myocardial infarction in mice through secreting the anti-inflammatory protein tumour factor-stimulated gene-6 (TSG-6) (Lee et al., 2009b).

3. Cell types used in stroke

Stem cells are defined as undifferentiated cells capable of self-renewal and differentiation. Truly totipotent stem cells can only be found in the embryo and these are capable of producing a new individual upon implantation. Depending on their origin, stem cells are classified as pluripotent (i.e., embryonic) or multipotent (i.e., fetal and adult) stem cells, referring also to their differentiation capacity. Intracerebral transplantation is the primary delivery route for embryonic stem cells (ESC), induced pluripotent stem cells and fetal stem cells in experimental stroke. Thus only intravascular delivery of adult stem/progenitor cells and genetically modified cells will be discussed in the following chapters.

3.1 Adult stem/progenitor cells

The majority of systemic transplantation studies in stroke have used non-neural cells; cells from bone marrow (BM), umbilical cord blood (UCB), adipose tissue, or peripheral blood. These are all typically defined as adult stem/progenitor cells and represent a group of heterogeneous cell types. Usually many cell types are present in the population, such as mesenchymal stem/stromal cells, hematopoietic progenitors and endothelial progenitors, as well as more mature cell types (Erices et al., 2000; Herzog et al., 2003; Harris et al., 2008). Typically either the whole cell population has been used or a subpopulation has been selected with, e.g., cell surface markers or culture conditions (like adherent MSCs). Adult stem cells lack the ethical controversies associated with embryonic or fetal cells and they are rather easily obtained from different clinical sources.

Different adult stem/progenitor cell populations have been reported to enhance functional recovery in experimental stroke models. When considering studies using human cells, mostly bone marrow stem/stromal cells (BM-MSC) (Li et al., 2002; Zhao et al., 2002; Chen et al., 2003; Zhang et al., 2004; Omori et al., 2008; Andrews et al., 2008; Mays et al., 2010; Yang et al., 2010; Bao et al., 2011) or UCBCs (Chen et al., 2001; Willing et al., 2003a; Borlongan et al., 2004; Vendrame et al., 2004; Xiao et al., 2005; Newcomb et al., 2006; Chen et al., 2006; Mäkinen et al., 2006; Zhang et al., 2011; Riegelsberger et al., 2011) have been used. Most studies have administered cells early (6-48 h) or at subacute phase (2-7 days) after stroke and only few comparisons have been made. Omori et al. (2008) compared multiple time points and found that the greatest functional benefit was achieved when BM-MSCs were injected 6 h after stroke compared to later time points, which is supported by the finding of Yang et al. (2010) that cells delivered 1 day have greater effect than those at 7 days. Instead, Mays et al. (2010) reported time window from 1 to 7 days post-stroke to be equally beneficial. For UCBCs, time window up to 30 days post-stroke was found to be therapeutically beneficial (Zhang et al., 2011).

In addition to BM and UCB cells, peripheral blood progenitor cells (Willing et al., 2003b), endothelial progenitors (Fan et al., 2010; Moubarik et al., 2011), CD34-positive progenitors from UCB (Taguchi et al., 2004; Boltze et al., 2005; Nystedt et al., 2006), CD133-positive cells from BM (Borlongan et al., 2005; Bakondi et al., 2009), as well as MSCs from placenta (Kranz et al., 2010) have provided therapeutic benefit in stroke. Also in these studies mostly early administration has been employed. For CD133 cells, delayed administration (7 d) was shown to improve graft survival but behavioral improvement was only apparent in immediate intravenous delivery (Borlongan et al., 2005).

MSCs from BM or UCB (or other tissues) are a particularly promising candidate for cell therapy in stroke. MSCs are defined as multipotent stem cells that are adherent and express CD73, CD90 and CD105. They show the potential to differentiate into bone, cartilage and fat, and also exhibit additional differentiation capacity (Dominici et al., 2006). MSCs can be highly expanded in culture with a minimal loss of multipotency and they show very little immunogenic activity. This is a major advantage, allowing them to be potentially used as allogeneic "off-the-shelf" products. They have already been explored in many experimental models and clinical trials for their beneficial effects to, e.g., regenerate damaged tissue, treat adverse immune reactions, promote angiogenesis, and increase tissue protection, and MSCs are generally considered safe (Malgieri et al., 2010).

Adult stem cells are particularly well suited for non-invasive vascular delivery, since they have been shown to target injured tissue and exert their therapeutic effect through secreted factors (Karp & Teo, 2008; Hess & Hill, 2011). The targeting of cells to the brain and especially their survival *in situ* have proven challenging, as in most studies very few cells are actually found in the brain. Interestingly, however, this may not be crucial, as intravenously administered cells may have a therapeutic effect on the brain by acting from peripheral organs as well, such as the spleen and the lung (Hess & Hill, 2011).

As a summary, adult stem cells have been shown to exert their positive effect through soluble factors that reduce apoptosis and promote neuroprotection, angiogenesis, brain plasticity, and/or endogenous progenitor proliferation. Some studies have shown differentiation towards neuronal phenotype, but the significance of this remains unclear.

3.2 Genetically modified cells

In addition to stem cells, several neural cell lines have been reported to enhance functional recovery after experimental stroke by intravenous delivery of cells (Jeong et al., 2003; Chu et al., 2004; Lee et al., 2008; Narantuya et al., 2010). These cell lines are immortalised and thus have the advantage of unlimited expansion in culture. However, there is a potential risk of malignant transformation (Newman et al., 2005).

One approach has been to use immortalised human MSCs to, e.g., expand the limited life-span of MSCs or include a gene for efficient *in vivo* tracking of cells. These cells have also shown positive effects in experimental stroke models when delivered intravenously (Honma et al., 2006; Wakabayashi et al., 2010). A critical aspect with these cells is that they should not lose their MSC phenotype upon modification.

Human BM-MSCs have also been genetically modified to express neuroprotective/ angiogenic growth factors, such as brain derived neurotrophic factor (BDNF) (Kurozumi et al., 2004; Nomura et al., 2005), placental growth factor (PlGF) (Liu et al., 2006), glial cell line-derived neurotrophic factor (GDNF) (Horita et al., 2006), erythropoietin (Cho et al., 2010), and vascular endothelial growth factor (VEGF) combined with angiopoietin-1 (Toyama et al., 2009). All these modified MSCs have shown their ability to improve functional recovery in ischemic rats, compared to unmodified MSCs, when delivered intravenously. GDNF-modified human UCB CD34+ cells have also shown similar positive effects *in vivo* supporting the combined gene and stem cell therapy for the treatment of stroke (Ou et al., 2010).

4. Cell modifications that improve the efficiency of cell therapy

The major problem with intravenous delivery is cell trapping within organs that filter the bloodstream. Previous studies have explored different strategies to minimize lung adhesion and improve homing of systemically introduced cells: use of vasodilators (Schrepfer et al., 2007), pre-bolus injection of MSCs (Fischer et al., 2009), reducing the number of injected cells (Lee et al., 2009b), blockade of $\alpha6$ and $\alpha4$ integrins (Bonig et al., 2007; Qian et al., 2006; Bonig et al., 2009; Fischer et al., 2009), heparin saturation of MSCs (Deak et al., 2010) or preincubation of cells with white blood cells (Chute, 2006). Some beneficial effects on lung adhesion have been concluded, but the major mechanism behind this profound phenomenon is still unsolved. Interestingly, glycosylation engineering of stem cell surfaces by enzymatic *ex vivo* cell surface fucosylation has improved the homing and engraftment capacity of cord blood-derived cells (Xia et al., 2004) and, interestingly, the homing of BM-MSCs to the bone marrow (Sackstein et al., 2008). One feasible approach to alter cell surface structures and migratory behavior is also through culture conditions. A recent preclinical study has shown that low passage and low-density cultures of BM-MSCs impact cell structures that favour *in vivo* targeting to the infarcted heart (Lee et al., 2009a). Culturing MSCs in low oxygen increases the levels of relevant cell surface chemokine and growth factor receptors, subsequently increasing the *in vitro* migratory behavior and the therapeutic potential of MSCs (Hung et al., 2007; Rosova et al., 2008). Cells are normally maintained in a 20% O_2 tension in culture, but a lower oxygen tension in culture is more akin to the physiological niche for the MSC in the bone marrow or placenta (2-7% O_2) and would facilitate in maintaining the authentic *in vivo* identity of the MSCs. Culturing MSCs without animal-derived reagents can produce beneficial changes in expression levels of important adhesion receptors and the secretion potential of trophic mediators, which might have an important impact on cell migratory behavior and therapeutic potential. To support this, Bieback et al. (2009) recently showed differential expression of the fibronectin receptor CD29 between MSCs cultured in fetal bovine serum versus human blood components. The impact of MSC xenofree culture conditions have not yet been studied or reported in preclinical stroke models.

5. Effect of administration route

The most effective transplantation route to deliver cells into the brain following cerebral ischemia remains to be addressed. Noninvasive intravascular administration of cells has perhaps the most immediate access for clinical applications. It provides a broad distribution of cells in close proximity to ischemic tissue, although the entry of intravenously injected cells into the central nervous system may not be required for therapeutic effects (Borlongan et al. 2004). However, a reliable estimation of cell numbers in the brain in relation to other organs is lacking.

Modern imaging methods such as single photon emission computed tomography (SPECT), positron emission computed tomography (PET), magnetic resonance imaging (MRI) or optical imaging can be used for the *in vivo* tracking of cells. Excellent reviews on the different imaging modalities are available (Sykova & Jendelova, 2007; Gera et al., 2010). SPECT imaging with indium oxine (^{111}In-oxine) offers an efficient method to study the whole body biodistribution of cells in stroke models (Figure 1). Firstly, the labeling of the cells is straightforward and relatively simple without significant loss of cell viability. The most common labels are ^{111}In-oxine or technetium-hexamethylpropyleneamine oxime (^{99m}Tc-HMPAO). Double labeling with ^{111}In and ^{131}I or ^{18}F- fluorodeoxyglucose (FDG) and ^{111}In is also possible (Blocklet et al., 2006; Stodilka et al., 2006). Secondly, the half-life of ^{111}In is optimal for several days follow-up after a single injection. Additional advantages of SPECT include high sensitivity, short scanning times (<5 min), and possible multimodal imaging (MRI, CT) with the same stereotaxic coordinates. More importantly, whole-body imaging provides an estimation of the proportion of injected cells that eventually enter the brain in relation to other organs, to help in the assessment of the functional value of transplantation. SPECT imaging is also truly translational and the same tracers can be used in human studies (Correa et al., 2007; Barbosa da Fonseca et al., 2010).

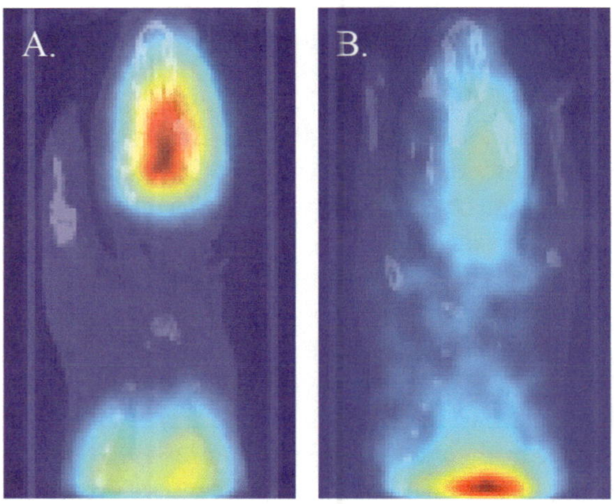

Fig. 1. Combined SPECT/CT images of ^{111}In-oxine labeled human bone marrow-derived mesenchymal stem cells (BM-MSCs) in a rat subjected to middle cerebral artery occlusion (MCAO). Images are taken 20 min (A) and 24 h (B) after intra-arterial administration of cells (4×10^5 cells; 3 MBq) 24 h after MCAO surgery. Please note the initial high signal in the brain followed by relocation of cells into the internal organs. A minor signal remains in the ischemic hemisphere.

Several studies have compared different administration routes. Willing et al. (2003a) concluded that intravenous administration of human UCBCs may be more effective that intracerebral transplantation. ^{111}In–oxine labeled human UCBCs have, however, been shown to localize primarily to the internal organs post intravenous injection in rats after middle cerebral artery occlusion (MCAO) (Mäkinen et al., 2006). Chen and co-workers also showed that after intravenous administration of human UCBCs in MCAO rats, only 1% of injected cells were detected in the brain (Chen et al., 2001). Undesirable biodistribution is most likely caused by the accumulation of cells in the trapping and filtering organs such as the lung, liver, and spleen, rather than due to the cell type injected or timing of administration. Thus, intra-arterial cell infusion may be a more efficient route to circumvent trapping in the internal organs and to target cells towards the ischemic brain (Lappalainen et al., 2008; Walczak et al., 2008; Li et al., 2010; Chua et al., 2011). Indeed, intra-arterial infusion resulted in minor engraftment of human ESCs into the ischemic hemisphere while no SPECT signal was detected after intravenous infusion (Lappalainen et al., 2008). Walczak et al. (2008) compared intravenous and intra-arterial delivery of MSCs in MCAO rats by using combined laser Doppler blood flow monitoring and MRI of iron labeled cells. The intra-arterial but not intravenous cell injection was shown to provide successful but variable cerebral engraftment, which was possibly due to microvascular occlusions. Engraftment was associated with high morbidity as also confirmed by Li et al. (2010). Later, it was shown that a modified injection technique with preserved flow in the carotid artery prevented decrease in cerebral blood flow and micro-occlusions (Chua et al., 2011). Intra-arterial over intravenous administration is also supported by the transplantation of mouse neural stem cells in a hypoxia-ischemia mouse model (Pendharkar et al., 2010). More importantly, a sustained presence (2 weeks) of transplanted cells in the brain was observed after intra-arterial administration. However, recently both intravenous and intra-arterial routes were shown to equally improve neurological recovery and provide neuroprotection (Gutierrez-Fernandez et al., 2011). In all above-mentioned studies, cell were administered within 24-48 h of ischemia.

Taken together, the delivery route seems to have an impact on the biodistribution of transplanted cells. Intra-arterial administration provides superior delivery of cells to the ischemic brain, although this depends on the type of cells and the experimental model employed.

6. Effective dose and therapeutic time window for cell transplantation

The effective cell dose needed for therapeutic effects in stroke animals is not well known. Intravenous infusion of 10^4 up to 5×10^7 human UCBCs improved behavioral deficits in MCAO rats in a dose-dependent manner, when administered at 24 h of ischemia (Vendrame et al., 2004). A dose of 10^6 cells was the threshold to promote functional recovery. Similarly, human umbilical tissue-derived cells at doses more than 3×10^6 have improved the behavioral outcome and enhanced several brain repair mechanisms (Zhang et al., 2011). A meta-analysis of 60 preclinical studies also found a dose-response association between the injected cell number and treatment effects (Janowski et al., 2010). Interestingly, the greatest therapeutic benefit is achieved following a single high cell dose injection of human MSCs (3.0×10^6) within 6 h of ischemia rather than multiple low dose injections (Omori et al., 2008). Thus, repeated dosing may not provide additional benefit. While the dose in preclinical studies is established to be around 10^6 cells per animal, it is more complicated to estimate the optimal dose for clinical studies. Dosing should be based on a dose-response curve and a maximum tolerated dose, as suggested by STEPS recommendations (STEPS Participant, 2009). The doses in early phase clinical trials were scaled to body weight and have varied from 5 x 10^7 (twice) (Bang et al., 2005) to 0.5 - 5 x 10^8 (Honmou et al., 2011) per patient.

In most of the experimental studies (67%), cells were given <24 h after ischemia (Hicks et al., 2009a). This is partly because of the opening of the blood-brain barrier after cerebral ischemia, which allows cells to enter the brain parenchyma (Belayev et al., 1996). Also, the expression of various chemotactic signals peaks at this time point and guides the cells towards ischemic areas (Imitola et al., 2004; Wang et al., 2008). However, while early cell transplantation may provide neuroprotection (Homna et al., 2006; Horita et al., 2006), the hostile environment endangers the long-term survival of transplanted cells. Transplantation at later time points may target against secondary neurodegeneration and promote enhancement of the brain's own repair mechanisms (Zhang & Chopp, 2009). Komatsu et al. (2010) showed that MCAO rats receiving MSCs up to 1 month after ischemia showed enhanced functional recovery and associated angiogenesis in cortical areas adjacent to the infarct. Shen et al. (2007) have also showed that cell transplantation 1 month after MCAO is effective by leading to long-lasting behavioral improvement. Behavioral and morphological evidence suggest that the post-stroke brain displays heightened sensitivity to rehabilitative treatment early after the stroke (1 wk), but declines with time (2-4 wk) (Biernaskie et al., 2004). Based on this, the time of cell transplantation could be extended to up to 7 days after ischemia.

7. Clinical perspectives and future directions of intravascular cell therapy for stroke

Promising experimental data have prompted early phase I/II patient studies. In these studies, either bone marrow mononuclear cells or bone marrow-derived mesenchymal stem cells have been used. Three phase I studies explored the use of bone marrow mononuclear (BM-MNC) cells for stroke (Barbosa da Fonseca et al., 2010; Battistella et al., 2011; Suarez-Monteagudo et al., 2009). The most important finding of these studies is that intra-arterial delivery of mononuclear cells directly to the infarcted hemisphere is safe. Interestingly, Barbosa da Fonseca et al. (2010) labeled the mononuclear cells with technetium-99m, and followed the distribution of the cells in six stroke patients. They were able to show that the cells remained at the site of the lesion for two hours, but then the signal disappeared on all but two patients after 24 hours. It is unclear whether this short time of action will be enough for any therapeutic benefit.

Bang and co-workers pioneered in the use of MSCs for ischemic stroke with two studies (Bang et al., 2005; Lee et al., 2010). In the first one, five patients with stroke received autologous MSCs as two intravenous infusions of 5 x 10^7 cells each. The outcome of the patients after one year was compared to 25 randomized controls in an open-label study. The second study followed the same protocol as the first one but included 16 patients and 36 controls, and the patients were followed for five years. Both studies concluded that intravenous infusion of MSCs in stroke patients was safe. There was no apparent increase in mortality, bovine spongiform encephalitis or other zoonoses, arrhythmias, seizures, or tumors. The true incidence of possible side effects of MSC therapy can, however, only be evaluated after much larger patient groups have been treated and followed. Interestingly, patients showed significant improvement in the Barthel Index (BI) in both studies, and a trend towards improvement in the modified Rankin Scale (mRS) in the first study. There was some concern as to whether the improved functional recovery was upheld with time (Bang et al., 2005), but later this was confirmed as improvement could still be measured at 3.5 years (Lee et al., 2010). Because of the time required to produce the autologous therapeutic cells used, the infusion of the cells occurred rather late, i.e. the first cell infusion was given at weeks 4-5 and the subsequent one at weeks 7-9 after the onset of the stroke. Honmou and colleagues (2011) studied the safety and feasibility of intravenous infusion into stroke patients of autologous MSCs that had been expanded in autologous serum. The cells were infused 33-133 days post-stroke, and the patients were followed and imaged at one year after. No adverse events were recorded.

Several important questions still need to be addressed in both preclinical and clinical testing. It is unclear which cell type is therapeutically the most beneficial. Both BM-MNCs and MSCs have shown promise in preclinical testing. Selecting the best route of therapeutic cell delivery is also a major issue. Presumably, the route yielding the most effective delivery of cells to the injured tissue might offer most therapeutic potential. In a recent study comparing intra-arterial and intravenous MSC delivery, infusion directly into the internal carotid artery resulted in more engrafted cells as well as a more widespread distribution of cells within the infarcted hemispheres of rats (Li et al. 2010). However, this mode of administration was also associated with high mortality as the MSCs were sequestered in the blood vessels of the treated hemisphere and formed micro-occlusions. Careful development of safe but effective modes of administration is required for the advancement of this technique towards the clinic. In contrast to MSCs, BM-MNCs can apparently be administered via the intra-arterial route without harmful effects.

Another important question is the preferred timing of the cell infusion. There is some evidence suggesting that MSCs infused early (within days of the infarct) have more therapeutic efficacy than those administered several weeks after the event (Zhou et al., 2011). Theoretical considerations support these findings: a major mode of action of the MSCs in stroke is attenuation of the post-infarct inflammatory milieu, which is at its strongest during the early days following the insult (Ohtaki et al., 2008). BM-MNCs have the advantage of having no need for cell expansion. Thus, the patient can be treated soon after stroke even if autologous cells are used.

Whether the patient should be given autologous or allogeneic MSCs is a major unanswered question. If the patient is given autologous cells, the therapy will necessarily take place several weeks after the infarct due to the time required to produce the cells. Allogeneic cells offer the critical advantage of being available off-the-shelf. It can be argued that allogeneic cells are quickly rejected and will rapidly lose their therapeutic efficacy, but if only short term action is required e.g., to modulate post-infarct inflammation, allogeneic cells may be an ideal therapeutic vehicle. In experimental animals, allogeneic cells appear to function as equally well as autologous MSCs (Li et al., 2006). Furthermore, risks of tumor formation are reduced, because the allogeneic cells are eventually rejected by the host's immune system (Poncelet et al., 2007).

Finally, the optimal characteristics of the therapeutic cells need to be determined. Special attention needs to be given to the conditions of cell production, because the administration of apoptotic or senescent cells is not only ineffective, but positively harmful (Modo et al., 2003; Prockop et al., 2010). Another important advance will be the adoption of xenofree culture methods for MSCs, reducing the risks of anaphylaxis and transmission of diseases (Horwitz et al., 2002). Modification of the cell surface prior to administration, to improve the delivery or therapeutic efficacy of the cells, is a promising strategy.

Taken together, there is already preliminary evidence for the safety of intravenously delivered BM-MNCs and MSCs for stroke patients. Further work is required to establish this safety profile. However, careful studies that also use an intra-arterial application of cells should proceed. Futhermore, the use of allogeneic MSCs should be explored, allowing treatment during the first week after infarct.

Currently, eight clinical phase I-II studies are underway, assessing the safety and possible efficacy of either fractionated BM cells or culture-expanded MSCs in patients with recent ischemic stroke (Table 2). Three studies will address the feasibility and safety of intravenous

Identifier/Sponsor*	**Cell type**	**Time window**	**Administration route**	**Comments**
NCT00859014/The University of Texas Health Science Center, Houston, USA	autologous BM-MNCs	24-72 h	i.v.	phase I, non-randomized
NCT01028794 National Cardiovascular Center, Japan	autologous BM-MNCs	day 7-10	i.v.	phase I-II, non-randomized
NCT00473057/Federal University of Rio de Janeiro, Brazil	autologous BM-MNCs	day 3 - 90	i.v./i.a.	phase I, completed
NCT761982/Hospital Universitario Central de Asturias, Spain	autologous BM CD34+ cells	day 5-9	i.a.	phase I-II, non-randomized, completed
NCT01273337/Aldagen, USA	autologous BM ALDH-positive cells	day 13-19	i.a.	phase II, randomized
NCT01297413/Stemedica Cell Technologies Inc, USA	allogeneic MSCs	> 6 mo	i.v.	phase I-II, non-randomized
NCT00875654/ University Hospital, Grenoble, France	autologous MSCs	within 6 wks	i.v.	phase II, randomized trial
NCT01389453/General Hospital of Chinese Armed Police Forces; China	allogeneic cord blood MSCs	day 7-14	i.v.	phase I, non-randomized

*from www.clinicaltrials.com

Table 2. Summary of ongoing clinical trials with intravascular administration of cell therapy in stroke. Only recruiting or completed studies are listed.

BM-MNCs, and one of these will compare intravenous with intra-arterial delivery. Two more trials will utilize BM cells that have been selected using stem cell markers (CD34 or aldehyde dehydrogenase). Three studies evaluate the use of MSCs for stroke, one of them using the patients' own and two using allogeneic cells. In most of these early studies, cells will be administered via the intravenous route. These studies, and others addressing the questions posed above, may help us in ameliorating the devastating consequences of ischemic stroke at the personal and societal level.

8. References

Allers, C., Sierralta, W.D., Neubauer, S., Rivera, F., Minguell, J.J., Conget, P.A. (2004). Dynamic of distribution of human bone marrow-derived mesenchymal stem cells after transplantation into adult unconditioned mice. *Transplantation,* 78, 503-8.

Andrews, E.M., Tsai, S.Y., Johnson, S.C., Farrer, J.R., Wagner, J.P., Kopen, G.C., Kartje, G.L. (2008). Human adult bone marrow-derived somatic cell therapy results in functional recovery and axonal plasticity following stroke in the rat. *Exp Neurol,* 211, 588-92.

Anjos-Afonso, F., Siapati, E.K., Bonnet, D. (2004). In vivo contribution of murine mesenchymal stem cells into multiple cell-types under minimal damage conditions. *J Cell Sci,* 117, 5655-64.

Bakondi, B., Shimada, I.S., Perry, A., Munoz, J.R., Ylostalo, J., Howard, A.B., Gregory, C.A., Spees, J.L. (2009). CD133 identifies a human bone marrow stem/progenitor cell sub-population with a repertoire of secreted factors that protect against stroke. *Mol Ther,* 17, 1938-47.

Banerjee, S., Williamson, D., Habib, N., Gordon, M., Chataway, J. (2011). Human stem cell therapy in ischaemic stroke: a review. *Age Ageing,* 40, 7-13.

Bang, O.Y., Lee, J.S., Lee, P.H., Lee, G. (2005). Autologous mesenchymal stem cell transplantation in stroke patients. *Ann Neurol,* 57, 874-82.

Bao, X., Wei, J., Feng, M., Lu, S., Li, G., Dou, W., Ma, W., Ma, S., An, Y., Qin, C., Zhao, R.C., Wang, R. (2011). Transplantation of human bone marrow-derived mesenchymal stem cells promotes behavioral recovery and endogenous neurogenesis after cerebral ischemia in rats. *Brain Res,* 1367, 103-13.

Barbash, I.M., Chouraqui, P., Baron, J., Feinberg, M.S., Etzion, S., Tessone, A., Miller, L., Guetta, E., Zipori, D., Kedes, L.H., Kloner, R.A., Leor, J. (2003). Systemic delivery of bone marrow-derived mesenchymal stem cells to the infarcted myocardium: feasibility, cell migration, and body distribution. *Circulation,* 108, 863-8.

Barbosa da Fonseca, L.M., Gutfilen, B., Rosado de Castro, P.H., Battistella, V., Goldenberg, R.C., Kasai-Brunswick, T., Chagas, C.L., Wajnberg, E., Maiolino, A., Salles Xavier, S., Andre, C., Mendez-Otero, R., de Freitas, G.R. (2010). Migration and homing of bone-marrow mononuclear cells in chronic ischemic stroke after intra-arterial injection. *Exp Neurol,* 221, 122-8.

Battistella, V., de Freitas, G.R., da Fonseca, L.M., Mercante, D., Gutfilen, B., Goldenberg, R.C., Dias, J.V., Kasai-Brunswick, T.H., Wajnberg, E., Rosado-de-Castro, P.H., Alves-Leon, S.V., Mendez-Otero, R., Andre, C. (2011). Safety of autologous bone marrow mononuclear cell transplantation in patients with nonacute ischemic stroke. *Regen Med*, 6, 45-52.

Belayev, L., Busto, R., Zhao, W., Ginsberg, M.D. (1996). Quantitative evaluation of blood-brain barrier permeability following middle cerebral artery occlusion in rats. *Brain Res*, 739, 88-96.

Bhalla, K.S., Folz, R.J. (2002). Idiopathic pneumonia syndrome after syngeneic bone marrow transplant in mice. *Am J Respir Crit Care Med*, 166, 1579-89.

Bieback, K., Hecker, A., Kocaömer, A., Lannert, H., Schallmoser, K., Strunk, D., Klüter, H. (2009). Human alternatives to fetal bovine serum for the expansion of mesenchymal stromal cells from bone marrow. *Stem Cells*, 27, 2331-41.

Biernaskie, J., Chernenko, G., Corbett, D. (2004). Efficacy of rehabilitative experience declines with time after focal ischemic brain injury. *J Neurosci*, 24, 1245-54.

Bliss, T., Guzman, R., Daadi, M., Steinberg, G.K. (2007). Cell transplantation therapy for stroke. *Stroke*, 38(2 Suppl), 817-26.

Bliss, T.M., Andres, R.H., Steinberg, G.K. (2010). Optimizing the success of cell transplantation therapy for stroke. *Neurobiol Dis*, 37, 275-83.

Blocklet, D., Toungouz, M., Berkenboom, G., Lambermont, M., Unger, P., Preumont, N., Stoupel, E., Egrise, D., Degaute, J.P., Goldman, M., Goldman, S. (2006). Myocardial homing of nonmobilized peripheral-blood CD34+ cells after intracoronary injection. *Stem Cells*, 24, 333-336.

Boltze, J., Kowalski, I., Geiger, K., Reich, D., Gunther, A., Buhrle, C., Egger, D., Kamprad, M., Emmrich, F. (2005). Experimental treatment of stroke in spontaneously hypertensive rats by CD34+ and CD34- cord blood cells. *Ger Med Sci*, 3, Doc 09.

Bonig, H., Priestley, G.V., Oehler, V., Papayannopoulou, T. (2007). Hematopoietic progenitor cells (HPC) from mobilized peripheral blood display enhanced migration and marrow homing compared to steady-state bone marrow HPC. *Exp Hematol*, 35, 326-34.

Bonig, H., Priestley, G.V., Wohlfahrt, M., Kiem, H.P., Papayannopoulou, T. (2009). Blockade of alpha6-integrin reveals diversity in homing patterns among human, baboon, and murine cells. *Stem Cells Dev*, 18, 839-44.

Bonita, R., Solomon, N., Broad, J.B. (1997). Prevalence of stroke and stroke-related disability. Estimates from the Auckland stroke studies. *Stroke*, 28, 898-902.

Borlongan, C.V., Hadman, M., Sanberg, C.D., Sanberg, P.R. (2004). Central nervous system entry of peripherally injected umbilical cord blood cells is not required for neuroprotection in stroke. *Stroke*, 35, 2385-9.

Borlongan, C.V., Evans, A., Yu, G., Hess, D.C. (2005). Limitations of intravenous human bone marrow CD133+ cell grafts in stroke rats. *Brain Res*, 1048, 116-22.

Chen, J., Sanberg, P.R., Li, Y., Wang, L., Lu, M., Willing, A.E., Sanchez-Ramos, J., Chopp, M. (2001). Intravenous administration of human umbilical cord blood reduces behavioral deficits after stroke in rats. *Stroke*, 32, 2682-8.

Chen, J., Zhang, Z.G., Li, Y., Wang, L., Xu, Y.X., Gautam, S.C., Lu, M., Zhu, Z., Chopp, M. (2003). Intravenous administration of human bone marrow stromal cells induces angiogenesis in the ischemic boundary zone after stroke in rats. *Circ Res*, 92, 692-9.

Chen, S.H., Chang, F.M., Tsai, Y.C., Huang, K.F., Lin, C.L., Lin, M.T. (2006). Infusion of human umbilical cord blood cells protect against cerebral ischemia and damage during heatstroke in the rat. *Exp Neurol*, 199, 67-76.

Cho, G.W., Koh, S.H., Kim, M.H., Yoo, A.R., Noh, M.Y., Oh, S., Kim, S.H. (2010). The neuroprotective effect of erythropoietin-transduced human mesenchymal stromal cells in an animal model of ischemic stroke. *Brain Res*, 1353, 1-13.

Chu, K., Kim, M., Park, K.I., Jeong, S.W., Park, H.K., Jung, K.H., Lee, S.T., Kang, L., Lee, K., Park, D.K., Kim, S.U., Roh, J.K. (2004). Human neural stem cells improve sensorimotor deficits in the adult rat brain with experimental focal ischemia. *Brain Res*, 1016, 145-53.

Chua, J.Y., Pendharkar, A.V., Wang, N., Choi, R., Andres, R.H., Gaeta, X., Zhang, J., Moseley, M.E., Guzman, R. (2011). Intra-arterial injection of neural stem cells using a microneedle technique does not cause microembolic strokes. *J Cereb Blood Flow Metab*, 31, 1263-71.

Chute, J.P. (2006). Stem cell homing. *Curr Opin Hematol*, 13, 399-406.

Correa, P.L., Mesquita, C.T., Felix, R.M., Azevedo, J.C., Barbirato, G.B., Falcão, C.H., Gonzalez, C., Mendonça, M.L., Manfrim, A., de Freitas, G., Oliveira, C.C., Silva, D., Avila, D., Borojevic, R., Alves, S., Oliveira, A.C. Jr, Dohmann, H.F. (2007). Assessment of intra-arterial injected autologous bone marrow mononuclear cell distribution by radioactive labeling in acute ischemic stroke. *Clin Nucl Med*, 32, 839-41.

Deak, E., Rüster, B., Keller, L., Eckert, K., Fichtner, I., Seifried, E., Henschler, R. (2010). Suspension medium influences interaction of mesenchymal stromal cells with endothelium and pulmonary toxicity after transplantation in mice. *Cytotherapy*, 12, 260-4.

Dominici, M., Le Blanc, K., Mueller, I., Slaper-Cortenbach, I., Marini, F., Krause, D., Deans, R., Keating, A., Prockop, Dj., Horwitz, E. (2006). Minimal criteria for defining multipotent mesenchymal stromal cells. The International Society for Cellular Therapy position statement. *Cytotherapy*, 8, 315-7.

Erices, A., Conget, P., Minguell, J.J. (2000). Mesenchymal progenitor cells in human umbilical cord blood. *Br J Haematol*, 109, 235-42.

Fan, Y., Shen, F., Frenzel, T., Zhu, W., Ye, J., Liu, J., Chen, Y., Su, H., Young, W.L., Yang, G.Y. (2010). Endothelial progenitor cell transplantation improves long-term stroke outcome in mice. *Ann Neurol*, 67, 488-97.

Fischer, U.M., Harting, M.T., Jimenez, F., Monzon-Posadas, W.O., Xue, H., Savitz, S.I., Laine, G.A., Cox, C.S. Jr. (2009). Pulmonary passage is a major obstacle for intravenous stem cell delivery: the pulmonary first-pass effect. *Stem Cells Dev*, 18, 683-92.

Gao, J., Dennis, J.E., Muzic, R.F., Lundberg, M., Caplan, A.I. (2001). The dynamic in vivo distribution of bone marrow-derived mesenchymal stem cells after infusion. *Cells Tissues Organs*, 169, 12-20.

Gera, A., Steinberg, G.K., Guzman, R. (2010). In vivo neural stem cell imaging: current modalities and future directions. *Regen Med* 5, 73-86.

Gutiérrez-Fernández, M., Rodríguez-Frutos, B., Alvarez-Grech, J., Vallejo-Cremades, M.T., Expósito-Alcaide, M., Merino, J., Roda, J.M., Díez-Tejedor, E. (2011). Functional recovery after hematic administration of allogenic mesenchymal stem cells in acute ischemic stroke in rats. *Neuroscience*, 175, 394-405.

Guzman, R., Choi, R., Gera, A., De Los Angeles, A., Andres, R.H., Steinberg, G.K. (2008). Intravascular cell replacement therapy for stroke. *Neurosurg Focus*, 24, E15.

Hakkarainen, T., Särkioja, M., Lehenkari, P., Miettinen, S., Ylikomi, T., Suuronen, R., Desmond, R.A., Kanerva, A., Hemminki, A. (2007). Human mesenchymal stem cells lack tumor tropism but enhance the antitumor activity of oncolytic adenoviruses in orthotopic lung and breast tumors. *Hum Gene Ther*, 18, 627-41.

Harris, D, T. (2008). Cord blood stem cells: a review of potential neurological applications. *Stem Cell Rev*, 4, 269-74.

Herzog, E.L., Chai, L., Krause, D.S. (2003). Plasticity of marrow-derived stem cells. *Blood*, 102, 3483-93.

Hess, D.C., Hill, W.D. (2011). Cell therapy for ischaemic stroke. *Cell Prolif*, 44 (Suppl 1),1-8.

Hicks, A., Jolkkonen, J. (2009). Challenges and possibilities of intravascular cell therapy in stroke. *Acta Neurobiol Exp (Wars)*, 69, 1-11.

Hicks, A., Schallert, T., Jolkkonen, J. (2009a). Cell-based therapies and functional outcome in experimental stroke. *Cell Stem Cell*, 5, 139-140.

Hicks, A.U., Lappalainen, R.S., Narkilahti, S., Suuronen, R., Corbett, D., Sivenius, J., Hovatta, O., Jolkkonen, J. (2009b). Transplantation of human embryonic stem cell (hESC)-derived neural precursor cells and enriched environment after cortical stroke in rats: cell survival and functional recovery. *Eur J Neurosci*, 29, 562-572.

Honma, T., Honmou, O., Iihoshi, S., Harada, K., Houkin, K., Hamada, H., Kocsis, J.D. (2006). Intravenous infusion of immortalized human mesenchymal stem cells protects against injury in a cerebral ischemia model in adult rat. *Exp Neurol*, 199, 56-66.

Honmou, O., Houkin, K., Matsunaga, T., Niitsu, Y., Ishiai, S., Onodera, R., Waxman, S.G., Kocsis, J.D. (2011). Intravenous administration of auto serum-expanded autologous mesenchymal stem cells in stroke. *Brain*, 134, 1790-807.

Horita, Y., Honmou, O., Harada, K., Houkin, K., Hamada, H., Kocsis, J.D. (2006). Intravenous administration of glial cell line-derived neurotrophic factor gene-modified human mesenchymal stem cells protects against injury in a cerebral ischemia model in the adult rat. *J Neurosci Res*, 84, 1495-504.

Horwitz, E.M., Gordon, P.L., Koo, W.K., Marx, J.C., Neel, M.D., McNall, R.Y., Muul, L., Hofmann, T. (2002). Isolated allogeneic bone marrow-derived mesenchymal cells engraft and stimulate growth in children with osteogenesis imperfecta: Implications for cell therapy of bone. *Proc Natl Acad Sci USA* 99, 8932-7.

Hung, S.C., Pochampally, R.R., Hsu, S.C., Sanchez, C., Chen, S.C., Spees, J., Prockop, D.J. (2007). Short-term exposure of multipotent stromal cells to low oxygen increases their expression of CX3CR1 and CXCR4 and their engraftment in vivo. *PLoS One*, 2, e416.

Imitola, J., Raddassi, K., Park, K.I., Mueller, F.J., Nieto, M., Teng, Y.D., Frenkel, D., Li, J., Sidman, R.L., Walsh, C.A., Snyder, E.Y., Khoury, S.J. (2004). Directed migration of neural stem cells to sites of CNS injury by the stromal cell-derived factor 1alpha/CXC chemokine receptor 4 pathway. *Proc Natl Acad Sci USA*, 101, 18117-22.

Janowski, M., Walczak, P., Date, I. (2010). Intravenous route of cell delivery for treatment of neurological disorders: a meta-analysis of preclinical results. *Stem Cells Dev*, 19, 5-16.

Jeong, S.W., Chu, K., Jung, K.H., Kim, S.U., Kim, M., Roh, J.K. (2003). Human neural stem cell transplantation promotes functional recovery in rats with experimental intracerebral hemorrhage. *Stroke*, 34, 2258-63.

Kang, W.J., Kang, H.J., Kim, H.S., Chung, J.K., Lee, M.C., Lee, D.S. (2006). Tissue distribution of 18F-FDG-labeled peripheral hematopoietic stem cells after intracoronary administration in patients with myocardial infarction. *J Nucl Med*, 47, 1295-301.

Karp, J.M., Leng Teo, G.S. (2006). Mesenchymal stem cell homing: the devil is in the details. *Cell Stem Cell*, 4, 206-16.

Komatsu, K., Honmou, O., Suzuki, J., Houkin, K., Hamada, H., Kocsis, J.D. (2010). Therapeutic time window of mesenchymal stem cells derived from bone marrow after cerebral ischemia. *Brain Res*, 1334, 84-92.

Kranz, A., Wagner, D.C., Kamprad, M., Scholz, M., Schmidt, U.R., Nitzsche, F., Aberman, Z., Emmrich, F., Riegelsberger, U.M., Boltze, J. (2010). Transplantation of placenta-derived mesenchymal stromal cells upon experimental stroke in rats. *Brain Res*, 1315, 128-36.

Kurozumi, K., Nakamura, K., Tamiya, T., Kawano, Y., Kobune, M., Hirai, S., Uchida, H., Sasaki, K., Ito, Y., Kato, K., Honmou, O., Houkin, K., Date, I., Hamada, H. (2004). BDNF gene-modified mesenchymal stem cells promote functional recovery and reduce infarct size in the rat middle cerebral artery occlusion model. *Mol Ther*, 9, 189-97.

Lappalainen, R., Narkilahti, S., Huhtala, T., Liimatainen, T., Suuronen, T., Närvänen, A., Suuronen, R., Hovatta, O., Jolkkonen, J. (2008). SPECT imaging shows accumulation of stem cells into internal organs after systemic administration in middle cerebral artery occlusion rats. *Neurosci Lett*, 440, 246-50.

Lee, J.S., Hong, J.M., Moon, G.J., Lee, P.H., Ahn, Y.H., Bang, O.Y.; STARTING collaborators. (2010). A long-term follow-up study of intravenous autologous mesenchymal stem cell transplantation in patients with ischemic stroke. *Stem Cells*, 28, 1099-106.

Lee, R.H., Seo, M.J., Pulin, A.A., Gregory, C.A., Ylostalo, J., Prockop, D.J. (2009a). The CD34-like protein PODXL and alpha6-integrin (CD49f) identify early progenitor MSCs with increased clonogenicity and migration to infarcted heart in mice. *Blood*, 113, 816-26.

Lee, R.H., Pulin, A.A., Seo, M.J., Kota, D.J., Ylostalo, J., Larson, B.L., Semprun-Prieto, L., Delafontaine, P., Prockop, D.J. (2009b). Intravenous hMSCs improve myocardial infarction in mice because cells embolized in lung are activated to secrete the anti-inflammatory protein TSG-6. *Cell Stem Cell*, 5, 54-63.

Lee, S.T., Chu, K., Jung, K.H., Kim, S.J., Kim, D.H., Kang, K.M., Hong, N.H., Kim, J.H., Ban, J.J., Park, H.K., Kim, S.U., Park, C.G., Lee, S.K., Kim, M., Roh, J.K. (2008). Anti-inflammatory mechanism of intravascular neural stem cell transplantation in haemorrhagic stroke. *Brain*, 131, 616-29.

Li, L., Jiang, Q., Ding, G., Zhang, L., Zhang, Z.G., Li, Q., Panda, S., Lu, M., Ewing, J.R., Chopp, M. (2010). Effects of administration route on migration and distribution of neural progenitor cells transplanted into rats with focal cerebral ischemia, an MRI study. *J Cereb Blood Flow Metab*, 30, 653-62.

Li, Y., Chen, J., Chen, X.G., Wang, L., Gautam, S.C., Xu, Y.X., Katakowski, M., Zhang, L.J., Lu, M., Janakiraman, N., Chopp, M. (2002). Human marrow stromal cell therapy for stroke in rat: neurotrophins and functional recovery. *Neurology*, 59, 514-23.

Li, Y., McIntosh, K., Chen, J., Zhang, C., Gao, Q., Borneman, J., Raginski, K., Mitchell, J., Shen, L., Zhang, J., Lu, D., Chopp, M. (2006). Allogeneic bone marrow stromal cells promote glial-axonal remodeling without immunologic sensitization after stroke in rats. *Exp Neurol* 198, 313-325.

Lindvall, O., Kokaia, Z. (2011). Stem cell research in stroke: how far from the clinic? *Stroke*, 42, 2369-75.

Liu, H., Honmou, O., Harada, K., Nakamura, K., Houkin, K., Hamada, H., Kocsis, J.D. (2006). Neuroprotection by PIGF gene-modified human mesenchymal stem cells after cerebral ischaemia. *Brain*, 129, 2734-45.

Malgieri, A., Kantzari, E., Patrizi, M.P., Gambardella, S. (2010). Bone marrow and umbilical cord blood human mesenchymal stem cells: state of the art. *Int J Clin Exp Med*, 3, 248-69.

Mays, R.W., Borlongan, C.V., Yasuhara, T., Hara, K., Mak, M., Carrol, J.E., Deans, R.J., Hess, D.C. (2010). Development of an allogeneic adherent stem cell therapy for treatment of ischemic stroke. *J Exp Stroke Transl Med*, 3, 34-46

Meyerrose, T.E., De Ugarte, D.A., Hofling, A.A., Herrbrich, P.E., Cordonnier, T.D., Shultz, L.D., Eagon, J.C., Wirthlin, L., Sands, M.S., Hedrick, M.A., Nolta, J.A. (2007). In vivo distribution of human adipose-derived mesenchymal stem cells in novel xenotransplantation models. *Stem Cells*, 25, 220-7.

Modo, M., Stroemer, R.P., Tang, E., Patel, S., Hodges, H. (2003). Effects of implantation site of dead stem cells in rats with stroke damage. *Neuroreport*, 14, 39-42.

Moubarik, C., Guillet, B., Youssef, B., Codaccioni, J.L., Piercecchi, M.D., Sabatier, F., Lionel, P., Dou, L., Foucault-Bertaud, A., Velly, L., Dignat-George, F., Pisano, P. (2011). Transplanted late outgrowth endothelial progenitor cells as cell therapy product for stroke. *Stem Cell Rev*, 7, 208-20.

Mäkinen, S., Kekarainen, T., Nystedt, J., Liimatainen, T., Huhtala, T., Närvanen, A., Laine, J., Jolkkonen, J. (2006). Human umbilical cord blood cells do not improve sensorimotor or cognitive outcome following transient middle cerebral artery occlusion in rats. *Brain Res*, 1123, 207-15.

Narantuya, D., Nagai, A., Sheikh, A.M., Masuda, J., Kobayashi, S., Yamaguchi, S., Kim, S.U. (2010). Human microglia transplanted in rat focal ischemia brain induce neuroprotection and behavioral improvement. PLoS One, 5, e11746.

Nemati, S., Hatami, M., Kiani, S., Hemmesi, K., Gourabi, H., Masoudi, N., Alaei, S., Baharvand, H. (2011). Long-term self-renewable feeder-free human induced pluripotent stem cell-derived neural progenitors. *Stem Cells Dev*, 20, 503-14.

Newcomb, J.D., Ajmo, C.T. Jr, Sanberg, C.D., Sanberg, P.R., Pennypacker, K.R., Willing, A.E. (2006). Timing of cord blood treatment after experimental stroke determines therapeutic efficacy. *Cell Transplant*, 15, 213-23.

Newman, M.B., Misiuta, I., Willing, A.E., Zigova, T., Karl, R.C., Borlongan, C.V., Sanberg, P.R. (2005). Tumorigenicity issues of embryonic carcinoma-derived stem cells: relevance to surgical trials using NT2 and hNT neural cells. *Stem Cells Dev*, 14, 29-43.

Nomura, T., Honmou, O., Harada, K., Houkin, K., Hamada, H., Kocsis, J.D. (2005). I.V. infusion of brain-derived neurotrophic factor gene-modified human mesenchymal stem cells protects against injury in a cerebral ischemia model in adult rat. *Neuroscience*, 136, 161-9.

Nystedt, J., Mäkinen, S., Laine, J., Jolkkonen, J. (2006). Human cord blood CD34+ cells and behavioral recovery following focal cerebral ischemia in rats. *Acta Neurobiol Exp (Wars)*, 66, 293-300.

Ohtaki, H., Ylostalo, J.H., Foraker, J.E., Robinson, A.P., Reger, R.L., Shioda, S., Prockop, D.J. (2008). Stem/progenitor cells from bone marrow decrease neuronal death in global ischemia by modulation of inflammatory/immune responses. *Proc Natl Acad Sci USA*, 105, 14638-14643.

Omori, Y., Honmou, O., Harada, K., Suzuki, J., Houkin, K., Kocsis, J.D. (2008). Optimization of a therapeutic protocol for intravenous injection of human mesenchymal stem cells after cerebral ischemia in adult rats. *Brain Res*,1236, 30-8.

Ou, Y., Yu, S., Kaneko, Y., Tajiri, N., Bae, E.C., Chheda, S.H., Stahl, C.E., Yang, T., Fang, L., Hu, K., Borlongan, C.V., Yu, G. (2010). Intravenous infusion of GDNF gene-modified human umbilical cord blood CD34+ cells protects against cerebral ischemic injury in spontaneously hypertensive rats. *Brain Res*, 1366, 217-25.

Pendharkar, A.V., Chua, J.Y., Andres, R.H., Wang, N., Gaeta, X., Wang, H., De, A., Choi, R., Chen, S., Rutt, B.K., Gambhir, S.S., Guzman, R. (2010). Biodistribution of neural stem cells after intravascular therapy for hypoxic-ischemia. *Stroke*, 41, 2064-70.

Poncelet, A.J., Vercruysse, J., Saliez, A., Gianello, P. (2007). Although pig allogeneic mesenchymal stem cells are not immunogenic in vitro, intracardiac injection elicits an immune response in vivo. *Transplantation*, 83, 783-790.

Prockop, D.J., Brenner, M., Fibbe, W.E., Horwitz, E., Le Blanc, K., Phinney, D.G., Simmons, P.J., Sensebe, L., Keating, A. (2010). Defining the risks of mesenchymal stromal cell therapy. *Cytotherapy*, 12, 576-8.

Qian, H., Tryggvason, K., Jacobsen, S.E., Ekblom, M. (2006). Contribution of alpha6 integrins to hematopoietic stem and progenitor cell homing to bone marrow and collaboration with alpha4 integrins. *Blood*, 107, 3503-10.

Riegelsberger, U.M., Deten, A., Pösel, C., Zille, M., Kranz, A., Boltze, J., Wagner, D.C. (2011). Intravenous human umbilical cord blood transplantation for stroke: impact on infarct volume and caspase-3-dependent cell death in spontaneously hypertensive rats. *Exp Neurol*, 227, 218-23.

Rosová, I., Dao, M., Capoccia, B., Link, D., Nolta, J.A. (2008). Hypoxic preconditioning results in increased motility and improved therapeutic potential of human mesenchymal stem cells. *Stem Cells*, 26, 2173-82.

Sackstein, R., Merzaban, J.S., Cain, D.W., Dagia, N.M., Spencer, J.A., Lin, C.P., Wohlgemuth, R. (2008). Ex vivo glycan engineering of CD44 programs human multipotent mesenchymal stromal cell trafficking to bone. *Nat Med*, 14, 181-7.

Savitz, S.I., Chopp, M., Deans, R., Carmichael, S.T., Phinney, D., Wechsler, L. (2011). Stem Cell Therapy as an Emerging Paradigm for Stroke (STEPS) II. *Stroke*, 42, 825-9.

Schrepfer, S., Deuse, T., Reichenspurner, H., Fischbein, M.P., Robbins, R.C., Pelletier, M.P. (2007). Stem cell transplantation: the lung barrier. *Transplant Proc*, 39, 573-6.

Shen, L.H., Li, Y., Chen, J., Zacharek, A., Gao, Q., Kapke, A., Lu, M., Raginski, K., Vanguri, P., Smith, A., Chopp, M. (2007). Therapeutic benefit of bone marrow stromal cells administered 1 month after stroke. *J Cereb Blood Flow Metab*, 27, 6-13.

Stem Cell Therapies as an Emerging Paradigm in Stroke Participants. (2009). Stem Cell Therapies as an Emerging Paradigm in Stroke (STEPS): bridging basic and clinical science for cellular and neurogenic factor therapy in treating stroke. *Stroke*, 40, 510-5.

Stodilka, R.Z., Blackwood, K.J., Prato, F.S. (2006). Tracking transplanted cells using dual-radionuclide SPECT. *Phys Med Biol*, 51, 2619-2632.

Suarez-Monteagudo, C., Hernandez-Ramirez, P., Alvarez-Gonzalez, L., Garcia-Maeso, I., de la Cuetara-Bernal, K., Castillo-Diaz, L., Bringas-Vega, M.L., Martinez-Aching, G., Morales-Chacon, L.M., Baez-Martin, M.M., Sánchez-Catasús, C., Carballo-Barreda, M., Rodríguez-Rojas, R., Gómez-Fernández, L., Alberti-Amador, E., Macías-Abraham, C., Balea, E.D., Rosales, L.C., Del Valle, Pérez, L., Ferrer, B.B., González, R.M., Bergado, J.A. (2009). Autologous bone marrow stem cell neurotransplantation in stroke patients. An open study. *Restor Neurol Neurosci*, 27, 151-161.

Sykova, E., Jendelova, P. (2007). In vivo tracking of stem cells in brain and spinal cord injury. *Prog Brain Res*, 161, 367-83.

Taguchi, A., Soma, T., Tanaka, H., Kanda, T., Nishimura, H., Yoshikawa, H., Tsukamoto, Y., Iso, H., Fujimori, Y., Stern, D.M., Naritomi, H., Matsuyama, T. (2004). Administration of CD34+ cells after stroke enhances neurogenesis via angiogenesis in a mouse model. *J Clin Invest*, 114, 330-8.

Tolar, J., O'shaughnessy, M.J., Panoskaltsis-Mortari, A., McElmurry, R.T., Bell, S., Riddle, M., McIvor, R.S., Yant, S.R., Kay, M.A., Krause, D., Verfaillie, C.M., Blazar, B.R. (2006). Host factors that impact the biodistribution and persistence of multipotent adult progenitor cells. *Blood*, 107, 4182-8.

Toyama, K., Honmou, O., Harada, K., Suzuki, J., Houkin, K., Hamada, H., Kocsis, J.D. (2009) Therapeutic benefits of angiogenetic gene-modified human mesenchymal stem cells after cerebral ischemia. *Exp Neurol*, 216, 47-55.

Vendrame, M., Cassady, J., Newcomb, J., Butler, T., Pennypacker, K.R., Zigova, T., Sanberg, C.D., Sanberg, P.R., Willing, A.E. (2004). Infusion of human umbilical cord blood cells in a rat model of stroke dose-dependently rescues behavioral deficits and reduces infarct volume. *Stroke*, 35, 2390-5.

Vilalta, M., Dégano, I.R., Bagó, J., Gould, D., Santos, M., García-Arranz, M., Ayats, R., Fuster, C., Chernajovsky, Y., García-Olmo, D., Rubio, N., Blanco, J. (2008). Biodistribution, long-term survival, and safety of human adipose tissue-derived mesenchymal stem cells transplanted in nude mice by high sensitivity non-invasive bioluminescence imaging. *Stem Cells Dev*, 17, 993-1003.

Wakabayashi, K., Nagai, A., Sheikh, A.M., Shiota, Y., Narantuya, D., Watanabe, T., Masuda, J., Kobayashi, S., Kim, S.U., Yamaguchi, S. (2010). Transplantation of human mesenchymal stem cells promotes functional improvement and increased expression of neurotrophic factors in a rat focal cerebral ischemia model. *J Neurosci Res*, 88, 1017-25.

Walczak, P., Zhang, J., Gilad, A.A., Kedziorek, D.A., Ruiz-Cabelo, J., Young, R.G., Pittenger, M.F., van Zijl, P.C.M., Huang, J., Bulte, J.W.M. (2008). Dual-modality monitoring of targeted intraarterial delivery of mesenchymal stem cells after transient ischemia. *Stroke* 39, 1569-74.

Wang, Y., Deng, Y., Zhou, G.Q. (2008). SDF-1alpha/CXCR4-mediated migration of systemically transplanted bone marrow stromal cells towards ischemic brain lesion in a rat model. *Brain Res*, 1195, 104-12.

Willing, A.E., Lixian, J., Milliken, M., Poulos, S., Zigova, T., Song, S., Hart, C., Sanchez-Ramos, J., Sanberg, P.R. (2003a). Intravenous versus intrastriatal cord blood administration in a rodent model of stroke. *J Neurosci Res*, 73, 296-307.

Willing, A.E., Vendrame, M., Mallery, J., Cassady, C.J., Davis, C.D., Sanchez-Ramos, J., Sanberg, P.R. (2003b). Mobilized peripheral blood cells administered intravenously produce functional recovery in stroke. *Cell Transplant*, 12, 449-54.

Xia, L., McDaniel, J.M., Yago, T., Doeden, A., McEver, R.P. (2004). Surface fucosylation of human cord blood cells augments binding to P-selectin and E-selectin and enhances engraftment in bone marrow. *Blood*, 104, 3091-6.

Xiao, J., Nan, Z., Motooka, Y., Low, W.C. (2005). Transplantation of a novel cell line population of umbilical cord blood stem cells ameliorates neurological deficits associated with ischemic brain injury. *Stem Cells Dev*, 14, 722-33.

Yang, M., Wei, X., Li, J., Heine, L.A., Rosenwasser, R., Iacovitti, L. (2010) Changes in host blood factors and brain glia accompanying the functional recovery after systemic administration of bone marrow stem cells in ischemic stroke rats. *Cell Transplant*, 19, 1073-84.

Zhang, J., Li, Y., Chen, J., Yang, M., Katakowski, M., Lu, M., Chopp, M. (2004). Expression of insulin-like growth factor 1 and receptor in ischemic rats treated with human marrow stromal cells. *Brain Res*, 1030, 19-27.

Zhang, Z.G., Chopp, M. (2009). Neurorestorative therapies for stroke: underlying mechanisms and translation to the clinic. *Lancet Neurol* 8, 491-500.

Zhang, L., Li, Y., Zhang, C., Chopp, M., Gosiewska, A., Hong, K. (2011). Delayed administration of human umbilical tissue-derived cells improved neurological functional recovery in a rodent model of focal ischemia. *Stroke*, 42, 1437-44.

Zhao, L.R., Duan, W.M., Reyes, M., Keene, C.D., Verfaillie, C.M., Low, W.C. (2002). Human bone marrow stem cells exhibit neural phenotypes and ameliorate neurological deficits after grafting into the ischemic brain of rats. *Exp Neurol*, 174, 11-20.

Zhou, J., Cheng, G., Kong, R., Gao, D.K., Zhang, X. (2011). The selective ablation of inflammation in an acute stage of ischemic stroke may be a new strategy to promote neurogenesis. *Med Hypotheses*, *76*, 1-3.

Part 3

Intravenous Thrombolysis and Intra-Arterial Procedures

9

Thrombolysis for Ischemic Stroke in Patients Aged 90 Years or Older

M. Balestrino, L. Dinia, M. Del Sette, B. Albano and C. Gandolfo
Department of Neuroscience, Ophthalmology and Genetics, University of Genova, Genova, Italy

1. Introduction

Currently, i.v. thrombolysis with recombinant tissue plasminogen activator (r-TPA) in acute ischemic stroke is approved in Europe only for patients 80 y.o. or younger, but data from the literature do suggest that patients older than 80 might also benefit from it (Sanossian and Ovbiagele, 2009). These older patients do have a worse outcome than younger ones, a finding that seems to be due to worse baseline prognostic factors, but they are not more prone to r-TPA -induced haemorrhage than younger ones (Engelter et al., 2006; Toni et al., 2008). Based on these considerations, a randomized clinical trial of i.v. r-TPA vs. usual care in acute stroke in patients older than 80 years is currently under way in Italy.

Even lesser data are available for stroke patients older than 90 years. Recently, it has been suggested that these patients may not benefit significantly from i.v. r-TPA, despite the fact that this treatment appeared reasonably safe in this age range, too (Mateen et al., 2009; Mishra et al., 2010).

We are reporting our experience with i.v. r-TPA in patients 90 year old or older in acute ischemic stroke.

2. Patients and results

We treated with i.v. r-TPA for acute stroke 6 patients (1male, 5 females). Median age was 92.5 y.o. (range 90-95). In all cases a CT scan had ruled out haemorrhagic stroke, and there were none of the exclusion criteria of the Safe Implementation of Thrombolysis in Stroke-Monitoring Study (SITS-MOST (Wahlgren et al., 2007)), apart from age and (in some cases, see below) elapsed time from onset of stroke. Specifically, all patient were in good neurological conditions before stroke. Modified Rankin Scale (mRS) before stroke was 0 or 1 in all cases, except in one 91 y.o. female where it was 3 due to severe hip arthrosis (no pre-existing neurological deficit). Risk factors were none (in 2 patients), hypertension (3 patients), previous minor stroke more than 3 months before the index one (2 patients), atrial fibrillation (2 patients), diabetes (1 patient). No patient had cognitive impairment.

In all cases a written informed consent was obtained either from the patient through a witness (n = 3) or from a first degree relative (n = 3).

The median National Institute of Health Stroke Scale (NIHSS) on presentation was 17.5 (range 3-21). It should be noted that that in the single case where NIHSS was less than 6, the symptoms included aphasia, a condition that is considered to be very disabling, thus worth the risk of thrombolysis (Kohrmann et al., 2009). All strokes were in the carotid arteries territory. Stroke was due to cardiac emboli in 3 cases and to large vessel disease in 2 cases; the remaining case was a lacunar stroke in the basal ganglia area. Median stroke-to-hospital time was 84.50′ (range 32′-102′). Upon their arrival into the Emergency Room the advanced age of these patients caused them to be initially excluded from thrombolytic treatment, therefore they were not handled in the rapid way that prospective r-TPA patients are usually treated. Only later the consulting stroke neurologist considered thrombolysis as an option. Thus, in-hospital times were rather slow. Median door-to-CT time was 82′ (range 39′-147′), median CT-to-treatment time was 49.5′ (range 15′-110′). Median door-to-treatment time was 150.5′ (range 80′-215′). Median time from stroke onset to thrombolysis ("stroke-to-needle" time) was 210′ (range 177′-273′).

Four patients received the standard r-TPA dose of 0.9 mg/Kg i.v., 10% of which was administered as an initial bolus, followed by the remaining 90% over 1h. One 94 y.o. patient received the lower dose of 0.67 mg/Kg i.v. (10% as a bolus, 90% in 1h) because in the Emergency Room she had already been given 500 mg acetyl-salicylic acid i.v. in the belief that she was not a candidate to thrombolysis. This same patient suffered shortly after thrombolysis a traumatic clavicle fracture, after which a major bleeding, requiring transfusion, occurred around the site of fracture. In another 91 y.o. patient r-TPA infusion was stopped after 25′ (when about 50% of the full dose had been administered) because of headache and a minor gingival bleeding. It should be noted that minor gingival bleeding occurred during drug infusion in another 91 y.o. patient who nevertheless went on to receive the full standard dose of r-TPA with no other harmful consequences. No bleeding was observed in the other patients.

No symptomatic intracranial haemorrhage occurred in any patient. In 1 patient a haemorrhagic transformation of the stroke (categorized HI2 according to ECASS III classification (Hacke et al., 2008)) was seen at the 24 hours follow-up CT scan, in all other cases no bleeding was observed at routine 24 hours follow-up CT scan.

In 3 patients transcranial ultrasound examination was carried out both before and immediately after thrombolysis. In all these cases occlusion of the middle cerebral artery was found, graded 0 (1 case) or 2 (2 cases) according to the TIBI classification (Demchuk et al., 2001). In all these cases improvement of flow to a TIBI 3 score was noted after thrombolysis. It should be noted that these are the 3 patients that showed good outcome at the 3-months follow up (mRS=0-1). In one of the 3 cases where ultrasound examination could not be carried out, a dense middle cerebral artery sign (Launes and Ketonen, 1987) was found upon non-contrast CT scan before thrombolysis. This patient showed an almost full recovery of motor function at the end of tPA infusion, that was followed by worsening 1 hour after. A CT scan right after the second worsening ruled out any intracranial bleeding. We interpret these findings as r-TPA -induced recanalization followed by early restenosis.

Figure 1 shows the NIHSS score before thrombolysis and at various times after it. As it can be seen, both patients with milder symptoms (NIHSS=3 and 6, respectively) had very good outcome (NIHSS=1 and 0, respectively, after 7 days). Of the 4 patients with more severe symptoms, 1 had very good improvement (NIHSS=1 after 7 days - this is the lady who received 0.67 mg/Kg because she had already received 500 mg ASA), the other 3 had minimal or no improvement. Of the latter 3 ones, 1 is the lady whose treatment was aborted because of gingival bleeding, and the other 2 are the patients who later died (see below).

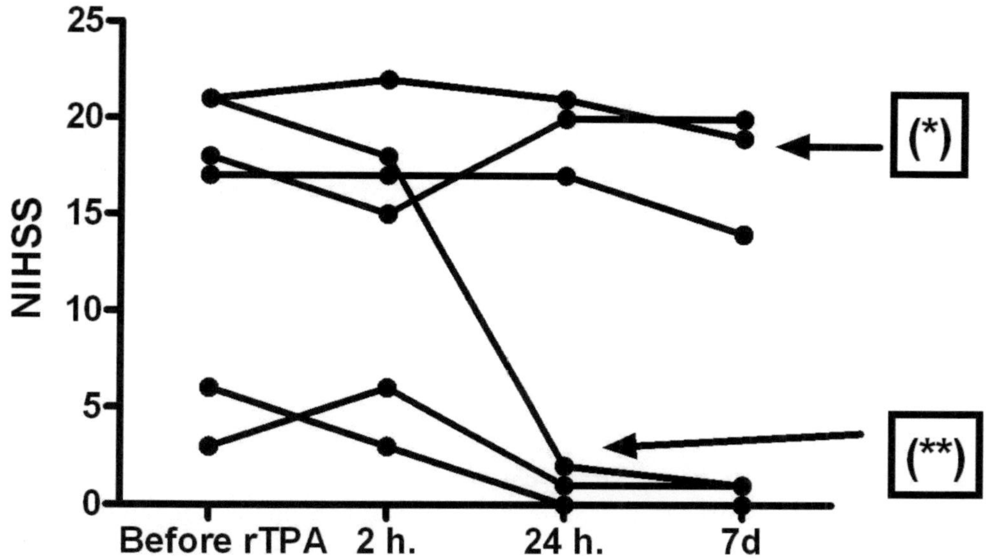

Fig. 1. NIHSS score before thrombolysis and 2 hours, 24 hours and 7 days after it in each patient. One asterisk indicates the patient in whom treatment was aborted before of gingival bleeding after 25', when approximately 50% of the full rTPA dose had been received. Two asterisks indicate the patient to whom the reduced rTPA dose of 0.67 mg/Kg was administered because she had already received 500 mg ASA i.v.

Figure 2 shows the score on the modified Rankin Scale (mRS) (Farrell et al., 1991) before stroke as well as 7 days and 3 months afterwards (note that mRS=6 was added to the original mRS, indicating death (Mateen et al., 2009)). As it can be seen, 2 patients died (14 and 16 days after thrombolysis, respectively), both of cardiac failure (acute pulmonary oedema). Neither one had significantly improved after thrombolysis (see above). One more patient was left with moderately severe disability (mRS=4) and 3 patients showed very good outcome (mRS=0-1).

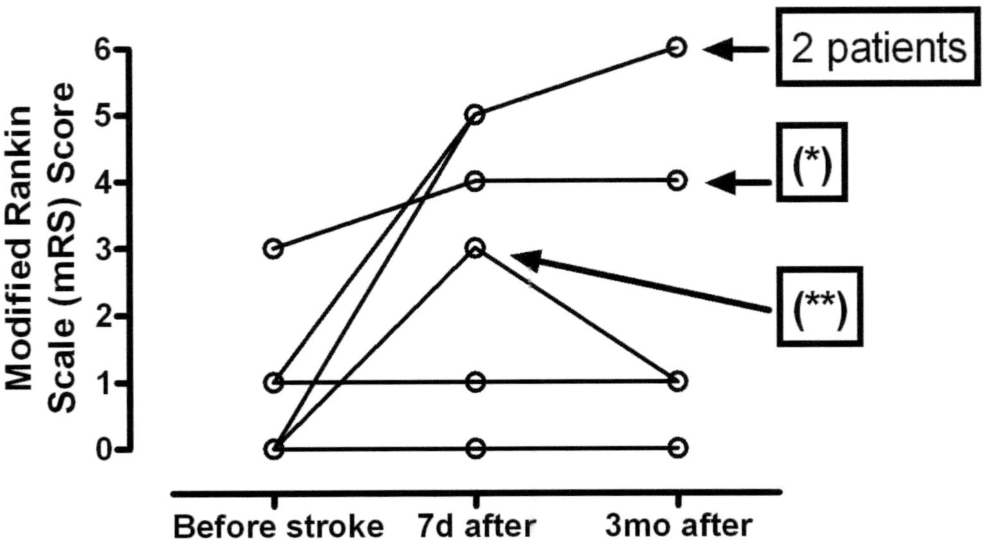

Fig. 2. mRS score before stroke, 7 days and 3 months after it in each patient. One asterisk indicates the patient in whom treatment was aborted before of gingival bleeding after 25′, when approximately 50% of the full rTPA dose had been received. Two asterisks indicate the patient to whom the reduced rTPA dose of 0.67 mg/Kg was administered because she had already received 500 mg ASA i.v.

3. Discussion

The present report has the obvious limitation of being retrospective, nevertheless we believe that it has some interest given the extreme paucity of literature reports on thrombolysis in 90 y.o. patients or older. To the best of our knowledge, so far only one such report exists in the literature (Mateen et al., 2009), while another recently published paper reports a comparison between two different databases (Mishra et al., 2010). Our findings are in agreement with those report, as far as they confirm that i.v. thrombolysis with r-TPA is reasonably safe in these patients. In our patients no treatment-related worsening occurred, and no intracranial symptomatic haemorrhage was observed. Retrospectively, abortion of therapy upon gingival bleeding may have been excessive (in the other case of minor gingival bleeding the full dose was still given with no harm). Death occurred in two of the three unimproved cases 14 and 16 days after thrombolysis because of cardiac failure, unrelated to treatment. Three out of six patients improved upon thrombolysis. Although this finding is obviously encouraging, randomized clinical trials are needed to reach firm conclusions that may be applied to the general population of 90 y.o. patients or older. Finally, we note that our patients were not handled rapidly by hospital personnel because it was initially believed that their old age should have ruled out thrombolysis. This resulted in a significant delay of treatment (see above), that has probably reduced the chances of good outcome in our patients (Lees et al., 2010). This finding strongly suggests that stroke therapy is a complex issue, one that, whenever possible, should be managed by stroke physicians having specific competence and experience.

4. References

Demchuk AM, Burgin WS, Christou I, Felberg RA, Barber PA, Hill MD, Alexandrov AV (Thrombolysis in Brain Ischemia (TIBI) Transcranial Doppler Flow Grades Predict Clinical Severity, Early Recovery, and Mortality in Patients Treated With Intravenous Tissue Plasminogen Activator. Stroke 32:89-93.2001).

Engelter ST, Bonati LH, Lyrer PA (Intravenous thrombolysis in stroke patients of >80 versus <80 years of age - A systematic review across cohort studies. Age and Ageing 35:572-580.2006).

Farrell B, Godwin J, Richards S, Warlow C (The United Kingdom transient ischaemic attack (UK-TIA) aspirin trial: final results. J Neurol Neurosurg Psychiatry 54:1044-1054.1991).

Hacke W, Kaste M, Bluhmki E, Brozman M, Davalos A, Guidetti D, Larrue V, Lees KR, Medeghri Z, Machnig T, Schneider D, von Kummer R, Wahlgren N, Toni D, the E, I (Thrombolysis with Alteplase 3 to 4.5 Hours after Acute Ischemic Stroke. N Engl J Med 359:1317-1329.2008).

Kohrmann M, Nowe T, Huttner HB, Engelhorn T, Struffert T, Kollmar R, Saake M, Doerfler A, Schwab S, Schellinger PD (Safety and outcome after thrombolysis in stroke patients with mild symptoms. Cerebrovasc Dis 27:160-166.2009).

Launes J, Ketonen L (Dense middle cerebral artery sign: an indicator of poor outcome in middle cerebral artery area infarction. J Neurol Neurosurg Psychiatry 50:1550-1552.1987).

Lees KR, Bluhmki E, von Kummer R, Brott TG, Toni D, Grotta JC, Albers GW, Kaste M, Marler JR, Hamilton SA, Tilley BC, DAVIS SM, Donnan GA, Hacke W (Time to treatment with intravenous alteplase and outcome in stroke: an updated pooled analysis of ECASS, ATLANTIS, NINDS, and EPITHET trials. The Lancet 375:1695-1703.2010).

Mateen FJ, Nasser M, Spencer BR, Freeman WD, Shuaib A, Demaerschalk BM, Wijdicks EFM (Outcomes of Intravenous Tissue Plasminogen Activator for Acute Ischemic Stroke in Patients Aged 90 Years or Older. Mayo Clinic Proceedings 84:334-338.2009).

Mishra NK, Ahmed N, Andersen G, Egido JA, Lindsberg PJ, Ringleb PA, Wahlgren NG, Lees KR (Thrombolysis in very elderly people: controlled comparison of SITS International Stroke Thrombolysis Registry and Virtual International Stroke Trials Archive. BMJ 341.2010).

Sanossian N, Ovbiagele B (Prevention and management of stroke in very elderly patients. The Lancet Neurology 8:1031-1041.2009).

Toni D, Lorenzano S, Agnelli G, Guidetti D, Orlandi G, Semplicini A, Toso V, Caso V, Malferrari G, Fanucchi S, Bartolomei L, Prencipe M (Intravenous thrombolysis with rt-PA in acute ischemic stroke patients aged older than 80 years in Italy. Cerebrovasc Dis 25:129-135.2008).

Wahlgren N, Ahmed N, Davalos A, Ford GA, Grond M, Hacke W, Hennerici MG, Kaste M, Kuelkens S, Larrue V, Lees KR, Roine RO, Soinne L, Toni D, Vanhooren G (Thrombolysis with alteplase for acute ischaemic stroke in the Safe Implementation of Thrombolysis in Stroke-Monitoring Study (SITS-MOST): an observational study. The Lancet 369:275-282.2007).

10

Mechanical Embolectomy

Jiří Lacman and František Charvát
Central Military Hospital Prague
Czech Republic

1. Introduction

After heart attack and cancer, cerebrovascular accident (CVA) is the third leading cause of death in the industrialized part of the world and the leading cause of long-term disability. Having suffered a stroke, 15% to 30% of patients are incapable of independent existence without help of others. After the first ischemic stroke, 10 to 15% of patients die within the first 30 days, 20% within 6 months, and this percentage increases to 25 - 30% within one year. CVA is a serious problem both in terms of significant disability of the population and very high costs of treatment. In addition to the high incidence, which increases with the increasing age of patients, also the fact that the cerebrovascular accidents affect younger age groups more and more often is serious. Thus, even accidents in patients between 30 - 40 years of age may occur.

The aim of treatment of ischemic CVA (iCVA) is to intervene as soon as possible and to achieve reperfusion of the affected area. Therefore, the time of recanalization plays a crucial role.

The only causal and potentially effective treatment for iCVA due to the occlusion of cerebral arteries is an attempt at recanalization. The first treatment introduced was intravenous administration of a thrombolytic agent, in the following years possibilities of intra-arterial recanalization using a thrombolytic agent or intra-arterial mechanical embolectomy have gradually been introduced and technologically improved.

2. IV thrombolysis

Experience with intravenous thrombolysis (IVT) of coronary and peripheral arteries have led to the introduction of intravenous thrombolysis also in acute occlusions of the major cerebral arteries. However, the method of IVT in the coronary and peripheral bloodstream has been abandoned with the passing because of the higher risk of overall complications and less pronounced local effect.

A good final effect of intravenously administered thrombolytic agent in acute CVA has been shown in the NINDS study (National Institute of Neurological Disorders and Stroke) (NINDS Stroke Study Group, 1995). This study randomized a set of 624 patients with occlusions of larger and smaller arteries not verified by angiography, the time interval from the CVA occurrence to the beginning of administration was 3 hours. Recombinant tissue plasminogen activator (rt-PA) was used, administered at a dose of 0.9 mg/kg up to a maximum dose of 90 mg, 10% bolus, the rest administered in course of 60 minutes. The study showed a better clinical outcome in the group treated with IVT (30% of patients had no or minor neurological deficit according to the modified Rankin scale mRS ≤ 1). Mortality after three months was 17% (versus 21% in placebo-treated patients). Based on the results of this study, the FDA in the U.S. approved in 1996 the use of IVT in clinical practice. In Europe, IVT with rt-PA was approved in 2002. Safety and efficacy of routinely administered rt-PA in clinical practice has been confirmed once more by the SITS-MOST monitoring study (SITS-MOnitoring STudy) based on the data from European registries (Wahlgren et al., 2007).

The ECASS I, II, and ATLANTIS studies have not shown significant benefit to the patient in case of use of an intravenous thrombolytic agent (rt-PA) in the time-window of 3 to 6 hours after the onset of symptoms. The number of patients with no or minimal neurological deficit (mRS ≤ 1) has not improved, but the number of disabled patients (mRS ≤ 2) has decreased (Clark et al., 1999; Hacke et al., 1995, 1998). The efficacy in the ECASS studies depended on the selection of patients with moderate to severe neurological deficit and without extended infarctsigns on the initial CT scan. These studies did not support use of intravenous thrombolysis after the 3-hour time window.

The recently published ECASS III study evaluating safety and efficacy of IVT in the time window of 3 - 4.5 hours has showed improvement of the clinical condition in patients after 3 months compared to placebo (mRS 0-1 52.4% vs. 45.2%). Intracranial hemorrhage occurred more often in the group treated with the thrombolytic agent: 27% vs. placebo 17.6%. Mortality in both groups did not significantly differ after 3 months (7.7% vs 8.4%) (Hacke et al., 2008). The results of this study have led to redefinition of the recommended procedure for IVT and the therapeutic window has increased to 4.5 hours.

The advantage of IVT is its widespread availability and the possibility of its early initiation. This treatment does not even require the specialized know-how that is needed to insert and navigate a catheter into the cerebral circulation. Unfortunately, experimental studies have shown that this treatment is less and less effective as the trombus grows larger and larger. (Zivin et al., 1985). More than 50% of patients treated with intravenous rtPA fail to achieve a favorable clinical outcome (Papadakis & Buchan, 2006; Qureshi et al., 2001).

3. IA thrombolysis

The basic study on intra-arterial thrombolysis (IAT) was a randomized multicenter study ProAct (Prolyse in Acute Cerebral thromboembolism), ongoing in phase I and II (del Zoppo et al., 1998; Furlan et al., 1999). In this study, recombinant pro-urokinase (r-proUK) was compared to placebo. As a placebo, intra-arterial administration of saline solution with heparin was used. The first 16 patients received a 100-IU/kg bolus followed by a 1000 IU/h constant infusion ("high heparin") for 4 hours. Thereafter, on the recommendation of the External Safety Committee, the heparin regimen was altered to a 2000 IU bolus and 500 IU/h infusion ("low heparin") for the remaining patients. In both groups, the end of the microcatheter was placed in the proximal third of the thrombus. Passing through the thrombus or mechanical action on it was not allowed. If partial recanalisation occured after 2 hours, the catheter was introduced more distally into the proximal third of the remaining thrombus. The time window was 6 hours, patients with middle cerebral artery occlusion (MCA) were indicated for the treatment.

In phase I of the PROACT study, 46 patients were randomized, the maximum total dose was 6 mg of r-proUK in course of 2 hours, recanalization occurred in 57.7% (compared to 14.3% in controls), whereas in the group with higher dose of heparin it was in 81.8% and in the group with lower dose of heparin it was in 40.0%. Serious cerebral bleeding in the first 24 hours after thrombolysis occurred on average in 15.4% compared to 7.1% in the control group, in the group with the higher administered dose of heparin it was more frequent (27.3%), after dose reduction it occurred only in 6.7%. Clinical condition was evaluated after 3 months according to mRS, 19.2% of patients were stage 0-2 (versus 7.1% of controls).

In phase II of the study, 180 patients were randomized, the total dose of r-proUK was increased to 9 mg, heparin was decreased to 2000 units. The percentage of recanalization was 66% in the r-proUK group compared to 18% in the placebo group for TIMI 2-3 and 19% in the r-proUK group compared to 2% in the placebo group for TIMI 3. The resulting morbidity after 3 months was more favorable by 60% compared to the control group (mRS 0-2 in 40% r-proUK treated patients in contrast to 25% in the control group), also mortality was slightly lower (24% compared to 27% in the control group); the number of symptomatic hemorrhage was significantly higher in the first 24 hours (35% versus 13% in the control group) but was not significantly different by 10 days (68% in the r-proUK group versus 57%, $p=0.23$); for comparison, in the NINDS study with IVT the ICH rate was 6.5%.

In phase II of the study with a higher dose of the thrombolytic agent and a lower dose of heparin, a higher number of recanalizations and a significant decrease in the number of hemorrhages was achieved, leading to better clinical results. However, FDA has not granted its approval to use IAT in clinical practice, and so IAT has remained reserved for use in studies only.

Also a number of nonrandomized studies with smaller numbers of patients, a number of case studies and meta-analyses have been published, which generally show a higher percentage of recanalizations and favorable clinical outcome after intra-arterial administration of a thrombolytic agent (Wardlaw et al., 1997; Gönner et al., 1998; Higashida et al., 2003).

In recent years, new antiplatelet medicines acting on the inhibitory IIb / IIIa platelet receptors, have been developed, which are highly effective and relatively safe even when administered intravenously. They act selectively on platelet thrombi, preventing their formation and growth. In the brain arteries they have been used so far mainly to treat thromboembolic complications of endovascular procedures, however, several papers on their use in acute iCVA has already been published (The Abciximab in Ischemic Stroke Investigators, 2000).

Compared to intravenous thrombolysis, intra-arterial thrombolysis has the advantage of achieving higher concentration of the thrombolytic agent at the site of the occlusion with lower total dose. This reduces the risk of systemic complications. Also results of a number of other studies showing a higher percentage of recanalization and better clinical results despite a slightly increased percentage of bleeding complications confirm this (Barnwell et al., 1994).

Opinion of the ASITN Executive Committee (American Society of Interventional and Therapeutic Neuroradiology), published in the AJNR in 2001, states that although the results of the PROACT study have not led the FDA to grant the approval, IAT is regarded, based on the experience gained so far, as a beneficial and acceptable method of treatment at the centers experienced in vascular diagnostic and interventional procedures. Indication for the IAT must be considered on an individual basis, with careful assessment of risks and benefits (Executive of the American Society of Interventional & Therapeutic Neuroradiology [ASITN], 2001).

In 2003, new guidelines of the ASA (American Stroke Association) and EUSI (European Stroke Initiative) were published, which indicate that EUSI acknowledges that the IAT performed within 6 hours after development of the CVA improves its course (Klijn & Hankey, 2003).

Thus, IAT is currently an appropriate method of iCVA treatment in patients within the time window of 3-6 hours (or 4.5 - 6 hours with regard to the results of the ECASS III study) in specialized centers. Intra-arterial thrombolysis may also be used to deal with acute embolic complications developed during endovascular procedures and last, but not least, an IA thrombolytic agent may be administered after unsuccessful response to its IV administration.

4. Combined IV / IA thrombolysis

Pursuit of a fast and simple administration of a thrombolytic agent together with angiographic monitoring of the treatment effect and the possibility of local intervention led to the IMS study (International Management of Stroke Trial) (IMS study investigators, 2004). Patients were treated with intravenous administration of a thrombolytic agent and in the case of angiographic confirmation of persisting arterial occlusion they were given the thrombolytic agent also locally. The treatment led to a partial or complete recanalization of the artery in 56% of patients, the percentage of bleeding was low (6.3%), as well as mortality was (16%). A subsequent study, IMS II (IMS II Trial Investigators, 2007; Tomsick et al., 2008), put together combined IVT and IAT with the use of a special micro-infusion catheter Ekos-MicroLysUS (Ekos, Washington, USA) or a standard microcatheter. The Ekos catheter allows simultaneous infusion of a thrombolytic agent and administration of low-energy ultrasound waves transmitted by a probe at the end of the micro-catheter. The median dose of the intravenously administered rt-PA in the study was 46.4 mg and that of the intra-arterial dose was 12 mg. The percentage of hemorrhage, compared to IMS I, increased to 9.9%, recanalization increased to 73% with 16% mortality. Favorable clinical outcome was achieved in 46% of patients. Currently, IMS III is being conducted, comparing the results of combined treatment with IVT, IAT, and the Ekos catheter with the MERCI apparatus (Khatri et al., 2008) and Synthesis-Expansion trial is comparing the IVT and IAT (Ciccone et al, 2011).

There are also other publications reporting improvement of the angiographic and clinical finding in different combinations of methods (Burns et al., 2008).

5. Mechanical embolectomy

After IAT has been introduced within the framework of clinical studies, first positive experience with mechanical disruption of the thrombus were reported, with the use of a guide wire or special instrumentation increasing the thrombus surface by means of its fragmentation and thus accelerating the effect of thrombolytic agent. For this purpose, micro-guide wire, extraction loops and baskets to capture foreign bodies were used (Sarimachi et al., 2004; Krajina et al., 2005).

Mechanical treatment reduces the need for a thrombolytic agent, reducing the risk of ICH (intracerebral hemorrhage) and has been used up to 8 hours after stroke, since it is still an experimental time window. Mechanical devices are able to recanalize the occluded artery faster and may be more effective in removing mature thrombi, cholesterol or calcified thrombi, and atherosclerotic plate emboli.

Mechanical embolectomy plays a role even in patients contraindicated for pharmacological treatment with a thrombolytic agent, usually because of recent surgery.

The disadvantage of mechanical embolectomy is often a difficult access to cerebral arteries, due to the anatomical relations of the arteries in the neck and brain, and the possibility of perforation or dissection of the artery or distal embolization of the parts of the thrombus into the so far unaffected bloodstream. However, the advantages of mechanical intervention outweigh the disadvantages and risks.

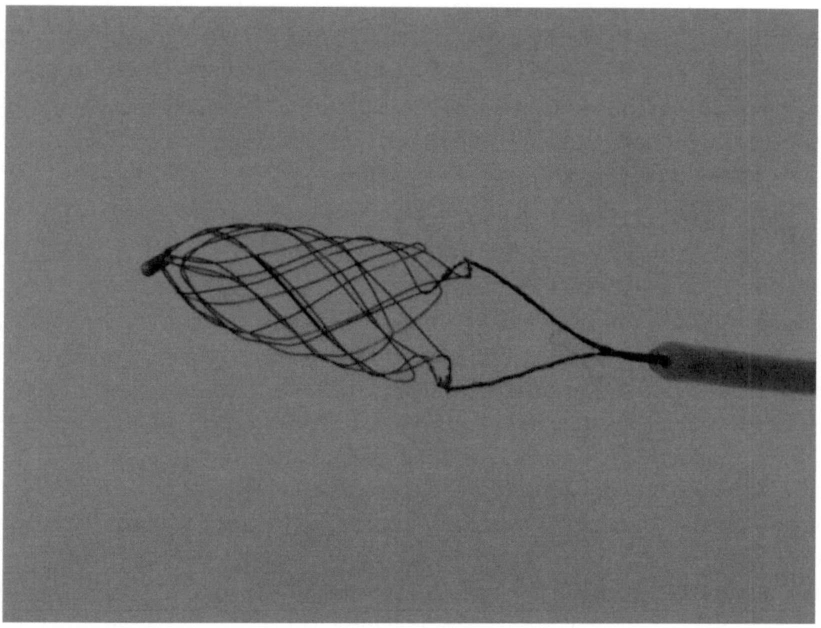

Fig. 1. Catch retriever

Mechanical devices vary depending on the site from where they act upon the thrombus. The approach is either a proximal one with aspiration of the thrombus or with a device "grasping" the thrombus or devices for distal access capturing the thrombus into a basket or snare. A study comparing the effectiveness of these approaches, using the Vasco35 micro-catheter and the Catch retriever (both by Balt Extrusion, France) has demonstrated that the use of the micro-catheter allows rapid and repeated use with a low percentage of thromboembolic events (3% versus 26%) but also significantly lower success rate as regards thrombus removal compared to the Catch retriever (39% vs. 83%). The percentage of embolic events may be reduced by the use of a proximal occlusion balloon (Gralla et al., 2006). In a recent retrospective case series, the Catch device again appears effective for recanalization and improving 90-day outcome in patients with acute ischemic stroke (Mourand et al., 2011).

5.1 Merci retriever

In 2004 the FDA approved the first extraction device for stroke treatment - Merci Retrieval system (by Concentric Medical, Calif., USA). It is a flexible nitinol guide wire, the distal end of which has shape memory and, being pushed out of the micro-catheter, it forms into a helix. Newer types have the end surrounded by a braided tangle of filaments to better capture the thrombus.

Using the micro-catheter, the system is introduced behind thrombus, it is expanded and then retracted under the vacuum in the balloon guide catheter. Merci Retrieval system was first used in a small multicenter study in 28 patients with NIHSS ≥ 10 with occlusion of a major cerebral artery documented by angiography (Gobin et al., 2004). Occlusion of intracranial ICA (internal carotid artery) was present in 5 patients (18%), occlusion of MCA (medial cerebral artery) in 18 patients (64%), common occlusion of both ICA and MCA in 3 patients (11%) and occlusion of basilar artery in 2 patients (7%). Successful recanalization (TIMI 2-3) was achieved when using the retriever alone in 12 patients (43%) and when combining it with IA rtPA in 18 patients (64%). In 12 patients, an asymptomatic ICH developed after the procedure. After one month, clinical improvement was achieved in 9 patients (32%). 10 patients died within 30 days (36%).

Another study investigating the effectiveness of the Merci Retrieval system was the MERCI study (Mechanical Embolus Removal in Cerebral Ischemia) (Smith et al., 2005). The aim of the MERCI study was to broaden the possibilities of treatment for patients contraindicated for IVT or for those who have passed the time window for IVT or in whom IVT was not successful. The therapeutic window was extended to 8 hours after the onset of complaints. IA thrombolytic agents were either not used at all or only in small doses. 151 patients were enrolled in the study, the average age was 67 years of age and the NIHSS score was 20. The inclusion criteria were: age over 18 years, the clinical picture of stroke with NIHSS ≥ 8, time window between 3 to 8 hours after the onset of complaints or 0 to 3 hours in the case of contraindicated IV rtPA, and artery occlusion confirmed by angiography. Patients with an ICH, significant mass effect with a midline shift, or with a hypodensity exceeding 1/3 of the MCA supply area were excluded from the study. In 90% of cases, the occlusion affected the anterior supply area. Recanalization was achieved in 48% of patients, in the case of combination with thrombolytic therapy it was in 60%. Symptomatic complications of the procedure, such as perforation, dissection, distal embolism occurred in 7.1%, symptomatic hemorrhage occurred in 7.8% and was partly due to artery perforation (4.3%). Conclusions of the first phase of the study showed good clinical results (mRS ≤ 0-2) in 27.7% of the total number of patients, i.e. in 46% of patients from the successful recanalization group.

In the MERCI multicenter study (Smith et al., 2008), the second generation of the retriever-type L5 was used. Inclusion and exclusion criteria did not differ from those in the previous study. 177 of patients were enrolled in the study, in 164 patients the Merci Retrieval system could be used. The average age was 68 years, and NIHSS score was 19. In 92% of cases, the occlusion affected the anterior supply area. 48 patients (29%) were given IV rtPA before the procedure, 57 patients received IA rtPA during the procedure. Recanalization was achieved using the Merci system alone in 57% of treated arteries, in the case of combination with IA thrombolytic and mechanical treatment it was in 69.5%. After 90 days, a good result was achieved in the whole group in 36% of patients (mRS 0-2), symptomatic hemorrhage occurred in 9.8%. 90-day mortality was 34%.

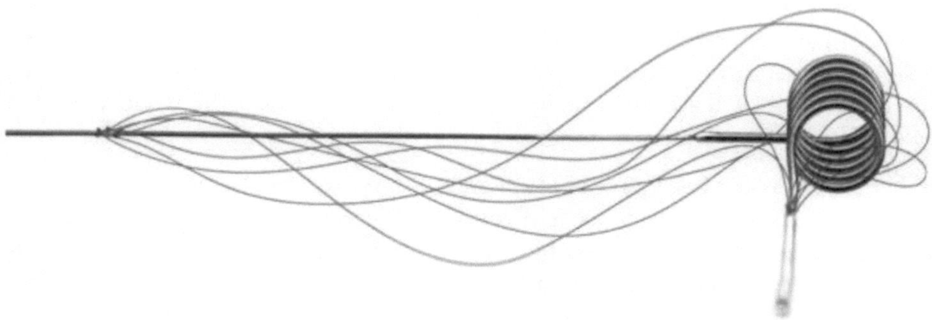

Fig. 2. Merci retriever, type L5

5.2 Neuronet

Neuronet (Abbott Vascular, Calif., USA) is a nitinol basket-shaped device attached to a micro-guide wire. The basket is opened distal to the embolus and the embolus is captured in it when the basket is pulled back.

Neuronet was successfully used for the removal of emboli in the cerebral arteries (Mayer et al., 2002a; Nesbit et al., 2004) and prospectively evaluated in a small study Neuronet Evaluation in Embolic Stroke Disease (Mayer et al., 2002b). This study evaluated 5 patients, treated with the Neuronet device. The device was used without flow control in the first 2 patients and with flow reversal in the basilar artery in the last 3 patients during retraction by transiently blocking both vertebral arteries or the left subclavian artery or both vertebral arteries with coaxial catheters. Patients 1, 2, and 4 were also administered recombinant tissue plasminogen activator (rtPA) intra-arterially. In the first 2 patients the deployment of the device was unsuccessfull. Two of the remaining patients were recanalised using mechanical thrombolysis only and one patient required additional fibrinolysis.

5.3 Penumbra system

Another device for thrombus removal is the Penumbra system (Penumbra, Calif., USA), which combines mechanical disruption of the thrombus with aspiration of its fragments. The system consists of a reperfusion micro-catheter connected to an aspiration pump and of a drop-like separator at the end of the micro-guide wire. The thrombus is fragmented and aspirated by the reperfusion micro-catheter, whereas the separator also fragments the thrombus and prevents micro-catheter clogging. This mechanical device is capable to restore the passage through the vessel without the use of a thrombolytic agent, the entire system is designed as a minimally invasive one, acting on the embolus from its proximal end. High flexibility and a wide spectrum of sizes (0.26 - 0.54 inch) of reperfusion micro-catheters enable their successful use even in smaller branches, such as M2 and A2.

The Penumbra system was first used in a European study in 20 patients (21 treated arteries) with ICA occlusion in 7 patients, MCA occlusion in 5 patients, and BA occlusion in 9 patients (Bose et al., 2008). The average NIHSS score was 21, patients were treated within 8 hours after the onset of complaints. Successful recanalization (TIMI 2 and 3) was achieved in 100%. Nine patients received IA rtPA after the embolectomy. Clinical improvement (mRS ≤ 2) within 30 days occurred in 45% of patients. Mortality reached 45%. In 8 patients, ICH occurred, two of these were symptomatic. The use of IA rtPA resulted in a higher incidence of bleeding.

In another multicenter prospective study (Penumbra Stroke Trial) 125 patients were treated in 24 centers in Europe and the USA (Penumbra Pivotal StrokeTrial Investigators, 2009). Inclusion and exclusion criteria were the same as in the MERCI and Multi MERCI studies. The average age of patients was 63 years, average NIHSS score was 17, median time from the onset of symptoms until the procedure was started was 4.1 hours. The occluded arteries were: ICA in 18%, M1 or M2 in 70%, vertebrobasilar artery in 9% and others in 3%. Complete or partial recanalization (TIMI 2 to 3) occurred in 81.6% of occluded arteries. Major complication of the procedure occurred in 4 patients. Symptomatic ICH occurred in 14 patients (11.2%) and asymptomatic ICH in 21 patients (16.8%). Good clinical outcome (mRS ≤ 2) after 90 days was achieved in 25% of patients. 90-day mortality was 32.8%. Just as in the MERCI and Multi MERCI studies, there was a marked improvement of neurological status (29% vs. 9%) and mortality (29% vs. 48%) in patients with successful recanalization. Based on this study, FDA consented to the use of this system in the treatment of ischemic stroke in 2007.

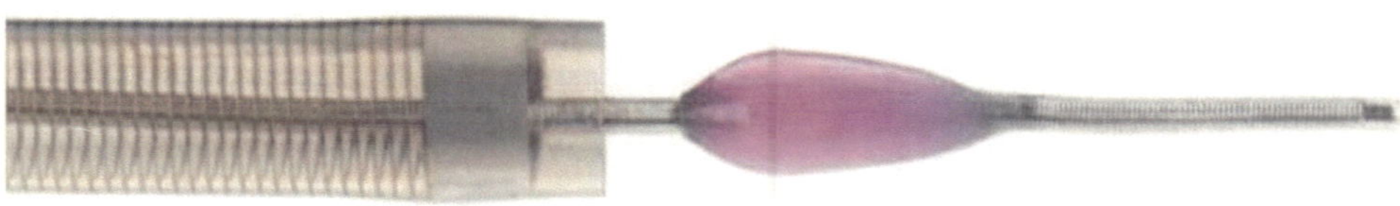

Fig. 3. Penumbra system

5.4 Stent solitaire

Stent Solitaire (Solitaire AB, EV3, USA) is a demountable retractable stent originally designed for intracranial bloodstream vessels where it served as a remodeling stent during coiling of wide neck aneurysms. It is a self-expanding closed cell nitinol stent, which allows for its pulling completely back into the microcatheter after having it fully inserted in the vessel. Using the device for demounting GDC embolization spirals, the stent may be electrolytically detached by the same mechanism as the GDC coils (Stryker, USA) and left at the site of the artery occlusion to help restore brain perfusion. Stent diameter expanded at maximum is 4 mm, the length is 15 mm or 20 mm. The stent is introduced through a microcatheter with an internal lumen of 0.021 inches. Its advantageous use even as a mechanical embolectomy device was demonstrated by a prospective study in 20 patients with an acute ischemic stroke in the time window up to 8 hours (Castaño et al., 2010). In 18 patients (90%) the thrombus was successfully removed by the stent and the affected brain tissue revascularized (TICI 2b-3); after 90 days, 45% of patients achieved the final mRS ≤ 2.

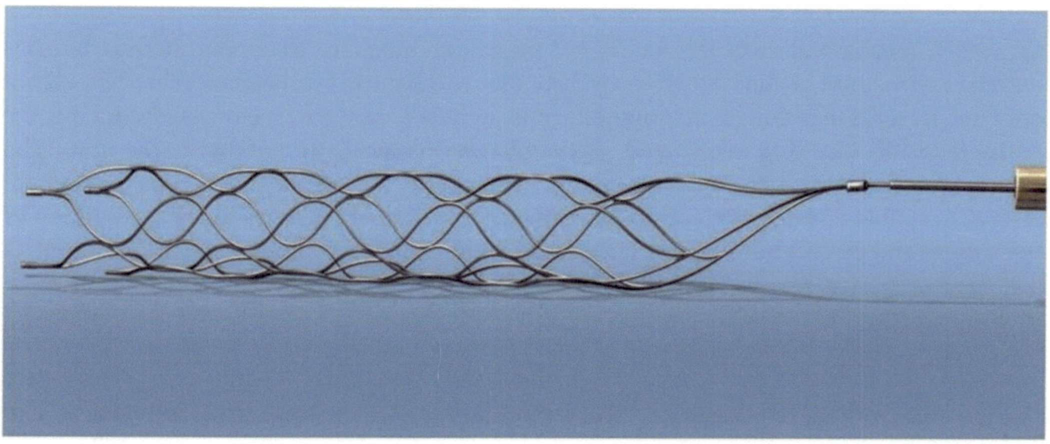

Fig. 4. Solitaire stent

5.5 Bonnet retriever

It is a novelty by Phenox company (Bochum, Germany), which launched Phenox Clot Retriever to the market already several years ago. It is a "self-expanding nitinol braiding with polyamide filaments." Larger surface helps better fixation of the thrombus and increases the chance for its extraction. Retriever is firmly connected to the micro-guide wire, it has contrast marks on both ends. It may be introduced distal to the thrombus or opened in the thrombus. By pulling it back slowly, the thrombus is captured in the braiding and extracted from the artery.

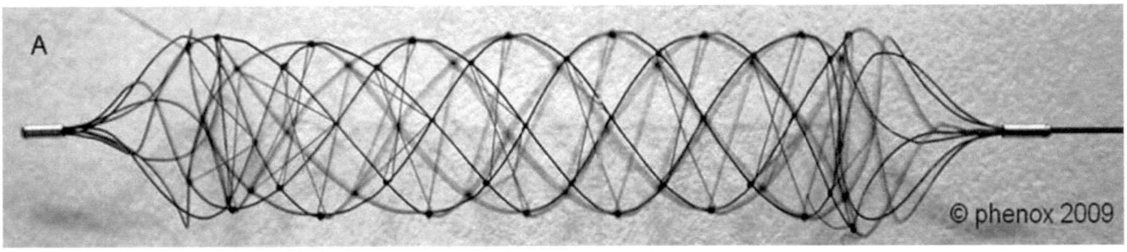

Fig. 5. Bonnet retriever

6. Conclusion

Endovascular treatment of ischemic cerebral strokes is still developing. There are advances both in the field of pharmacology and in the field of mechanical devices. New medicines improve the safety of the procedure and together with more and more perfect and efficient devices markedly improve the clinical condition of patients.

Currently, mechanical embolectomy is not merely an auxiliary treatment method but has become the method of choice in given indications. Mechanical recanalization has helped not only to investigate a longer therapeutic time window, but may also help reducing the percentage of bleeding associated with pharmacological thrombolytic therapy. The indicated option of mechanical embolectomy in patients with occlusion of a major cerebral artery is capable of improving brain perfusion and thus ameliorate the clinical condition of the patient and protect him/her from a disabling affliction.

7. References

Barnwell, SL., Clark, WM., Nguyen, TT., O'Neill, OR., Wynn ML, & Coull BM. (1994) Safety and efficacy of delayed intraarterial urokinase therapy with mechanical clot disruption for thromboembolic stroke. *AJNR*, Vol. 15, No. 10, pp. 1817–1822, ISSN: 0195-6108

Bose, A., Henkes, H., Alfke, K., Reith, W., Mayer, TE., Berlis, A., Branca, V., Sit, SP., & Penumbra Phase 1 Stroke Trial Investigators. (2008) The Penumbra System: a mechanical device for the treatment of acute stroke due to thromboembolism. *AJNR*, Vol. 29, No. 7, pp. 1409-1413, ISSN: 0195-6108

Burns, TC., Rodriguez, GJ., Patel, S., Hussein, HM., Georgiadis, AL., Lakshminarayan, K., & Qureshi, AI. (2008) Endovascular Interventions following Intravenous Thrombolysis May Improve Survival and Recovery in Patients with Acute Ischemic Stroke: A Case-Control Study. *AJNR*, Vol. 29, No. 10, pp. 1918 - 1924, ISSN: 0195-6108

Castaño, C., Dorado, L., Guerrero, C., Millán, M., Gomis, M., Perez de la Ossa, N., Castellanos, M., García, MR., Domenech, S., & Dávalos, A. (2010) Mechanical thrombectomy with the Solitaire AB device in large artery occlusions of the anterior circulation: a pilot study. *Stroke*, Vol. 41, No. 8, pp. 1836–1840, ISSN: 1747-4930

Ciccone, A, Valvassori, L, Nichellatti, M, SYNTHESIS Expansion investigators. (2011) SYNTHESIS Expansion: design of a nonprofit, pragmatic, randomized, controlled trial on the best fast-track endovascular treatment vs. standard intravenous alteplase for acute ischemic stroke. *Int J Stroke*, Vol. 6, No. 3, pp.259-265, ISSN: 1747-4949

Clark, WM., Wissman, S., Albers, GW., Jhamandas, JH., Madden, KP., & Hamilton, S. (1999) For the ATLANTIS Study Investigators. Recombinant Tissue-Type Plasminogen Activator (Alteplase) for Ischemic Stroke 3 to 5 Hours After Symptom Onset. The ATLANTIS Study: A Randomized Controlled Trial. *JAMA*, Vol. 282, No. 21, pp. 2019-2026, ISSN: 0098-7484

del Zoppo, GJ., Higashida, RT., Furlan, AJ., Pessin, MS., Rowley, HA., & Gent, M. (1998) PROACT: A Phase II Randomized Trial of Recombinant Pro-Urokinase by Direct Arterial Delivery in Acute Middle Cerebral Artery Stroke. *Stroke*, Vol. 29, No. 1, pp. 4-11, ISSN 0039-2499

Executive of the ASITN. (2001) Intraarterial thrombolysis: Ready for Prime Time? *AJNR*, Vol. 22, pp. 55-58, ISSN: 0195-6108

Furlan, A., Higashida, R., Wechsler, L., Gent, M., Rowley, H., Kase, C., Pessin, M., Ahuja, A., Callahan, F., Clark, WM., Silver, F., & Rivera F. (1999) Intra-arterial prourokinase for acute ischemic stroke. The PROACT II study: a randomised controlled trial. Prolyse in Acute Cerebral Thromboembolism. *JAMA*, Vol. 282, No. 21, pp. 2003-2011, ISSN: 0098-7484

Gobin, YP., Starkman, S., Duckwiler, GR., Grobelny, T., Kidwell, CS., Jahan, R., Pile-Spellman, J., Segal, A., Vinuela, F., & Saver, JL. (2004) MERCI 1: a phase 1 study of Mechanical Embolus Removal in Cerebral Ischemia. *Stroke*, Vol. 35, No. 12, pp. 2848–2854, ISSN: 1747-4930

Gönner, F., Remonda, L., Mattle, H., Sturzenegger, M., Ozdoba, C., Lövblad, KO., Baumgartner, R., Bassetti, C., & Schroth, G. (1998) Local intra-arterial thrombolysis in acute ischaemic stroke. *Stroke*, Vol. 29, No. 9, pp. 1894 - 1900, ISSN 0039-2499

Gralla, J., Schroth, G., Remonda, L., Nedeltchev, K., Slotboom, J., & Brekenfeld, C. (2006) Mechanical Thrombectomy for Acute Ischemic Stroke: Thrombus-Device Interaction, Efficiency, and Complications In Vivo. *Stroke*, Vol. 37, No. 12, pp. 3019–3024, ISSN: 1747-4930

Hacke, W., Kaste, M., Fieschi, C., Toni, D., Lesaffre, E., von Kummer, R., Boysen, G., Bluhmki, E., Höxter, G., & Mahagne, MH. (1995) Intravenous thrombolysis with recombinant tissue plasminogen acivator for acute hemispheric stroke. The European Cooperative Acute Stroke Study (ECASS). *JAMA*, Vol. 274, No. 13, pp. 1017-1025, ISSN: 0098-7484

Hacke, W., Kaste, M., Fieschi, C., von Kummer, R., Davalos, A., Meier, D., Larrue, V., Bluhmki, E., Davis, S., Donnan, G., Schneider, D., Diez-Tejedor, E., & Trouillas, P. (1998) Randomised double-blind placebo-controlled trial of thrombolytic therapy with intravenous alteplase in acute ischemic stroke (ECASS II). Second European-

Australasian Acute Stroke Study Investigators. *Lancet*, Vol. 352, No. 9136, pp. 1245-1251, ISSN:0140-6736

Hacke, W., Kaste, M., Bluhmki, E., Brozman, M., Dávalos, A., Guidetti, D., Larrue, V., Lees, KR., Medeghri, Z., Machnig, T., Schneider, D., von Kummer, R., Wahlgren, N., Toni, D., & ECASS Investigators. (2008) Trombolysis with alteplase 3 to 4.5 hours after acute ischemic stroke. *N Engl J Med*, Vol. 359, No. 13, pp. 1317-1329, ISSN 0028-4793

Higashida, RT., Furlan, AJ., Roberts, H., Tomsick, T., Connors, B., Barr, J., Dillon, W., Warach, S., Broderick, J., Tilley, B., Sacks, D., Technology Assessment Committee of the American Society of Interventional and Therapeutic Neuroradiology., & Technology Assessment Committee of the Society of Interventional Radiology. (2003) Trial Design and Reporting Standards for Intra-Arterial Cerebral Thrombolysis for Acute Ischemic Stroke. *Stroke*, Vol. 34, No. 8, pp. e109-e137, ISSN 0039-2499

IMS study investigators. (2004) Combined intravenous and intra-arterial recanalization for acute ischemic stroke: the Interventional Management of Stroke study. *Stroke*, Vol. 35, No. 4, pp. 904-911, ISSN 0039-2499

IMS II The IMS II Trial Investigators. (2007) The Interventional Management of Stroke (IMS) II Study. *Stroke*, Vol. 38, No. 7, pp. 2127-2135, ISSN 0039-2499

Khatri, P., Hill, MD., Palesch, YY., Spilker, J., Jauch, EC., Carrozzella, JA., Demchuk, AM., Martin, R., Mauldin, P., Dillon, C., Ryckborst, KJ., Janis, S., Tomsick, TA., Broderick, JP., & Interventional Management of Stroke III Investigators. (2008) Methodology of the Interventional Management of Stroke III Trial. *Int J Stroke*, Vol. 3, No. 2, pp. 130-137, ISSN: 1747-4930

Klijn, CJM. & Hankey, GJ. (2003) Management of acute ischemic stroke: new guidelines from American Stroke Assotiation and European Stroke Iniciative. *Lancet Neurol*, Vol. 2, No. 11, pp. 698-701, ISSN: 1474-4422

Krajina, A., Krajickova, D., Sprinar, Z., Nikolov, DH., Parizkova, R., & Lojik M. (2005) Percutaneous mechanical extraction of emboli in acute ischemic stroke: a case report and literature survey. *Cesko Slov Neurol N*, Vol. 65, No. 101, pp. 51-57, ISSN 1802-4041

Mayer, TE., Hamann, GF., & Brueckmann, H. (2002a) Mechanical extraction of a basilar-artery embolus with the use of flow reversal and a microbasket. *N Engl J Med*, Vol. 347, No. 10, pp. 769–770, ISSN 0028-4793

Mayer, TE, Hamann, GF, & Brueckmann, H. (2002b) Treatment of Basilar Artery Embolism With a Mechanical Extraction Device. Necessity of Flow Reversal. *Stroke*, Vol. 33, No. 9, pp. 2232-2235, ISSN: 0039-2499

Mourand, I., Brunel, H., Constalat, V., Riquelme, C., Lobotesis, K., Milhaud, D., Héroum, C., Arquizan, C., Moynier, M., & Bonafé, A. (2011) Mechanical Thrombectomy in Acute Ischemic Stroke: Catch Device. *AJNR*, Vol. 32, No. 8, pp. 1381-1385, ISSN: 0195-6108

Nesbit, GM., Luh, G., Tien, R., & Barnwell, SL. (2004) New and future endovascular treatment strategies for acute ischemic stroke. *J Vasc Interv Radiol*, Vol. 15, pp. 103–110, ISSN: 1051-0443

NINDS rt-PA Stroke Study Group. (1995). Tissue plasminogen activator for acute ischemic stroke. *N Engl J Med,* Vol. 333, No. 24, pp. 1581–1587, ISSN 0028-4793

Papadakis, M. & Buchan, AM. (2001) Translational vehicles for neuroprotection. *Biochem Soc Trans,* Vol. 34, pp. 1318–1322, ISSN 0300-5127

Penumbra Pivotal StrokeTrial Investigators. (2009) The penumbra pivotal stroke trial: safety and effectivenessof a new generation of mechanical devices for clot removal in intracranial large vessel occlusive disease. *Stroke,* Vol. 40, No. 8, pp. 2761-2768, ISSN: 1747-4930

Qureshi, AI., Ali, Z., Suri, MF., Kim, SH., Shatla, AA., Ringer, AJ., Lopes, DK., Guterman, LR., & Hopkins LN. (2001) Intra-arterial third generation recombinant tissue plasminogen activator (reteplase) for acute ischemic stroke. *Neurosurgery,* Vol. 49, No. 1, pp. 41–48, discussion 48–50, ISSN 0148-396X

Sarimachi, T., Fujii, Y., Tsuchiya, N., Nashimoto, T., Harada, A., Ito, Y., & Tanaka, R. (2004) Recanalization by mechanical embolus disruption during intraarterial trombolysis in the carotid territory. *AJNR,* Vol. 25, No. 8, pp. 1391-1402, ISSN: 0195-6108

Smith, WS., Sung, G., Starkman, S., Saver, JL., Kidwell, CS., Gobin, YP., Lutsep, HL., Nesbit, GM., Grobelny, T., Rymer, MM., Silverman, IE., Higashida, RT., Budzik, RF., Marks, MP., & MERCI Trial Investigators. (2005) Safety and efficacy of mechanical embolectomy in acute ischemic stroke: results of MERCI trial. *Stroke,* Vol. 36, No. 7, pp. 1432-1438, ISSN: 1747-4930

Smith, WS., Sung, G., Saver, J., Budzik, R., Duckwiler, G., Liebeskind, DS., Lutsep, HL., Rymer, MM., Higashida, RT., Starkman, S., Gobin, YP., Multi MERCI Investigators, Frei, D., Grobelny, T., Hellinger, F., Huddle, D., Kidwell, C., Koroshetz, W., Marks, M., Nesbit, G., & Silverman, IE. (2008) Mechanical thrombectomy for acute ischemic stroke: final results of the Multi MERCI trial. *Stroke,* Vol. 39, No. 4, pp. 1205-1212, ISSN: 1747-4930

The Abciximab in Ischemic Stroke Investigators. (2000) Abciximab in acute ischemic stroke: a randomized, double-blind, placebo-controlled, dose-escalation study. *Stroke,* Vol. 31, No. 3, pp. 601-609, ISSN 0039-2499

Tomsick, T., Broderick, J., Carrozella, J., Khatri, P., Hill, M., Palesch, Y., Khoury, J., & Interventional Management of Stroke II Investigators. (2008) Revascularization Results in the Interventional Management of Stroke II Trial. *AJNR,* Vol. 29, pp. 582-587, ISSN: 0195-6108

Wahlgren, N., Ahmed, N., Davalos, A., Ford, GA., Grond, M., Hacke, W., Hennerici, MG., Kaste, M., Kuelkens, S., Larrue, V., Lees, KR., Roine, RO., Soinne, L., Toni, D., Vanhooren, G., & SITS-MOST investigators. (2007) Trombolysis with alteplase for acute ischemic stroke in the Safe Implementation of Thrombolysis in Stroke-Monitoring Study(SITS-MOST): an observational study. *Lancet,* Vol. 369, No. 9558, pp. 275-282, ISSN:0140-6736

Wardlaw, JM., Warlow, CP., & Counsell C. (1997) Systemic review of evidence on thrombolytic therapy for acute ischaemic stroke. *Lancet,* Vol. 350, No. 9078, pp. 607-614, ISSN:0140-6736

Zivin, JA., Fisher, M., De Girolami, U., Hemenway, CC., & Stashak JA. (1985) Tissue plasminogen activator reduces neurological damage after cerebral embolism. *Science*, Vol. 230, No. 4731, pp. 1289–1292, ISSN 0036-8075

11

Decreased Cerebral Perfusion in Carotid Artery Stenosis, Carotid Angioplasty and Its Effects on Cerebral Circulation

Antenor Tavares and José Guilherme Caldas
Universidade de São Paulo; São Paulo, Brazil

1. Introduction

Cervical carotid artery stenosis frequently causes ischemia by embolic phenomena, but stenosis can also induce ischemia via a reduction in blood flow. Cerebral perfusion failure can be caused by reductions in blood flow arising from a stenosis or an occlusive vascular lesion (Bokkers et al., 2008; Wilkinson et al., 2003).

Two major mechanisms are involved in cerebral ischemia (Mohr et al., 1997). Cerebral perfusion failure may occur as the result of flow reduction mediated by stenosis or occlusive vascular damage such as that arising from atheroma plaques that affect the carotid bulb. Perfusion failure affects the distal territories, and it is known as a watershed infarction. The second mechanism of cerebral ischemia is artery-to-artery embolism, which is considered to be the main cause of ischemia associated with plaques in the larger arteries. There are thrombosis in situ and embolic infarctions associated with these plaques, which usually involve the occlusion of intracranial distal vessels.

Ischemia is the physiological term indicating a blood flow that is insufficient for the normal functioning of cells, whereas infarction is the pathological term denoting permanent tissue damage caused by ischemia (Jensen et al., 2005). Classically, atherosclerotic infarctions, particularly those in major arteries, are caused by the formation of thrombosis at the atheroma plaque site or from an embolism that originates from the same plaque. The most common location of plaques in the cerebrovascular circulation is at the carotid bifurcation involving the distal common carotid artery and the first 2 cm of the internal carotid artery.

Treatment options for atherosclerotic carotid artery stenosis include drug therapy and surgical treatment (Mas et al., 2006; NASCET, 1991; Ringleb et al., 2006; Yadav et al., 2004). Clinical treatment includes antiplatelet aggregation agents and statins, along with the control of risk factors such as arterial hypertension, dyslipidemia, hyperglycemia and tobacco use (Bates et al., 2007). Two relevant studies comparing clinical and surgical treatments showed a significant reduction of cerebral infarction risk in the group of patients selected for surgical intervention. In the NASCET (North American Symptomatic Carotid Endarterectomy Trial) study, the stroke risk was significantly reduced by surgery (9% in the surgical group versus 26% in the clinical treatment group) (NASCET, 1991). In another classic trial with asymptomatic patients, the prognosis also improved with surgical treatment (4.8% versus 10.6% with drug treatment) (ACAS Trial investigators, 1995).

Endovascular surgical treatment consists of carotid angioplasty and stenting (CAS), a procedure in which the stenosis is approached through a natural passage inside the vessels themselves. In this technique, a flexible guidewire and catheter are inserted into the arterial system through a peripheral puncture site. The guidewire and catheter are then directed to the site of the stenosis, where the stent and the balloon open the narrowing.

CAS was designed as a less invasive method to treat carotid artery stenosis, and the Carotid Revascularization Endarterectomy Versus Stenting Trial (CREST) showed that angioplasty was as effective as endarterectomy for the prevention of stroke (Bates et al., 2007; Caldas, 2006; Chaturvedi & Yadav, 2006; Gurm et al., 2008; Menon & Stafinski, 2006). CAS is also advantageous because it most often uses a local anesthetic, maintains a constant blood flow to the brain and allows an earlier hospital discharge (Menon & Stafinski, 2006).

Initially, the "status" of cerebral perfusion in patients with carotid artery stenosis, or the changes that occur after angioplasty, was not well known. Research suggesting that the modification of cerebral blood flow following angioplasty possibly induces cerebral reperfusion syndrome (Fukuda et al., 2007) is relatively recent. The course of other diseases can be potentially altered after carotid angioplasty.

Recent publications show cognitive dysfunction in some patients with carotid artery stenosis (Takaiwa et al., 2009; Takaiwa et al., 2006; Turk et al., 2008). Takaiwa et al. observed cognitive improvement in elderly patients after CAS (Takaiwa et al., 2009).

2. Cerebral ischemia and carotid artery stenosis

Carotid artery stenosis is present in 7% of men and 5% of women aged 65 years or older (O'Leary et al., 1992). Severe carotid artery stenosis is a leading cause of cerebral infarction and transitory ischemic attack. The estimated risk of ipsilateral carotid artery infarction in a 5-year period is 4% in the population without carotid artery stenosis, 18% in asymptomatic patients with stenosis of over 75% and 27% in symptomatic patients with stenosis of over 75% (Inzitari et al., 2000).

3. Poor cerebral perfusion in carotid artery stenosis

Cerebral infarction resulting from hypotension, which is the result of cerebral blood flow that is insufficient to comply with metabolic demands, represents about 0.7 to 3.2% of cerebral infarctions. The terms "border-zone" and "watershed" are used synonymously to describe this condition in everyday clinical practice. The appearance of these infarcts varies but can include wedge-shaped lesions with their base on the surface of the cortex, exhibiting a border-zone irrigation topography. There may be a string-of-pearls presentation of the infarcts where there are three or more lesions in the deep white matter within the centrum semiovale, with a linear orientation parallel to the lateral ventricle. These ischemic lesions are identified on magnetic resonance imaging (MRI) studies (Ogata et al., 2005) and on CT scan (Del Sette et al., 2010). This appearance can be unilateral when there is ipsilateral carotid artery stenosis or bilateral when there is a severe and global hemodynamic event or bilateral arterial stenosis (Hamilton, 2005). Regarding pathology, infarctions arising from hypotension occur in border zones and result in encephalomalacia and ulegyria (Hamilton, 2005).

The etiology of infarctions in watershed areas is varied and includes global cerebral injury resulting from perfusion or oxygenation disorders such as prolonged severe hypotension, cardiac arrest with resuscitation, asphyxia and carbon monoxide inhalation. Carotid artery stenosis or occlusion is another etiology that predisposes subjects to border-zone infarctions during episodes of hemodynamic compromise, often with rosary-like patterns. Carotid artery disease infarctions can occur in the cortical and watershed areas as well (Krogias et al., 2010; Momjian-Mayor & Baron, 2005). Although embolisms can occur in the border zones, they are difficult to identify with precision by imaging (Hamilton, 2005; Marks, 2002; Yamada et al., 2010).

In addition to Perfusion-Weighted Imaging (PWI) by MRI, brain perfusion can be evaluated by several techniques including parenchymography, perfusion by helicoidal tomography, perfusion by multislice tomography, transcranial Doppler, intracarotid xenon injection and MRI volume flow quantification analysis. Transcranial doppler studies have the advantage of being non-invasive, but the technique is usually limited to an analysis of a single vessel, usually the middle cerebral artery ipsilateral to the carotid artery stenosis. Doppler studies can also be impaired by cardiac arrhythmias, which make the recorded velocity on the spectral trace unreliable (Kleiser & Widder, 1992; Niesen et al., 2004). Perfusion by tomography has attained a level of technical refinement; however, its application is still restricted to small slices or only one slice per examination (Jongen et al., 2010; Roberts et al., 2000; Waaijer et al., 2007). Evaluation by intracarotid xenon has been limited to research (Bando et al., 2001).

3.1 Digital subtraction angiography

Parenchymography, as described by Theron et al., is an innovative and simple method to evaluate cerebral perfusion. The research on parenchymography included an analysis by angiography with modified windowing to allow the preferential study of the brain blood supply (Theron et al., 1996). Angiography has the capacity to directly assess the degree of carotid artery stenosis and, with the addition of parenchymography, the hemodynamic effects between the brain and its blood supply (Theron et al., 1996).

During the angiographic study of many of our patients, we observed (Figure 1 A) a poor opacification of the anterior and middle cerebral arteries when ipsilateral severe cervical carotid artery stenosis was present.

3.2 Cerebral perfusion by MRI (PWI)

Although perfusion evaluation by MRI is semi-quantitative, its advantages are that MRI is a non-invasive method of evaluating cerebral blood flow that is available in most large centers, avoids ionizing radiation and retains good spatial resolution and good correlation with SPECT (Single-photon emission computed tomography) (Bokkers et al., 2008; Martin, A. J. et al., 2005; Rasmussen et al., 2010; Rohl et al., 2001; Tavares et al., 2010; Van Laar et al., 2007).

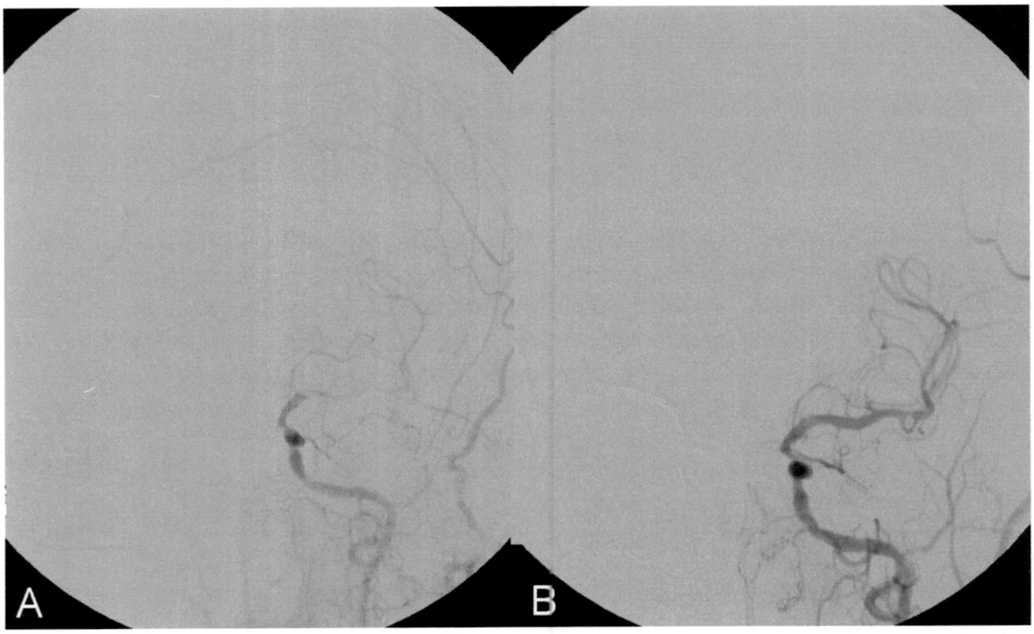

Fig. 1. A 61-year-old patient with severe left cervical carotid artery stenosis. Angiography of the left carotid artery in PA position before angioplasty with stenting shows the slight opacification of the middle cerebral artery (A). Angiogrpahy of the left carotid artery in PA after angioplasty with stenting shows the normal opacification of the left middle cerebral artery (B).

A recent study of PWI by MRI in patients with carotid artery stenosis investigated the relative values of cerebral blood volume (CBV), mean transit time (MTT) and time to peak (TTP). A reduction in the CBV and a time-to-peak delay in the middle cerebral artery territory ipsilateral to the cervical carotid artery stenosis were noted (Tavares et al., 2010). The intracranial blood supply was shown to be compromised by a timing delay (increase in the TTP) of the contrast peak in the territory fed by the middle cerebral artery ipsilateral to the cervical carotid artery stenosis. There was less signal loss (negative wave) in the area analyzed (reduction in CBV), which indicated a lower regional blood volume in the middle cerebral artery territory (Tavares et al., 2010).

The TTP delays in the middle cerebral artery ipsilateral to the cervical stenosis compared with the contralateral territory have also been observed by Gauvrit et al. (2004).

Teng et al. showed that analysis of the TTP is a valuable tool for the assessment of the hemodynamic changes in carotid artery stenosis that correlates well with the mean transit time (MTT) parameter (Teng et al., 2001).

An earlier study of cerebral perfusion in patients with carotid artery stenosis showed MTT prolongation ipsilateral to the carotid artery stenosis and stated that this finding was due to hemodynamic compensation between the cerebral hemispheres (Bozzao et al., 2002).

Even when the perfusion damage is limited, the timing parameters (TTP and MTT) are delayed, but these ischemic alterations are not sufficiently significant to be detected by diffusion MRI. Delayed perfusion parameters indicate infarction risk, whereas alterations in diffusion indicate an already-established lesion.

4. Modification of the cerebral flow after carotid angioplasty with stenting

4.1 Digital subtraction angiography

The analysis of cerebral perfusion by parenchymography has revealed an improvement in cerebral hemodynamic flow achieved by carotid angioplasty (Theron 1996). We observed restored intradural artery opacification (Figure 1 B) and a significant improvement in parenchymography results following CAS.

4.2 PWI

Several authors observed cerebral blood flow modifications after angioplasty (Bozzao et al., 2002; Martin, A. J. et al., 2005; Van Laar et al., 2007; van Laar et al., 2006)

A recent publication by our team that assessed patients with carotid artery stenosis used PWI by MRI to show that there was a discrete increase in rCBV ($p=0.940$) and important reductions in dMTT ($p<0.001$) resulting from the reduction in MTT at the site of the angioplasty.

However, during the evaluation before CAS, we observed a delay in the TTP in the middle cerebral artery (MCA) territory ipsilateral to the carotid stenosis when compared to the contralateral artery. After CAS, there was a remarkable reduction in dTTP ($p=0.019$)and the TPP was earlier at the site of the stent than in the contralateral territory (Tavares et al., 2010).

Timing parameters such as the MTT proved to be the most sensitive and the most reproducible in measuring cerebral hemodynamic alterations (Soinne et al., 2003; Waaijer et al., 2007; Wintermark et al., 2006).

The variations in relative CBV, MTT and TTP are illustrated in Figure 2, where patient improvement is shown through the imaging parameters of PWI by MRI after CAS (Figure 2).

Laar et al. found an increase in cerebral blood flow (CBF) on the side of carotid stenosis, after treatment of the stenosis with CAS or EAC; this increased rate of CBF became similar value to the CBF of the control group (Van Laar et al., 2007).

In our study, we observed an increase in CBV in the cerebral territory ipsilateral to the CAS as well as in the contralateral territory (Tavares et al., 2010). Other authors have documented this same phenomenon in the intracranial territory on the healthy side (Hino et al., 2005; Ko et al., 2005). The treatment of the carotid artery stenosis cannot explain this phenomenon based solely on hemodynamics (Ko et al., 2005). Other unknown factors must play an important role in the perfusion modifications before and after the treatment of carotid artery stenosis. This increased flow in the contralateral hemisphere may occur via collateral circulation through the anterior communicating artery (Tavares et al., 2010) or from the leptomeningeal anastomosis (Matsubara et al., 2009).

Thus, the benefits and eventual risks resulting from an increase in cerebral blood flow may occur in both the ipsilateral and contralateral hemispheres during the treatment of carotid artery stenosis.

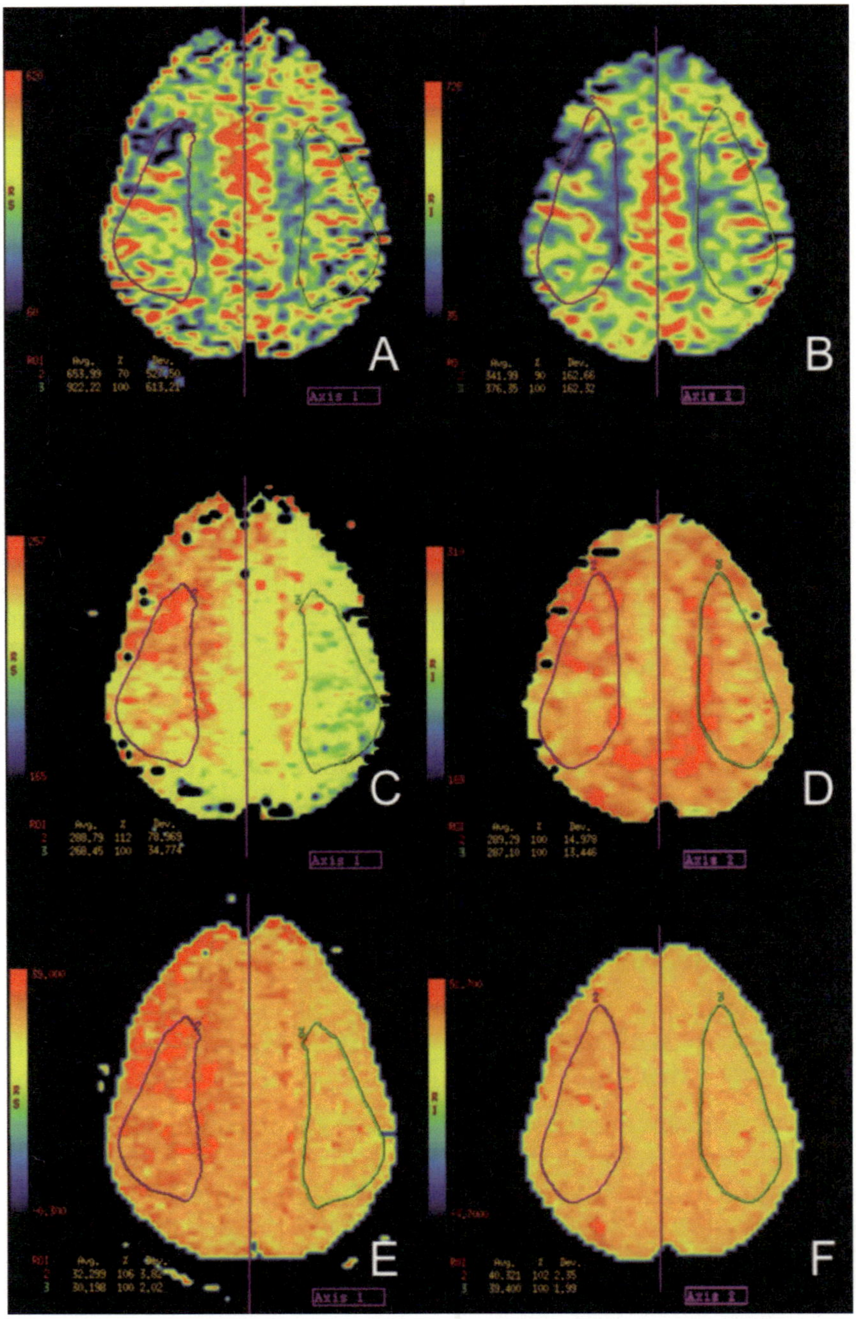

Fig. 2. A 56-year-old patient with severe right cervical carotid artery stenosis evaluated by digital angiography (not shown). Perfusion by MRI was performed in a section above the lateral ventricles. Before carotid angioplasty, the cerebral blood volume (CBV) of the right ACM territory was reduced by 30% compared with the contralateral (**A**), the mean transit time (MTT) was delayed by 12% (**C**) and the time to peak (TTP) was delayed by 6% (**E**). In the same axial plane after CAS, perfusion studies show improved parameters: the right side CBV

became less reduced (10%) compared with the contralateral (**B**), there was no difference in the MTT compared with the left side (**D**) and the TTP delay was lessened to 2% (**F**).

The improvement in cerebral perfusion after CAS, as documented by imaging technologies (Ko et al., 2005; Martin, A. J. et al., 2005; Tavares et al., 2010; Van Laar et al., 2007), has also been documented through questionnaires for dementia that show clinical improvement in scores after CAS(Moftakhar et al., 2005; Takaiwa et al., 2009; Turk et al., 2008).

According to Takaiwa, patients with carotid artery stenosis who scored below average on the questionnaire prior to carotid angioplasty and endarterectomy showed an improvement in cognitive function after the procedures (Takaiwa et al., 2009). Scores were decreased temporarily at 1 week after carotid endarterectomy (CEA), but not after CAS (Takaiwa et al., 2009).

Moftakhar concluded that the improvement in the perfusion parameters shown by MRI is predictive of cognitive improvement after angioplasty (Moftakhar et al., 2005).

These data together suggest that CAS not only prevents stroke but also improves cognitive function (Takaiwa et al., 2009).

In addition to carotid angioplasty, intracranial angioplasty is also associated with perfusion improvement as shown by MRI (Bendok et al., 2010).

According to Takaiwa, rapid improvement in cognitive function occurs following CAS (Takaiwa et al., 2009). Some studies, however, showed a worsening of cognitive function following CEA, which may be associated with hypoperfusion during the cross-clamping of the internal carotid artery (Costin et al., 2002; Heyer et al., 1998; Takaiwa et al., 2009).

Endarterectomy may interrupt the cerebral blood flow at several moments during the procedure, and it certainly does so during the construction of the bypass. However, the treatment of carotid artery stenosis by CAS with cerebral protection filters does not interrupt the cerebral blood flow, which avoids momentary deficits in cerebral perfusion.

Complex self-regulatory mechanisms may be set in motion by severe stenosis and may include microcirculatory vasodilatation or a direct response from the central nervous system. Therefore, it is theoretically possible that the microcirculatory condition that results from long-standing severe proximal stenosis might excite a state of microcirculatory vasodilatation after CAS, causing some blood volume reduction in the ipsilateral territory. This tends to occur during a reduction in the MTT. The equation CBF=CBV/MTT expresses the flow equilibrium (Kluytmans et al., 1998).

5. The risks of the immediate increase in cerebral perfusion following carotid angioplasty

5.1.1 Reperfusion syndrome

Because CAS induces changes in cerebral perfusion, it has the inherent although low risk of complications related to increased cerebral blood flow such as reperfusion syndrome (Matsubara et al., 2009; Morrish et al., 2000; Tavares et al., 2010) and even fatal hemorrhagic complications (Hartmann et al., 2004; Morrish et al., 2000). Patients who receive endarterectomies are exposed to similar risks (Bodenant et al., 2010).

Signal alteration, especially as seen with FLAIR sequence imaging, has been reported and is located in the spaces of the cortical convolutions that occur unilaterally after CAS. This finding may be related to modifications in perfusion that are clinically difficult to correlate (Martin, A. J. et al., 2005; Michel et al., 2001; Wilkinson et al., 2003).

Through MRI studies before and after conventional angioplasty with stenting, we observed hypersignal areas in the subarachnoid space of some of our patients in the FLAIR sequence after CAS. These areas showed gadolinium enhancement in the T1-weighted sequence and did not show signs of ischemia or neurological symptoms upon diffusion (Figure 3).

Wilkinson et al. called this finding "unilateral leptomeningeal enhancement" (Wilkinson et al., 2003). Although they stated that there was no definite cause, it is possible that there is an increase in flow to the territory of the MCA ipsilateral to the stent, and this greater flow might contribute to the leptomeningeal enhancement, especially because the enhancement occurred in the areas fed by the MCA that experienced transit time reductions (Wilkinson et al., 2003). Martin et al. showed similar images ipsilateral to the CAS in the liquor space of the ipsilateral cerebral sulcus in the watershed area (Martin, A. J. et al., 2005).

This asymmetric appearance is not fully understood and may be related to factors such as the leakage of the contrast agent or the possible alteration of the partial pressure of oxygen following treatment (Braga et al., 2003; Michel et al., 2001).

Patients who showed a signal increase on FLAIR in the subarachnoid space ipsilateral to the CAS experienced isolated symptoms (headache, transitory neurological deficit or mental confusion) in the series of Grunwald et al. (2009) These authors also reported that the signal increase on FLAIR is temporary and disappears within 3 to 5 days.

Thurley et al. (2009) described a similar unilateral leptomeningeal enhancement in a patient exhibiting headache, vomiting, seizure and hypertension after endarterectomy.

We believe that the finding described by Grunwald et al. (2009) and Thurley et al. (2009) may be related to a clinical syndrome of hyperperfusion that is not complicated by intracranial hemorrhage. Our patients with leptomeningeal enhancement after CAS (Figure 3) did not exhibit the classical symptoms of cerebral hyperperfusion, such as transitory focal deficit, unilateral migraine and seizures.

Nevertheless, we believe that areas with chronic ischemia have increased perfusion after CAS and that there is some degree of poor hematoencephalic barrier regulation. This notion is grounded on the pattern of dural/sulcal enhancement in border zones.

However, because of the reduction in the timing parameters of perfusion following CAS (Tavares et al., 2010) without a reduction in CBV, cerebral hyperflow with possible complications such as hyperperfusion syndrome can theoretically exist (Ko et al., 2005; Niesen et al., 2004; Wilkinson et al., 2003). This concept follows the equation of hemodynamic equilibrium, where cerebral blood flow is proportional to CBV and inversely proportional to the timing parameter MTT (Kluytmans et al., 1998; Waaijer et al., 2007; Wilkinson et al., 2003).

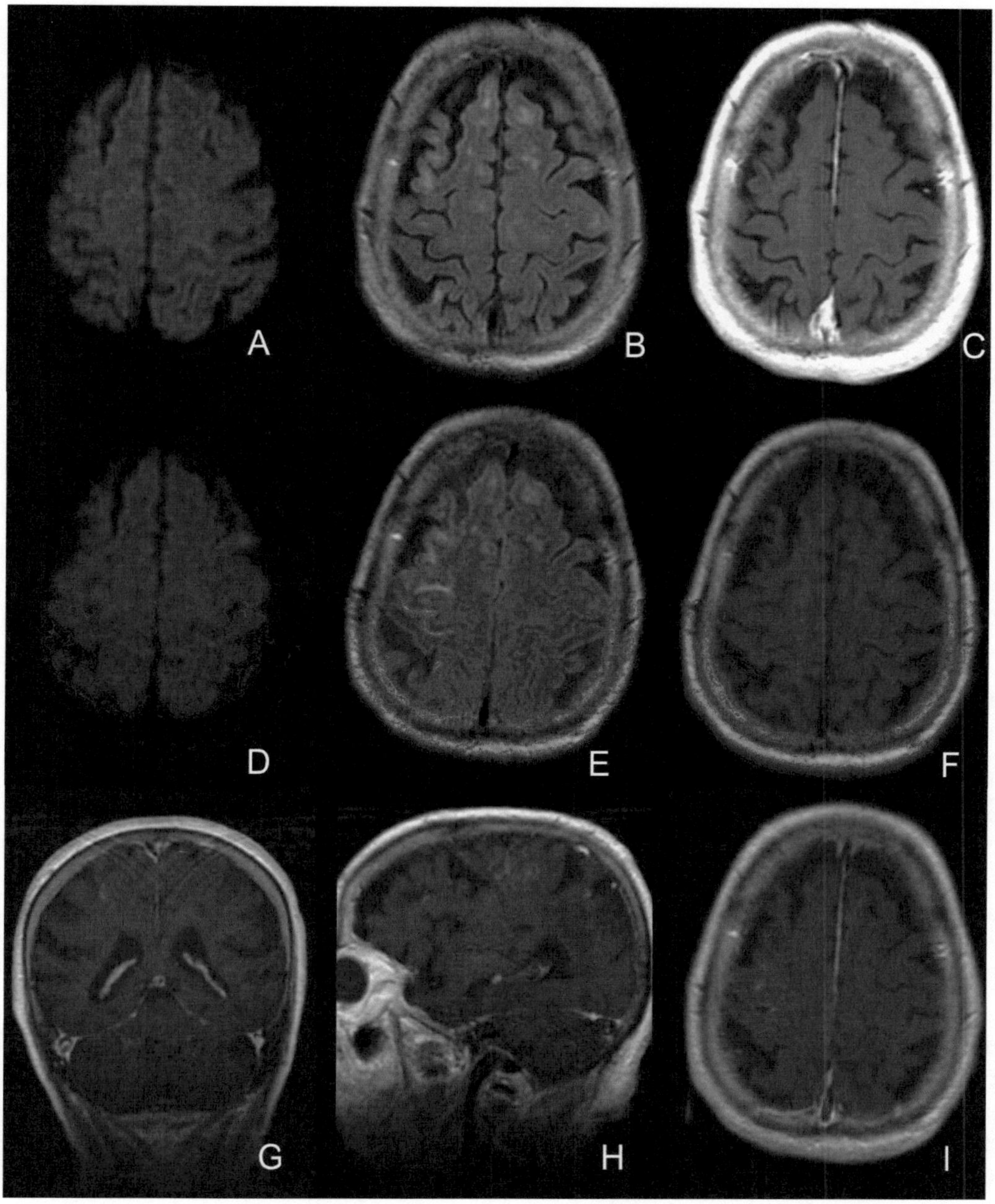

Fig. 3. An 87-year-old patient with right carotid artery stenosis. MRI in diffusion sequence before CAS: DWI (A), FLAIR (B) and T1 with gadolinium (C). Conventional angioplasty of the carotid artery was performed (not shown), and the patient was asymptomatic after endovascular treatment. A new MRI study was performed 12 hours after CAS (D,E,F,G,H,I): note the hypersignal in the central and precentral sulcus of the FLAIR sequence (E) without expression in DWI (D) or T1 without gadolinium (F). T1 sequences with coronal (G), sagittal (H) and axial (I) contrast after CAS: enhancement in the sulcal (pial) space in the right posterior frontal area.

5.1.2 Carotid stenting without postdilatation

Some authors hold that cerebral circulation is somehow altered when there is severe carotid artery stenosis by a "plegic vasodilatation" induced by chronic ischemia (Abou-Chebl et al., 2004; Hartmann et al., 2004; Matsubara et al., 2009).

The study of CBF by intracarotid xenon injection or photon emission tomography showed that patients with a two-fold increase in CBF after endarterectomy were at a greater risk of developing hyperperfusion syndrome (Bando et al., 2001; Henderson et al., 2001).

According to Hirooka et al. (2008), cerebral hyperperfusion after CEA is defined as a 100% increase in CBF in PWI by MRI.

Hosoda et al. based on SPECT data obtained before and after CEA, noted that the abrupt restoration of cerebral perfusion pressure immediately after the surgical correction of a tight carotid stenosis could not be quickly compensated for by vasoconstriction (Hosoda et al., 2001). However, the arterioles presumably contracted to normal size over the long term, thus reducing the CBV.

Matsubara et al. (2009) postulated that after the treatment of carotid stenosis, if the vascular walls of the arterioles were damaged and permeability increased, cerebral edema and convulsions could occur.

Furthermore, rapidly increased CBF and CBV (Matsubara et al., 2009) and a reduction in MTT by CAS (Tavares et al., 2010) and CEA may be risk factors for reperfusion syndrome after the treatment of carotid stenosis (Fukuda et al., 2007).

The technique of CAS has not been standardized, and there are technical variations among different institutions and neurointerventionists. The CAS technique that we call conventional has already been described in the literature (Caldas, 2006; Tavares et al., 2010). Briefly, the procedure is performed in an angiography suite and includes transfemoral access under local anesthesia, sedation and full heparinization (70 U/kg). All patients receive double anti-aggregation agents 5 days before CAS. The activated coagulation time (ACT) is not routinely checked. We initially establish access up to the common carotid artery with a 6F wired sheath or 8F guiding catheter. Next, a cerebral protection system is delicately placed in the internal carotid artery distal to the stenosis. We usually use filter protection because it affords intraprocedural safety against emboli without flow interruption. Most often, the protection filter is advanced without predilatation. Balloon dilatation before stenting is only employed when the filter or the stent cannot pass by the stenosis. Next, a self-expandable stent is advanced over the filter guide and released so as to cover the atheroma plaque. At this point, the stent is able to perform the partial or the subtotal opening of most stenoses and to reduce the friction of the blood flow against the plaque. Thus, the stent protects against emboli. The next step is to utilize balloon angioplasty within the stent to mold and widen the internal diameter of the stent or to continue to reduce the size of the residual stenosis to less than 30%.

Despite the lack of clear standards, we believe that the consensus is to achieve a post-stenting stenosis of less than 30%. The CREST study, which compared CAS and endarterectomy, required a final stenosis of less than 30% at the end of each procedure.

The CAS variation that we employ is a simplification of the technique described above, and it does not employ any pre-stenting or post-stenting balloons. By not fully opening the stent at the procedure, we seek to avoid the instantaneous increase of carotid flux that occurs during conventional angioplasty.

Because carotid angioplasty is most often indicated to avoid atheroembolic ischemic events rather than to correct hemodynamic effects, there is no need for complete stent expansion immediately during angioplasty (Bussiere et al., 2008). The restoration of a normal lumen diameter should be considered a secondary goal.

To avoid an abrupt increase in cerebral perfusion as well as the risks of reperfusion syndrome with its serious complications, we started a case series in which the release of the self-expandable stent was performed without the use of the balloon to completely open the stent (Figure 4). In this group, intra-stent angioplasty (post-dilatation balloon) was used only when the residual stenosis post-stenting did not change compared with the degree of stenosis before the stent placement; intra-stent angioplasty was also used when the residual stenosis was greater than 30%. The force of an self-expanding stent alone can dilate severely stenosed carotid (Lownie et al., 2005), and as the dilation is gradual, it may function protectively against hemorrhagic reperfusion (Maynar et al., 2007).

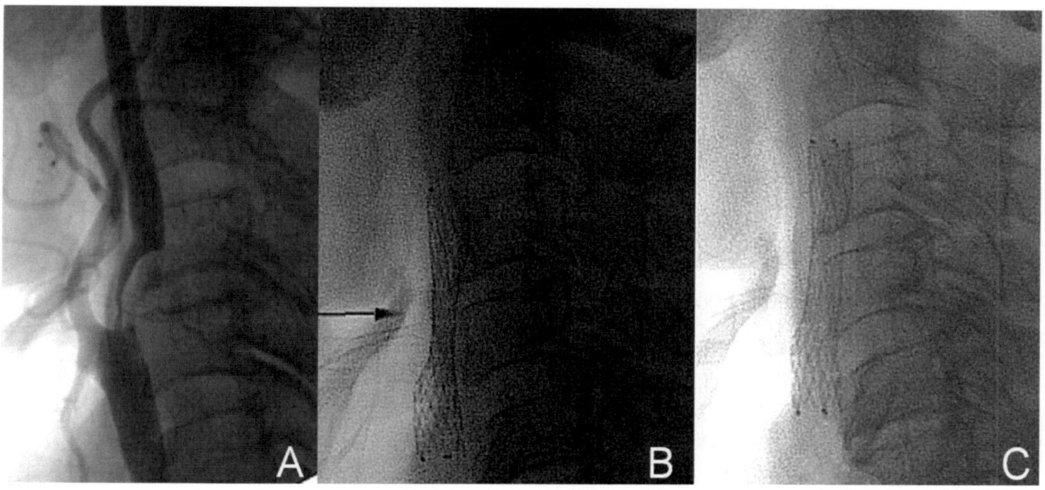

Fig. 4. A 78-year-old patient with severe aortic valve insufficiency and chronic obstructive pulmonary disease who exhibited an episode of amaurosis fugax of the right eye. Angiography showed severe (80%) right internal carotid artery stenosis (A). A 7 X 40 mm self-expandable stent (arrow) (Protégé®, eV3 Inc., Plymouth, Minnesota,USA) for stenting without balloon angioplasty; the immediate result after placement of the stent showed a small residual stenosis of about 30% (b). Lateral radiography 1 month later (C) showed complete expansion of the stent.

In a manner similar to that employed by Baldi et al. (2011), our group followed patients after the placement of the self-expandable stent with cervical radiographic studies at 3, 6 and 12 months. These studies showed a tendency for the stents to completely open. Baldi et al. showed an average reduction of stenosis from 82% to 30% immediately after stenting. In follow-up, however, 14.8% of the patients exhibited restenosis (Baldi et al., 2011).

In the patients who received CAS without balloon inflation, we also noticed the absence of intraprocedural bradycardia and hypotension induced by the compression of the balloon on the carotid baroreceptors. This effect was also observed by Rebellino et al. (Rabellino et al., 2010). This factor may contribute to shorter hospitalizations and increased periprocedural safety.

Each step of an angioplasty procedure includes the risk of thromboembolic events. The procedure with the highest risk of embolism is balloon angioplasty (Martin, J. B. et al., 2001; Men et al., 2002; Vitek et al., 2000). The use of the simplified technique (without balloon dilatation) may reduce such periprocedural risks and improve the clinical results (Jin et al., 2010; Men et al., 2002). The simplified technique may work especially well in cases with soft plaques that can easily rupture and form a large number of emboli.

We consider accentuated calcification, as shown by ultrasound or other imaging methods before angioplasty, to be a relative contraindication for the angioplasty without balloon that we used in our case series.

Other technical variations of CAS have been described in the literature. Maynar et al. (2007) published a series of 100 cases where neither the protection system nor balloon angioplasty was used. They demonstrated an average reduction of stenosis from 79% to 21% immediately after stent release, and in some cases, there was an almost complete opening of the stent. In most cases, however, the stent opening occurred gradually over time with an average final stenosis of 15% at the 6-month radiographic evaluation (Maynar et al., 2007).

Bussière published the results of a series of endovascular treatments of carotid stenosis in which 79% of the patients received a stent-only procedure (Bussiere et al., 2008). The risk of intra-stent restenosis (mostly asymptomatic) was higher than in conventional angioplasty, and residual stenosis below 30% occurred in only 25% of the cases at the end of the procedure. Bussière stated that plaques with circumferential calcification or very severe stenosis (>90%) must be treated with conventional CAS (with postdilatation).

Jin et al. used a 3-to-5-mm balloon after placement of the protection system (Jin et al., 2010). They used predilatation balloons with diameters similar to the luminal caliber of the internal carotid artery distal to the stenosis. After the placement of the stent, the arterial lumen exhibited minimal residual stenosis in about 70% of cases, which indicates that the use of the postdilatation balloon was unnecessary. We find the technique described by Jin et al. (2010) to be inadequate because we believe that the stent has a stabilizing action on the plaque that reduces embolic phenomena. Because the balloon has the potential to induce plaque rupture, it seems contradictory to use the balloon before stenting.

None of the protection systems now in use can perfectly prevent the occurrence of embolism. Emboli have been detected by diffusion MRI when using most of the protection systems described in the literature (Angelini et al., 2002; du Mesnil de Rochemont et al., 2006; Kastrup et al., 2008; Maleux et al., 2006; Muller-Hulsbeck et al., 2003; Schnaudigel et al., 2008; Vos et al., 2005; Yamada et al., 2010).

Indeed, other risk factors for reperfusion syndrome that might be involved include hypertension, severe carotid stenosis (>90%), severe stenosis of the contralateral carotid artery and recent cerebral ischemic or infarction events. The use of anticoagulants and antiplatelet agents may also play a role in this syndrome (Abou-Chebl et al., 2004).

Thus, new studies are required to establish whether or not CAS without postdilatation is a positive factor in clinical results.

5.2.1 The association of carotid stenosis with cerebral aneurysms

Cerebral aneurysms are incidentally found in 5% of adults by autopsy (McCormick & Schochet, 1976). The development of neuroimaging techniques has allowed to increasingly detect unruptured aneurysms in vivo. The overall risk of hemorrhage from incidental aneurysms was initially described as 1% per year by the ISUIA study (ISUIA, 1998). The incidence of unruptured, intracranial aneurysms that are present along with carotid artery stenosis is 3 to 5% (Suh et al., 2011). In most cases, this association involves unruptured aneurysms (Navaneethan et al., 2006). Some risk factors, such as hypertension and tobacco use, influence the development of both carotid artery stenosis and cerebral aneurysms (NASCET Trial Investigators, 1991; Wiebers et al., 2003).

Endarterectomy and conventional carotid artery angioplasty can induce an abrupt increase in cerebral perfusion. Additionally, any angioplasty technique that employs antiplatelet agents and heparin makes subsequent use of surgical correction by craniotomy and clipping immediately after CAS impossible. A further complication is that patients with carotid artery stenosis are usually greater than 50 years of age and frequently exhibit hypertension and diabetes, which are complicating factors in craniotomy and clipping according to the results of the ISUIA trial and other studies (Chen et al., 2004; Cowan et al., 2007; Wiebers et al., 2003; Xu et al., 2011).

Some studies have reported on the stability of small unruptured aneurysms after stenting or endarterectomy (Ladowski et al., 1984; Stern et al., 1979; Suh et al., 2011). These studies, however, do not describe with any uniformity the dimensions of the aneurysms (some studies describe aneurysms as large as 4 mm or 7 mm) or the treatments employed to control hypertension and to achieve hemodynamic stability; furthermore, the periods of observation are short. These studies have a relatively small sample size and include only a single center (Kang et al., 2007; Suh et al., 2011). Because only 3 to 5% of the patients have both lesions (Kappelle et al., 2000; Suh et al., 2011), 2,000 to 3,334 cases must be included to reach the statistically significant number of approximately 100 cases in which patients have both carotid artery stenosis and a cerebral aneurysm.

Stern et al. (1979) reported on one case of a fatal rupture of a cerebral aneurysm after EAC. They suggested that the risk of subarachnoid hemorrhage during endarterectomy might be increased in patients with aneurysms (Stern et al., 1979).

The analysis of the NASCET trial data (1991) by Kapelle et al. (2000) identified one case of a fatal subarachnoid hemorrhage that occurred 6 days after endarterectomy among 25 patients, which, according to the authors, increased the risk of rupture to 4% in this population (Kappelle et al., 2000).

Al-Mubarak et al. (2001) reported on the case of one patient with 90% symptomatic right carotid artery stenosis who developed a massive and fatal intracerebral and subarachnoid hemorrhage (arising from the left sylvian fissure) that occurred immediately after CAS. In this case, there was an occlusion of the left carotid artery.

Cheung et al. (2003) reported a case of a fatal subarachnoid hemorrhage after endarterectomy. The patient had two 4-mm ipsilateral intracranial aneurysms. The authors reported, however, that the autopsy failed to identify areas of rupture in these aneurysms.

Hartmann et al. (2004) reported a case of a fatal subarachnoid hemorrhage after the treatment of carotid artery stenosis. Their report did not include the complete angiography of the four main cerebral arteries. However, we believe that CAS increases the cerebral blood flow in the treated territory and in the other intracranial territories via the polygon collateral circulation. Hartmann et al. (2004) further reported that necropsy could not definitively rule out saccular aneurysms.

Riphagen et al. reported a case of one patient with an occlusion of the left carotid artery and symptomatic stenosis of the right carotid artery that was treated with endarterectomy. This patient developed a 6-mm aneurysm in the anterior communicating artery 10 days after endarterectomy with subarachnoid and intraventricular hemorrhage. The patient underwent a craniotomy and clipping (Riphagen & Bernsen, 2009). The annual risk of rupture from incidental aneurysms smaller than 7 mm in the anterior circulation is close to 0% in the ISUIA 2003 study (Wiebers et al., 2003). The case described by Riphagen et al. thus represents a negative shift in the natural course of this type of aneurysm.

Although hypertension is a classic factor that contributes to the rupture of cerebral aneurysms (Obray et al., 2003; Vlak et al., 2011), there is no consensus concerning the effects of CAS on the course of unruptured intracranial aneurysms. However, intracranial hemorrhages are associated with high morbidity and high mortality during the treatment of carotid artery stenosis (Cheung et al., 2003; Hartmann et al., 2004). We therefore conclude that there should be no delay in the treatment of cerebral aneurysms in hypertensive patients during the treatment of carotid stenosis. Theoretically, conditions such as hypertension and increased cerebral blood flow may contribute to the rupture of aneurysms. Other factors inherent to CAS can also worsen the course of hemorrhage, such as the use of antiplatelet agents and the non-reversal of heparin after the CAS.

Badruddin et al. (2010) suggested that intracranial aneurysms might be partially protected by the reduced flow and pressure in the intracranial arteries when there is carotid artery stenosis (Badruddin et al., 2010). Another author suggested that the intentional partial occlusion of the carotid artery might protect cerebral aneurysms, and a reduction of intracranial aneurysms has been observed (Cronqvist et al., 1964)

5.2.2 Minimally invasive treatments for carotid stenosis and cerebral aneurysm

Because of the potential risk for aneurysm rupture after carotid revascularization surgery (endarterectomy or stenting) arising from the increase in cerebral blood flow (Hartmann et al., 2004; Suh et al., 2011) and because there is a minimally invasive and safe technique for endovascular treatment (coiling) of cerebral aneurysms (Cowan et al., 2007; Higashida et al., 2007; Molyneux et al., 2009; Raja et al., 2008; Spelle & Pierot, 2008; van Rooij et al., 2006; van Rooij & Sluzewski, 2006), we believe that patients must be immediately protected by performing embolization (coiling) consecutively after CAS.

In patients for whom carotid artery stenosis is initially diagnosed and further investigation reveals an incidental aneurysm, the treatment of the aneurysm is based upon its rupture risk, which is calculated by the diameter and localization of the aneurysm and the age of each patient (Wiebers et al., 2003). However, most neuroradiology interventionists believe that aneurysms that are 3 mm or larger must be treated by coiling when it can be safely performed (Fang 2009). The decision to treat aneurysms is also based on the geometry of the lesion (including the neck-dome ratio) and on the availability of experienced specialists in interventionist neuroradiology.

In patients where a cerebral aneurysm was initially diagnosed and further investigation revealed asymptomatic severe carotid stenosis, the ACAS trial showed that treatment of asymptomatic carotid stenosis is beneficial (ACAS Trial investigators, 1995).

We prefer to treat these patients during a single procedure. We treat the carotid artery stenosis before the cerebral aneurysm to avoid the crossing of the stenosis by the endovascular materials that are necessary for the treatment of the aneurysm and that cause the iatrogenic embolism of parts of the cervical plaque (Figure 5). In such patients, we use antiplatelet agents by the oral route (or nasogastric tube) just one hour before angioplasty (clopidogrel 300 mg and aspirin 300 mg). We employ intravenous heparin to maintain an ACT between 200 and 300 seconds.

We perform the angioplasty with a closed-cell stent, since access of catheters to the intracranial carotid seems to us easier. We do not routinely use a post-dilatation balloon (intra-stent before removing the embolic protection device) because we want to limit the increase of the blood flow towards the intracranial territory. However, if the arterial lumen is too small for catheters up to 6F (3 mm) to pass, we perform a stent dilatation with a 4- or 5-mm balloon. After completion of the angioplasty, a 6F guide catheter is placed in the common carotid artery below the stent followed by a microcatheter and micro-guide that are carefully advanced through the stent and into the internal carotid artery distal to the stent. The microcatheter and micro-guide are then advanced up to the aneurysm, and coiling is performed according to the conventional embolization technique. Alternatively, the guide catheter can be carefully advanced over the microcatheter through the stent to the internal carotid distal to the stent. In cases where we employed CAS immediately followed by coiling, the aneurysm was on the same side as the carotid stenosis.

Aneurysms located outside of the circulation subjected to CAS also received increased blood flow after CAS, probably via the Willis polygon (Tavares et al., 2010). Theoretically, then, these aneurysms have an increased risk of rupture and can also be treated by CAS and coiling together.

Navaneethan et al. (2006) reported a case of bilateral carotid stenosis and a 25-mm aneurysm in the left middle cerebral artery. They performed an angioplasty of the left carotid artery and coiling in one single procedure. They believe that craniotomy and clipping of the aneurysm may reduce the cerebral blood flow and thus worsen the eventual hypoflow resulting from the concomitant carotid artery stenosis (Navaneethan et al., 2006).

Iwata et al. reported the case of a patient with 60% asymptomatic stenosis of the left cervical carotid artery with a 6-mm unruptured aneurysm in the ipsilateral intracranial carotid artery. They treated the aneurysm first by coiling, and 1 month later, they performed the CAS (Iwata 2008). A disadvantage of this approach to treatment arises from the potential risks due to the general anesthesia required for the cooling of the patient to reduce the

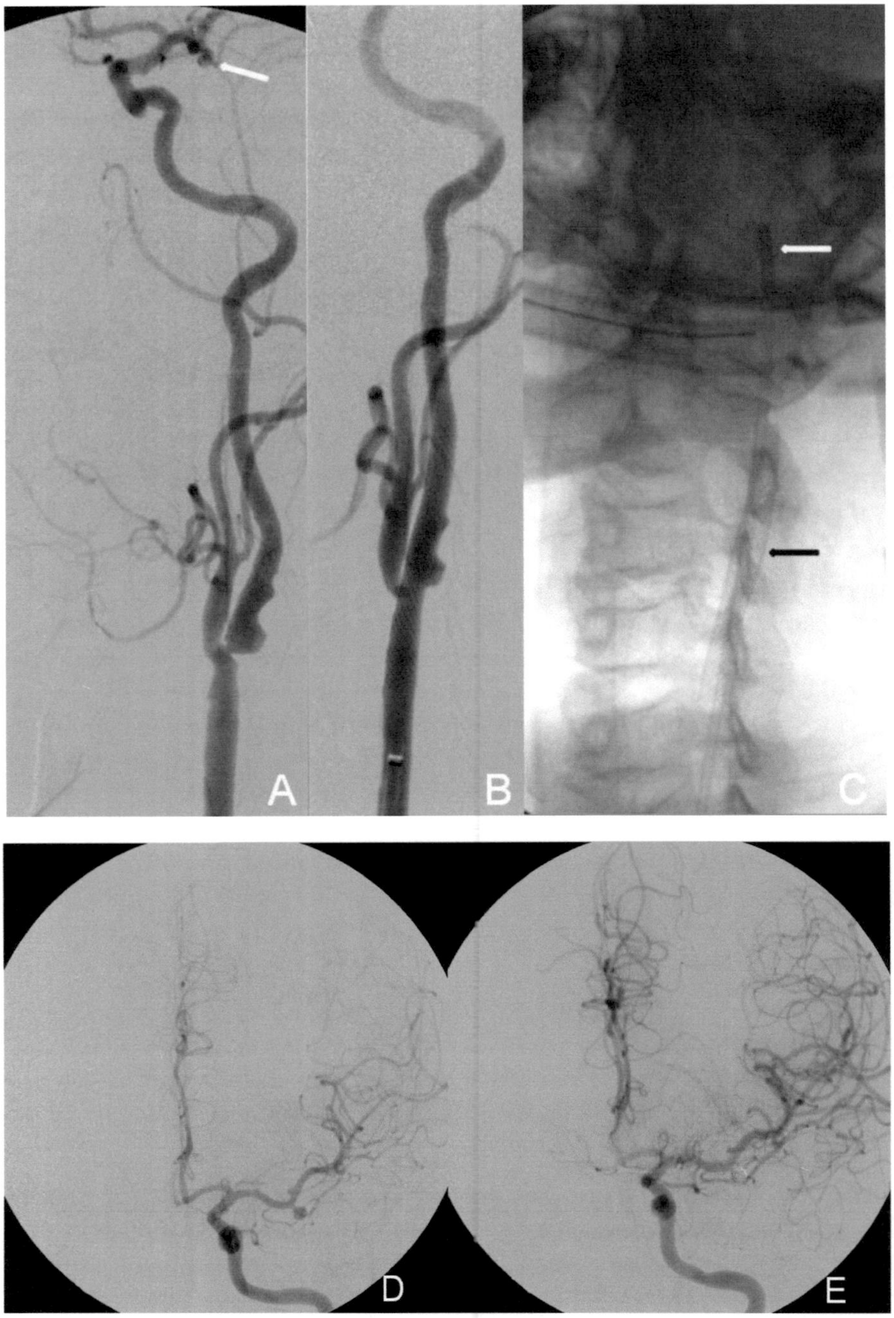

Fig. 5. A 52-year-old patient with a family history of deaths (brother and mother) from subarachnoid hemorrhage presented with an unruptured saccular aneurysm with a narrow neck in the left middle cerebral artery and severe ipsilateral cervical carotid artery stenosis. Left oblique incidence angiography showed (A) severe (70%) left cervical carotid artery stenosis by plaque with ulcers and an aneurysm in the same intracranial territory (arrow). Treatment with a self-expandable stent under anesthetic sedation employing a 6F sheath

resulted in the satisfactory opening of the stenosis (B). During the same procedure, the patent was subjected to general anesthesia, and a 6F guide catheter (white arrow) was advanced through the stent (black arrow, C) to the petrous segment of the carotid artery. Full embolization of the aneurysm was completed (F).

cerebral blood flow (Kappelle et al., 2000; Ladowski et al., 1984). This approach to treatment decreases the stenotic lumen generated by the atheroma plaque during embolization because of the passage of embolization materials through the stenosis, which can hinder the cerebral blood flow. We further note that during coiling, the patient was unnecessarily exposed to the risk of embolism because of the crossing of embolization materials over the cervical atheroma plaque. The authors employed antiplatelet agents, but the additional use of statins might afford higher protection.

The risk of hematoma in the inguinal area of coiling is avoided by use the puncture of the femoral artery for CAS. The single femoral approach is safe, and it decreases the risks inherent to each separate procedure.

The technique we described above has some limitations. The use of antiplatelet agents in carotid stenting might increase the risk if accidental rupture of the aneurysm occurs during coiling. Because the treatment of the carotid artery stenosis is performed first, a sudden increase in cerebral perfusion might theoretically increase the risk of aneurysm rupture; therefore, special efforts are required to diminish the interval between CAS and cooling. For these reasons, we believe that CAS and coiling should be performed exclusively by experienced neuroradiologists. A further risk of the combined CAS and coiling technique is the increase in exposure to ionizing radiation and contrast. However, experienced physicians are agile and employ less contrast, thus reducing the exposure to radiation and the risk of kidney damage.

Finally, in addition to aneurysms, angiographically hidden micro MAVs may theoretically cause intracranial hemorrhages for unknown reasons (Berker et al., 2003; Cheung et al., 2003; Hartmann et al., 2004) after treatment of the carotid artery stenosis. These patients may benefit from the treatment of carotid stenosis without the angioplasty balloon to prevent intracranial hemorrhages.

6. Conclusion

Although angioplasty was initially designed as a minimally invasive therapeutic approach to prevent ischemic events, several reports mention its ability to restore cerebral perfusion (Tavares et al., 2010).

The endovascular technique for the treatment of carotid artery stenosis is still being developed. Stent placement without post-dilatation by balloon allows the gradual dilatation of the stenosis and may reduce the risk of reperfusion syndrome and possibly also reduce the embolic load to the protection system. Confirmation of this hypothesis requires further studies.

Although the relationship with the possible rupture of aneurysms after treatment of carotid stenosis is not clear, it seems safe to us to suggest that cases with simultaneous carotid stenosis and cerebral aneurysm should be treated by endovascular techniques (CAS and coiling) in a single procedure to avoid the increase in cerebral flow, which is a risk factor for aneurysm bleeding. Again, further studies are required to confirm this hypothesis.

7. Acronyms

ACT - achieve activated clotting time

CAS - carotid angioplasty and stenting

CBF- regional cerebral blood flow

CBV - regional cerebral blood volume

CEA - carotid endarterectomy

Coiling – endovascular treatment of cerebral aneurysms

CREST - Carotid Revascularization Endarterectomy Versus Stenting Trial

DWI - Diffusion -Weighted Imaging

FLAIR - Fluid Attenuated Inversion Recovery

MCA - middle cerebral artery

MRI – magnetic resonance imaging

MTT - regional mean transit time

NASCET - North American Symptomatic Carotid Endarterectomy Trial

PWI - perfusion weighted images or cerebral perfusion by MRI

PWI - Perfusion-Weighted Imaging

SPECT - Single-photon emission computed tomography

TTP- regional time to peak

8. References

Abou-Chebl, A., Yadav, J.S., Reginelli, J.P., Bazer, C., Bhatt, D. & Krieger, D.W. (2004). Intracranial Hemorrhage and Hyperperfusion Syndrome Following Carotid Artery Stenting: Risk Factors, Prevention, and Treatment. *Journal of the American College of Cardiology*, Vol. 43, No. 9, (Epub 2004/05/04), pp. 1596-1601, ISSN 0735-1097 (Print) 0735-1097 (Linking).

ACAS Trial investigators. (1995). Endarterectomy for Asymptomatic Carotid Artery Stenosis. Executive Committee for the Asymptomatic Carotid Atherosclerosis Study. *JAMA : the journal of the American Medical Association*, Vol. 273, No. 18, (Epub 1995/05/10), pp. 1421-1428, ISSN 0098-7484 (Print) 0098-7484 (Linking).

Angelini, A., Reimers, B., Della Barbera, M., Sacca, S., Pasquetto, G., Cernetti, C., Valente, M., Pascotto, P. & Thiene, G. (2002). Cerebral Protection During Carotid Artery Stenting: Collection and Histopathologic Analysis of Embolized Debris. *Stroke; a journal of cerebral circulation*, Vol. 33, No. 2, (Epub 2002/02/02), pp. 456-461, ISSN 1524-4628 (Electronic) 0039-2499 (Linking).

Badruddin, A., Teleb, M.S., Abraham, M.G., Taqi, M.A. & Zaidat, O.O. (2010). Safety and Feasibility of Simultaneous Ipsilateral Proximal Carotid Artery Stenting and Cerebral Aneurysm Coiling. *Frontiers in neurology*, Vol. 1, No. (Epub 2011/01/06), pp. 120, ISSN 1664-2295 (Electronic) 1664-2295 (Linking).

Baldi, S., Zander, T., Rabellino, M., Gonzalez, G. & Maynar, M. (2011). Carotid Artery Stenting without Angioplasty and Cerebral Protection: A Single-Center Experience with up to 7 Years' Follow-Up. *AJNR. American journal of neuroradiology*, Vol. 32, No. 4, (Epub 2011/02/26), pp. 759-763, ISSN 1936-959X (Electronic) 0195-6108 (Linking).

Bando, K., Satoh, K., Matsubara, S., Nakatani, M. & Nagahiro, S. (2001). Hyperperfusion Phenomenon after Percutaneous Transluminal Angioplasty for Atherosclerotic Stenosis of the Intracranial Vertebral Artery. Case Report. *Journal of neurosurgery*, Vol. 94, No. 5, (Epub 2001/05/17), pp. 826-830, ISSN 0022-3085 (Print) 0022-3085 (Linking).

Bates, E.R., Babb, J.D., Casey, D.E., Jr., Cates, C.U., Duckwiler, G.R., Feldman, T.E., Gray, W.A., Ouriel, K., Peterson, E.D., Rosenfield, K., Rundback, J.H., Safian, R.D., Sloan, M.A. & White, C.J. (2007). Accf/Scai/Svmb/Sir/Asitn 2007 Clinical Expert Consensus Document on Carotid Stenting: A Report of the American College of Cardiology Foundation Task Force on Clinical Expert Consensus Documents (Accf/Scai/Svmb/Sir/Asitn Clinical Expert Consensus Document Committee on Carotid Stenting). *Journal of the American College of Cardiology*, Vol. 49, No. 1, (Epub 2007/01/09), pp. 126-170, ISSN 1558-3597 (Electronic) 0735-1097 (Linking).

Bendok, B.R., Sherma, A.K., Hage, Z.A., Das, S., Naidech, A.M., Surdell, D.L., Adel, J.G., Shaibani, A., Batjer, H.H., Carroll, T.J. & Walker, M. (2010). Periprocedural MRI Perfusion Imaging to Assess and Monitor the Hemodynamic Impact of Intracranial Angioplasty and Stenting for Symptomatic Atherosclerotic Stenosis. *Journal of Clinical Neuroscience: Official Journal of the Neurosurgical Society of Australasia*, Vol. 17, No. 1, (Epub 2009/12/17), pp. 54-58, ISSN 1532-2653 (Electronic) 0967-5868 (Linking).

Berker, M., Ulus, A., Palaoglu, S., Soylemezoglu, F., Ay, H. & Cekirge, S. (2003). Intracranial Haemorrhage Probably Due to an Angiographically Occult Avm after Carotid Stenting. A Case Report. *Interventional neuroradiology : journal of peritherapeutic neuroradiology, surgical procedures and related neurosciences*, Vol. 9, No. 3, (Epub 2003/09/30), pp. 315-320, ISSN 1591-0199 (Print) 1591-0199 (Linking).

Bodenant, M., Leys, D. & Lucas, C. (2010). Isolated Subarachnoidal Hemorrhage Following Carotid Endarterectomy. *Case reports in neurology*, Vol. 2, No. 2, (Epub 2010/07/31), pp. 80-84, ISSN 1662-680X (Electronic) 1662-680X (Linking).

Bokkers, R.P., van Laar, P.J., van de Ven, K.C., Kapelle, L.J., Klijn, C.J. & Hendrikse, J. (2008). Arterial Spin-Labeling Mr Imaging Measurements of Timing Parameters in Patients with a Carotid Artery Occlusion. *AJNR. American journal of neuroradiology*, Vol. 29, No. 9, (Epub 2008/08/15), pp. 1698-1703, ISSN 1936-959X (Electronic) 0195-6108 (Linking).

Bozzao, A., Floris, R., Gaudiello, F., Finocchi, V., Fantozzi, L.M. & Simonetti, G. (2002). Hemodynamic Modifications in Patients with Symptomatic Unilateral Stenosis of the Internal Carotid Artery: Evaluation with Mr Imaging Perfusion Sequences. *AJNR. American journal of neuroradiology*, Vol. 23, No. 8, (Epub 2002/09/12), pp. 1342-1345, ISSN 0195-6108 (Print) 0195-6108 (Linking).

Braga, F.T., da Rocha, A.J., Hernandez Filho, G., Arikawa, R.K., Ribeiro, I.M. & Fonseca, R.B. (2003). Relationship between the Concentration of Supplemental Oxygen and Signal Intensity of Csf Depicted by Fluid-Attenuated Inversion Recovery Imaging. *AJNR. American journal of neuroradiology*, Vol. 24, No. 9, (Epub 2003/10/17), pp. 1863-1868, ISSN 0195-6108 (Print) 0195-6108 (Linking).

Bussiere, M., Pelz, D.M., Kalapos, P., Lee, D., Gulka, I., Leung, A. & Lownie, S.P. (2008). Results Using a Self-Expanding Stent Alone in the Treatment of Severe Symptomatic Carotid Bifurcation Stenosis. *Journal of neurosurgery*, Vol. 109, No. 3, (Epub 2008/09/02), pp. 454-460, ISSN 0022-3085 (Print) 0022-3085 (Linking).

Caldas, J.G. (2006). Angioplastie Carotidienne Avec Stent Et Protection Cérébrale. *e-mémoires de l'Académie Nationale de Chirurgie*, Vol. 5, No. 4, pp. 1-4.

Chaturvedi, S. & Yadav, J.S. (2006). The Role of Antiplatelet Therapy in Carotid Stenting for Ischemic Stroke Prevention. *Stroke; a journal of cerebral circulation*, Vol. 37, No. 6, (Epub 2006/04/22), pp. 1572-1577, ISSN 1524-4628 (Electronic) 0039-2499 (Linking).

Chen, P.R., Frerichs, K. & Spetzler, R. (2004). Current Treatment Options for Unruptured Intracranial Aneurysms. *Neurosurgical focus*, Vol. 17, No. 5, (Epub 2005/01/07), pp. E5, ISSN 1092-0684 (Electronic) 1092-0684 (Linking).

Cheung, R.T., Eliasziw, M., Meldrum, H.E., Fox, A.J. & Barnett, H.J. (2003). Risk, Types, and Severity of Intracranial Hemorrhage in Patients with Symptomatic Carotid Artery Stenosis. *Stroke; a journal of cerebral circulation*, Vol. 34, No. 8, (Epub 2003/06/28), pp. 1847-1851, ISSN 1524-4628 (Electronic) 0039-2499 (Linking).

Costin, M., Rampersad, A., Solomon, R.A., Connolly, E.S. & Heyer, E.J. (2002). Cerebral Injury Predicted by Transcranial Doppler Ultrasonography but Not Electroencephalography During Carotid Endarterectomy. *Journal of neurosurgical anesthesiology*, Vol. 14, No. 4, (Epub 2002/10/03), pp. 287-292, ISSN 0898-4921 (Print) 0898-4921 (Linking).

Cowan, J.A., Jr., Ziewacz, J., Dimick, J.B., Upchurch, G.R., Jr. & Thompson, B.G. (2007). Use of Endovascular Coil Embolization and Surgical Clip Occlusion for Cerebral Artery Aneurysms. *Journal of neurosurgery*, Vol. 107, No. 3, (Epub 2007/09/25), pp. 530-535, ISSN 0022-3085 (Print) 0022-3085 (Linking).

Cronqvist, S., Lundberg, N. & Troupp, H. (1964). Temporary or Incomplete Occlusion of the Carotid Artery in the Neck for the Treatment of Intracranial Arterial Aneurysms. *Neurochirurgia*, Vol. 7, No. (Epub 1964/08/01), pp. 146-151, ISSN 0028-3819 (Print) 0028-3819 (Linking).

Del Sette M., Eliasziw M., Streifler J.Y., Hachinski V.C., Fox A.J., Barnett H.J. (2000). Internal borderzone infarction: a marker for severe stenosis in patients with symptomatic internal carotid artery disease. For the North American Symptomatic Carotid Endarterectomy (NASCET) Group. *Stroke*, Vol.31, No 3, (Epub 1999/28/12) pp. 631-636, ISSN 0039-2499 (Print) 1524-4628 (Linking).

du Mesnil de Rochemont, R., Schneider, S., Yan, B., Lehr, A., Sitzer, M. & Berkefeld, J. (2006). Diffusion-Weighted Mr Imaging Lesions after Filter-Protected Stenting of High-Grade Symptomatic Carotid Artery Stenoses. *AJNR. American journal of neuroradiology*, Vol. 27, No. 6, (Epub 2006/06/16), pp. 1321-1325, ISSN 0195-6108 (Print) 0195-6108 (Linking).

Fukuda, T., Ogasawara, K., Kobayashi, M., Komoribayashi, N., Endo, H., Inoue, T., Kuzu, Y., Nishimoto, H., Terasaki, K. & Ogawa, A. (2007). Prediction of Cerebral Hyperperfusion after Carotid Endarterectomy Using Cerebral Blood Volume Measured by Perfusion-Weighted Mr Imaging Compared with Single-Photon Emission Ct. *AJNR. American journal of neuroradiology*, Vol. 28, No. 4, (Epub 2007/04/10), pp. 737-742, ISSN 0195-6108 (Print) 0195-6108 (Linking).

Gauvrit, J.Y., Delmaire, C., Henon, H., Debette, S., al Koussa, M., Leys, D., Pruvo, J.P. & Leclerc, X. (2004). Diffusion/Perfusion-Weighted Magnetic Resonance Imaging after Carotid Angioplasty and Stenting. *Journal of neurology*, Vol. 251, No. 9, (Epub 2004/09/17), pp. 1060-1067, ISSN 0340-5354 (Print) 0340-5354 (Linking).

Grunwald, I.Q., Politi, M., Reith, W., Krick, C., Karp, K., Zimmer, A., Struffert, T., Roth, C., Kuhn, A.L., Haass, A. & Papanagiotou, P. (2009). Hyperperfusion Syndrome after Carotid Stent Angioplasty. *Neuroradiology*, Vol. 51, No. 3, (Epub 2008/12/24), pp. 169-174, ISSN 1432-1920 (Electronic) 0028-3940 (Linking).

Gurm, H.S., Nallamothu, B.K. & Yadav, J. (2008). Safety of Carotid Artery Stenting for Symptomatic Carotid Artery Disease: A Meta-Analysis. *European heart journal*, Vol. 29, No. 1, (Epub 2007/09/21), pp. 113-119, ISSN 0195-668X (Print) 0195-668X (Linking).

Hamilton, E.B. (2005). Intracranial Atherosclerosis, Atherosclerosis, Acute Cerebral Ischemia- Infarction, Hypotensive Cerebral Infarction, In: *Diagnostic Imaging Brain*, A.G. Osborn, S. Blaser and K. Salzman, pp. I-4-24 to I-24-35 and I-24-76 to I-24-99, Amirsys / Elservier Saunders,

Hartmann, M., Weber, R., Zoubaa, S., Schranz, C. & Knauth, M. (2004). Fatal Subarachnoid Hemorrhage after Carotid Stenting. *Journal of neuroradiology. Journal de neuroradiologie*, Vol. 31, No. 1, (Epub 2004/03/18), pp. 63-66, ISSN 0150-9861 (Print) 0150-9861 (Linking).

Henderson, R.D., Phan, T.G., Piepgras, D.G. & Wijdicks, E.F. (2001). Mechanisms of Intracerebral Hemorrhage after Carotid Endarterectomy. *Journal of neurosurgery*, Vol. 95, No. 6, (Epub 2002/01/05), pp. 964-969, ISSN 0022-3085 (Print) 0022-3085 (Linking).

Heyer, E.J., Adams, D.C., Solomon, R.A., Todd, G.J., Quest, D.O., McMahon, D.J., Steneck, S.D., Choudhri, T.F. & Connolly, E.S. (1998). Neuropsychometric Changes in Patients after Carotid Endarterectomy. *Stroke; a journal of cerebral circulation*, Vol. 29, No. 6, (Epub 1998/06/17), pp. 1110-1115, ISSN 0039-2499 (Print) 0039-2499 (Linking).

Higashida, R.T., Lahue, B.J., Torbey, M.T., Hopkins, L.N., Leip, E. & Hanley, D.F. (2007). Treatment of Unruptured Intracranial Aneurysms: A Nationwide Assessment of Effectiveness. *AJNR. American journal of neuroradiology*, Vol. 28, No. 1, (Epub 2007/01/11), pp. 146-151, ISSN 0195-6108 (Print) 0195-6108 (Linking).

Hino, A., Tenjin, H., Horikawa, Y., Fujimoto, M. & Imahori, Y. (2005). Hemodynamic and Metabolic Changes after Carotid Endarterectomy in Patients with High-Degree Carotid Artery Stenosis. *Journal of stroke and cerebrovascular diseases : the official journal of National Stroke Association*, Vol. 14, No. 6, (Epub 2007/10/02), pp. 234-238, ISSN 1532-8511 (Electronic) 1052-3057 (Linking).

Hirooka, R., Ogasawara, K., Sasaki, M., Yamadate, K., Kobayashi, M., Suga, Y., Yoshida, K., Otawara, Y., Inoue, T. & Ogawa, A. (2008). Magnetic Resonance Imaging in Patients with Cerebral Hyperperfusion and Cognitive Impairment after Carotid Endarterectomy. *Journal of neurosurgery*, Vol. 108, No. 6, (Epub 2008/06/04), pp. 1178-1183, ISSN 0022-3085 (Print) 0022-3085 (Linking).

Hosoda, K., Kawaguchi, T., Shibata, Y., Kamei, M., Kidoguchi, K., Koyama, J., Fujita, S. & Tamaki, N. (2001). Cerebral Vasoreactivity and Internal Carotid Artery Flow Help to Identify Patients at Risk for Hyperperfusion after Carotid Endarterectomy. *Stroke; a journal of cerebral circulation*, Vol. 32, No. 7, (Epub 2001/07/07), pp. 1567-1573, ISSN 1524-4628 (Electronic) 0039-2499 (Linking).

Inzitari, D., Eliasziw, M., Gates, P., Sharpe, B.L., Chan, R.K., Meldrum, H.E. & Barnett, H.J. (2000). The Causes and Risk of Stroke in Patients with Asymptomatic Internal-Carotid-Artery Stenosis. North American Symptomatic Carotid Endarterectomy Trial Collaborators. *The New England journal of medicine*, Vol. 342, No. 23, (Epub 2000/06/08), pp. 1693-1700, ISSN 0028-4793 (Print) 0028-4793 (Linking).

ISUIA. (1998). Unruptured Intracranial Aneurysms--Risk of Rupture and Risks of Surgical Intervention. International Study of Unruptured Intracranial Aneurysms Investigators. *The New England journal of medicine*, Vol. 339, No. 24, (Epub 1998/12/29), pp. 1725-1733, ISSN 0028-4793 (Print) 0028-4793 (Linking).

Jensen, M.C., Brant-Zawadki, M.N. & Jacobs, B.C. (2005). Isquemia, In: *Ressonância Magnética*, D.D. Stark and W.G. Bradley Jr, pp. 1255-1276, Revinter, Rio de Janeiro.

Jin, S.C., Kwon, O.K., Oh, C.W., Jung, C., Han, M.G., Bae, H.J., Lee, S.H., Jung, Y.S., Han, M.H. & Kang, H.S. (2010). A Technical Strategy for Carotid Artery Stenting: Suboptimal Prestent Balloon Angioplasty without Poststenting Balloon Dilatation. *Neurosurgery*, Vol. 67, No. 5, (Epub 2010/09/28), pp. 1438-1442; discussion 1442-1433, ISSN 1524-4040 (Electronic) 0148-396X (Linking).

Jongen, L.M., Hendrikse, J., Moll, F.L., Mali, W.P. & van der Worp, H.B. (2010). Cerebral Perfusion Affects the Risk of Ischemia During Carotid Artery Stenting. *Cerebrovascular diseases*, Vol. 29, No. 6, (Epub 2010/04/09), pp. 538-545, ISSN 1421-9786 (Electronic) 1015-9770 (Linking).

Kang, H.S., Han, M.H., Kwon, O.K., Kwon, B.J., Kim, S.H. & Oh, C.W. (2007). Intracranial Hemorrhage after Carotid Angioplasty: A Pooled Analysis. *Journal of endovascular therapy : an official journal of the International Society of Endovascular Specialists*, Vol. 14, No. 1, (Epub 2007/02/13), pp. 77-85, ISSN 1526-6028 (Print) 1526-6028 (Linking).

Kappelle, L.J., Eliasziw, M., Fox, A.J. & Barnett, H.J. (2000). Small, Unruptured Intracranial Aneurysms and Management of Symptomatic Carotid Artery Stenosis. North American Symptomatic Carotid Endarterectomy Trial Group. *Neurology*, Vol. 55, No. 2, (Epub 2000/07/26), pp. 307-309, ISSN 0028-3878 (Print) 0028-3878 (Linking).

Kastrup, A., Groschel, K., Nagele, T., Riecker, A., Schmidt, F., Schnaudigel, S. & Ernemann, U. (2008). Effects of Age and Symptom Status on Silent Ischemic Lesions after Carotid Stenting with and without the Use of Distal Filter Devices. *AJNR. American journal of neuroradiology*, Vol. 29, No. 3, (Epub 2007/12/11), pp. 608-612, ISSN 1936-959X (Electronic) 0195-6108 (Linking).

Kleiser, B. & Widder, B. (1992). Course of Carotid Artery Occlusions with Impaired Cerebrovascular Reactivity. *Stroke; a journal of cerebral circulation*, Vol. 23, No. 2, (Epub 1992/02/01), pp. 171-174, ISSN 0039-2499 (Print) 0039-2499 (Linking).

Kluytmans, M., van der Grond, J., Eikelboom, B.C. & Viergever, M.A. (1998). Long-Term Hemodynamic Effects of Carotid Endarterectomy. *Stroke; a journal of cerebral circulation*, Vol. 29, No. 8, (Epub 1998/08/26), pp. 1567-1572, ISSN 0039-2499 (Print) 0039-2499 (Linking).

Ko, N.U., Achrol, A.S., Chopra, M., Saha, M., Gupta, D., Smith, W.S., Higashida, R.T. & Young, W.L. (2005). Cerebral Blood Flow Changes after Endovascular Treatment of Cerebrovascular Stenoses. *AJNR. American journal of neuroradiology*, Vol. 26, No. 3, (Epub 2005/03/12), pp. 538-542, ISSN 0195-6108 (Print) 0195-6108 (Linking).

Krogias, C., Hennebohl, C., Geier, B., Hansen, C., Hummel, T., Meves, S., Lukas, C. & Eyding, J. (2010). Transcranial Ultrasound Perfusion Imaging and Perfusion-MRI--a Pilot Study on the Evaluation of Cerebral Perfusion in Severe Carotid Artery Stenosis. *Ultrasound in medicine & biology*, Vol. 36, No. 12, (Epub 2010/10/19), pp. 1973-1980, ISSN 1879-291X (Electronic) 0301-5629 (Linking).

Ladowski, J.S., Webster, M.W., Yonas, H.O. & Steed, D.L. (1984). Carotid Endarterectomy in Patients with Asymptomatic Intracranial Aneurysm. *Annals of surgery*, Vol. 200, No. 1, (Epub 1984/07/01), pp. 70-73, ISSN 0003-4932 (Print) 0003-4932 (Linking).

Lownie, S.P., Pelz, D.M., Lee, D.H., Men, S., Gulka, I. & Kalapos, P. (2005). Efficacy of Treatment of Severe Carotid Bifurcation Stenosis by Using Self-Expanding Stents without Deliberate Use of Angioplasty Balloons. *AJNR. American journal of neuroradiology*, Vol. 26, No. 5, (Epub 2005/05/14), pp. 1241-1248, ISSN 0195-6108 (Print) 0195-6108 (Linking).

Maleux, G., Demaerel, P., Verbeken, E., Daenens, K., Heye, S., Van Sonhoven, F., Nevelsteen, A. & Wilms, G. (2006). Cerebral Ischemia after Filter-Protected Carotid Artery Stenting Is Common and Cannot Be Predicted by the Presence of Substantial Amount of Debris Captured by the Filter Device. *AJNR. American journal of neuroradiology*, Vol. 27, No. 9, (Epub 2006/10/13), pp. 1830-1833, ISSN 0195-6108 (Print) 0195-6108 (Linking).

Marks, M.P. (2002). Cerebral Ischemia and Infarction, In: *Magnetic Resonance Imaging of the Brain and Spine*, S.W. Atlas, pp. 919-980, Lippincott Williams & Wilkins, Philadelphia.

Martin, A.J., Saloner, D.A., Roberts, T.P., Roberts, H., Weber, O.M., Dillon, W., Cullen, S., Halbach, V., Dowd, C.F. & Higashida, R.T. (2005). Carotid Stent Delivery in an Xmr Suite: Immediate Assessment of the Physiologic Impact of Extracranial Revascularization. *AJNR. American journal of neuroradiology*, Vol. 26, No. 3, (Epub 2005/03/12), pp. 531-537, ISSN 0195-6108 (Print) 0195-6108 (Linking).

Martin, J.B., Pache, J.C., Treggiari-Venzi, M., Murphy, K.J., Gailloud, P., Puget, E., Pizzolato, G., Sugiu, K., Guimaraens, L., Theron, J. & Rufenacht, D.A. (2001). Role of the Distal Balloon Protection Technique in the Prevention of Cerebral Embolic Events During Carotid Stent Placement. *Stroke; a journal of cerebral circulation*, Vol. 32, No. 2, (Epub 2001/02/07), pp. 479-484, ISSN 1524-4628 (Electronic) 0039-2499 (Linking).

Mas, J.L., Chatellier, G., Beyssen, B., Branchereau, A., Moulin, T., Becquemin, J.P., Larrue, V., Lievre, M., Leys, D., Bonneville, J.F., Watelet, J., Pruvo, J.P., Albucher, J.F., Viguier, A., Piquet, P., Garnier, P., Viader, F., Touze, E., Giroud, M., Hosseini, H., Pillet, J.C., Favrole, P., Neau, J.P. & Ducrocq, X. (2006). Endarterectomy Versus Stenting in Patients with Symptomatic Severe Carotid Stenosis. *The New England journal of medicine*, Vol. 355, No. 16, (Epub 2006/10/20), pp. 1660-1671, ISSN 1533-4406 (Electronic) 0028-4793 (Linking).

Matsubara, S., Moroi, J., Suzuki, A., Sasaki, M., Nagata, K., Kanno, I. & Miura, S. (2009). Analysis of Cerebral Perfusion and Metabolism Assessed with Positron Emission Tomography before and after Carotid Artery Stenting. Clinical Article. *Journal of neurosurgery*, Vol. 111, No. 1, (Epub 2009/03/24), pp. 28-36, ISSN 0022-3085 (Print) 0022-3085 (Linking).

Maynar, M., Baldi, S., Rostagno, R., Zander, T., Rabellino, M., Llorens, R., Alvarez, J. & Barajas, F. (2007). Carotid Stenting without Use of Balloon Angioplasty and Distal Protection Devices: Preliminary Experience in 100 Cases. *AJNR. American journal of neuroradiology*, Vol. 28, No. 7, (Epub 2007/08/19), pp. 1378-1383, ISSN 0195-6108 (Print) 0195-6108 (Linking).

McCormick, W.F. & Schochet, S.S. (1976). *Atlas of Cerebrovascular Disease*, Saunders, ISBN 0721658962, Philadelphia.

Men, S., Lownie, S.P. & Pelz, D.M. (2002). Carotid Stenting without Angioplasty. *The Canadian journal of neurological sciences. Le journal canadien des sciences neurologiques*, Vol. 29, No. 2, (Epub 2002/05/31), pp. 175-179, ISSN 0317-1671 (Print) 0317-1671 (Linking).

Menon, D. & Stafinski, T. (2006). Cerebral Protection Devices for Use During Carotid Artery Angioplasty with Stenting: A Health Technology Assessment. *International journal of technology assessment in health care*, Vol. 22, No. 1, (Epub 2006/05/06), pp. 119-129, ISSN 0266-4623 (Print) 0266-4623 (Linking).

Michel, E., Liu, H., Remley, K.B., Martin, A.J., Madison, M.T., Kucharczyk, J. & Truwit, C.L. (2001). Perfusion Mr Neuroimaging in Patients Undergoing Balloon Test Occlusion of the Internal Carotid Artery. *AJNR. American journal of neuroradiology*, Vol. 22, No. 8, (Epub 2001/09/18), pp. 1590-1596, ISSN 0195-6108 (Print) 0195-6108 (Linking).

Moftakhar, R., Turk, A.S., Niemann, D.B., Hussain, S., Rajpal, S., Cook, T., Geraghty, M., Aagaard-Kienitz, B., Turski, P.A. & Newman, G.C. (2005). Effects of Carotid or Vertebrobasilar Stent Placement on Cerebral Perfusion and Cognition. *AJNR. American journal of neuroradiology*, Vol. 26, No. 7, (Epub 2005/08/11), pp. 1772-1780, ISSN 0195-6108 (Print) 0195-6108 (Linking).

Mohr, J.P., Albers, G.W., Amarenco, P., Babikian, V.L., Biller, J., Brey, R.L., Coull, B., Easton, J.D., Gomez, C.R., Helgason, C.M., Kase, C.S., Pullicino, P.M. & Turpie, A.G. (1997). American Heart Association Prevention Conference. Iv. Prevention and

Rehabilitation of Stroke. Etiology of Stroke. *Stroke; a journal of cerebral circulation*, Vol. 28, No. 7, (Epub 1997/07/01), pp. 1501-1506, ISSN 0039-2499 (Print) 0039-2499 (Linking).

Molyneux, A.J., Kerr, R.S., Birks, J., Ramzi, N., Yarnold, J., Sneade, M. & Rischmiller, J. (2009). Risk of Recurrent Subarachnoid Haemorrhage, Death, or Dependence and Standardised Mortality Ratios after Clipping or Coiling of an Intracranial Aneurysm in the International Subarachnoid Aneurysm Trial (Isat): Long-Term Follow-Up. *Lancet neurology*, Vol. 8, No. 5, (Epub 2009/03/31), pp. 427-433, ISSN 1474-4422 (Print) 1474-4422 (Linking).

Momjian-Mayor, I. & Baron, J.C. (2005). The Pathophysiology of Watershed Infarction in Internal Carotid Artery Disease: Review of Cerebral Perfusion Studies. *Stroke; a journal of cerebral circulation*, Vol. 36, No. 3, (Epub 2005/02/05), pp. 567-577, ISSN 1524-4628 (Electronic) 0039-2499 (Linking).

Morrish, W., Grahovac, S., Douen, A., Cheung, G., Hu, W., Farb, R., Kalapos, P., Wee, R., Hudon, M., Agbi, C. & Richard, M. (2000). Intracranial Hemorrhage after Stenting and Angioplasty of Extracranial Carotid Stenosis. *AJNR. American journal of neuroradiology*, Vol. 21, No. 10, (Epub 2000/12/08), pp. 1911-1916, ISSN 0195-6108 (Print) 0195-6108 (Linking).

Muller-Hulsbeck, S., Jahnke, T., Liess, C., Glass, C., Grimm, J. & Heller, M. (2003). Comparison of Various Cerebral Protection Devices Used for Carotid Artery Stent Placement: An in Vitro Experiment. *Journal of vascular and interventional radiology : JVIR*, Vol. 14, No. 5, (Epub 2003/05/23), pp. 613-620, ISSN 1051-0443 (Print) 1051-0443 (Linking).

NASCET. (1991). Beneficial Effect of Carotid Endarterectomy in Symptomatic Patients with High-Grade Carotid Stenosis. North American Symptomatic Carotid Endarterectomy Trial Collaborators. *The New England journal of medicine*, Vol. 325, No. 7, (Epub 1991/08/15), pp. 445-453, ISSN 0028-4793 (Print) 0028-4793 (Linking).

NASCET Trial Investigators. (1991). Clinical Alert: Benefit of Carotid Endarterectomy for Patients with High-Grade Stenosis of the Internal Carotid Artery. National Institute of Neurological Disorders and Stroke Stroke and Trauma Division. North American Symptomatic Carotid Endarterectomy Trial (Nascet) Investigators. *Stroke; a journal of cerebral circulation*, Vol. 22, No. 6, (Epub 1991/06/01), pp. 816-817, ISSN 0039-2499 (Print) 0039-2499 (Linking).

Navaneethan, S.D., Kannan, V.S., Osowo, A., Shrivastava, R. & Singh, S. (2006). Concomitant Intracranial Aneurysm and Carotid Artery Stenosis: A Therapeutic Dilemma. *Southern medical journal*, Vol. 99, No. 7, (Epub 2006/07/27), pp. 757-758, ISSN 0038-4348 (Print) 0038-4348 (Linking).

Niesen, W.D., Rosenkranz, M., Eckert, B., MeISSNer, M., Weiller, C. & Sliwka, U. (2004). Hemodynamic Changes of the Cerebral Circulation after Stent-Protected Carotid Angioplasty. *AJNR. American journal of neuroradiology*, Vol. 25, No. 7, (Epub 2004/08/18), pp. 1162-1167, ISSN 0195-6108 (Print) 0195-6108 (Linking).

O'Leary, D.H., Polak, J.F., Kronmal, R.A., Kittner, S.J., Bond, M.G., Wolfson, S.K., Jr., Bommer, W., Price, T.R., Gardin, J.M. & Savage, P.J. (1992). Distribution and Correlates of Sonographically Detected Carotid Artery Disease in the

Cardiovascular Health Study. The Chs Collaborative Research Group. *Stroke; a journal of cerebral circulation*, Vol. 23, No. 12, (Epub 1992/12/11), pp. 1752-1760, ISSN 0039-2499 (Print) 0039-2499 (Linking).

Obray, R., Clatterbuck, R., Olvi, A., Tamargo, R., Murphy, K.J. & Gailloud, P. (2003). De Novo Aneurysm Formation 6 and 22 Months after Initial Presentation in Two Patients. *AJNR. American journal of neuroradiology*, Vol. 24, No. 9, (Epub 2003/10/17), pp. 1811-1813, ISSN 0195-6108 (Print) 0195-6108 (Linking).

Ogata, J., Yonemura, K., Kimura, K., Yutani, C. & Minematsu, K. (2005). Cerebral Infarction Associated with Essential Thrombocythemia: An Autopsy Case Study. *Cerebrovascular diseases*, Vol. 19, No. 3, (Epub 2005/02/11), pp. 201-205, ISSN 1015-9770 (Print) 1015-9770 (Linking).

Rabellino, M., Garcia-Nielsen, L., Baldi, S., Zander, T., Casasola, C., Estigarribia, A., Llorens, R. & Maynar, M. (2010). Non-Protected Carotid Artery Stent without Angioplasty in High-Risk Patients with Carotid and Coronary Artery Disease Undergoing Cardiac Surgery. *Minimally invasive therapy & allied technologies : MITAT : official journal of the Society for Minimally Invasive Therapy*, Vol. 19, No. 3, (Epub 2010/02/16), pp. 184-188, ISSN 1365-2931 (Electronic) 1364-5706 (Linking).

Raja, P.V., Huang, J., Germanwala, A.V., Gailloud, P., Murphy, K.P. & Tamargo, R.J. (2008). Microsurgical Clipping and Endovascular Coiling of Intracranial Aneurysms: A Critical Review of the Literature. *Neurosurgery*, Vol. 62, No. 6, (Epub 2008/10/01), pp. 1187-1202; discussion 1202-1183, ISSN 1524-4040 (Electronic) 0148-396X (Linking).

Rasmussen, M., Juul, N., Christensen, S.M., Jonsdottir, K.Y., Gyldensted, C., Vestergaard-Poulsen, P., Cold, G.E. & Ostergaard, L. (2010). Cerebral Blood Flow, Blood Volume, and Mean Transit Time Responses to Propofol and Indomethacin in Peritumor and Contralateral Brain Regions: Perioperative Perfusion-Weighted Magnetic Resonance Imaging in Patients with Brain Tumors. *Anesthesiology*, Vol. 112, No. 1, (Epub 2009/12/03), pp. 50-56, ISSN 1528-1175 (Electronic) 0003-3022 (Linking).

Ringleb, P.A., Allenberg, J., Bruckmann, H., Eckstein, H.H., Fraedrich, G., Hartmann, M., Hennerici, M., Jansen, O., Klein, G., Kunze, A., Marx, P., Niederkorn, K., Schmiedt, W., Solymosi, L., Stingele, R., Zeumer, H. & Hacke, W. (2006). 30 Day Results from the Space Trial of Stent-Protected Angioplasty Versus Carotid Endarterectomy in Symptomatic Patients: A Randomised Non-Inferiority Trial. *Lancet*, Vol. 368, No. 9543, (Epub 2006/10/10), pp. 1239-1247, ISSN 1474-547X (Electronic) 0140-6736 (Linking).

Riphagen, J.H. & Bernsen, H.J. (2009). Rupture of an Intracerebral Aneurysm after Carotid Endarterectomy: A Case Report. *Acta neurologica Belgica*, Vol. 109, No. 4, (Epub 2010/02/03), pp. 314-316, ISSN 0300-9009 (Print) 0300-9009 (Linking).

Roberts, H.C., Dillon, W.P. & Smith, W.S. (2000). Dynamic Ct Perfusion to Assess the Effect of Carotid Revascularization in Chronic Cerebral Ischemia. *AJNR. American journal of neuroradiology*, Vol. 21, No. 2, (Epub 2000/03/01), pp. 421-425, ISSN 0195-6108 (Print) 0195-6108 (Linking).

Rohl, L., Ostergaard, L., Simonsen, C.Z., Vestergaard-Poulsen, P., Andersen, G., Sakoh, M., Le Bihan, D. & Gyldensted, C. (2001). Viability Thresholds of Ischemic Penumbra of Hyperacute Stroke Defined by Perfusion-Weighted MRI and Apparent Diffusion Coefficient. *Stroke; a journal of cerebral circulation*, Vol. 32, No. 5, (Epub 2001/09/06), pp. 1140-1146, ISSN 1524-4628 (Electronic) 0039-2499 (Linking).

Schnaudigel, S., Groschel, K., Pilgram, S.M. & Kastrup, A. (2008). New Brain Lesions after Carotid Stenting Versus Carotid Endarterectomy: A Systematic Review of the Literature. *Stroke; a journal of cerebral circulation*, Vol. 39, No. 6, (Epub 2008/04/05), pp. 1911-1919, ISSN 1524-4628 (Electronic) 0039-2499 (Linking).

Soinne, L., Helenius, J., Tatlisumak, T., Saimanen, E., Salonen, O., Lindsberg, P.J. & Kaste, M. (2003). Cerebral Hemodynamics in Asymptomatic and Symptomatic Patients with High-Grade Carotid Stenosis Undergoing Carotid Endarterectomy. *Stroke; a journal of cerebral circulation*, Vol. 34, No. 7, (Epub 2003/06/14), pp. 1655-1661, ISSN 1524-4628 (Electronic) 0039-2499 (Linking).

Spelle, L. & Pierot, L. (2008). [Endovascular Treatment of Non-Ruptured Intracranial Aneurysms: Critical Analysis of the Literature]. *Journal of neuroradiology. Journal de neuroradiologie*, Vol. 35, No. 2, (Epub 2008/05/10), pp. 116-120, ISSN 0150-9861 (Print) 0150-9861 (Linking).

Stern, J., Whelan, M., Brisman, R. & Correll, J.W. (1979). Management of Extracranial Carotid Stenosis and Intracranial Aneurysms. *Journal of neurosurgery*, Vol. 51, No. 2, (Epub 1979/08/01), pp. 147-150, ISSN 0022-3085 (Print) 0022-3085 (Linking).

Suh, B.Y., Yun, W.S. & Kwun, W.H. (2011). Carotid Artery Revascularization in Patients with Concomitant Carotid Artery Stenosis and Asymptomatic Unruptured Intracranial Artery Aneurysm. *Annals of vascular surgery*, Vol. 25, No. 5, (Epub 2011/05/03), pp. 651-655, ISSN 1615-5947 (Electronic) 0890-5096 (Linking).

Takaiwa, A., Hayashi, N., Kuwayama, N., Akioka, N., Kubo, M. & Endo, S. (2009). Changes in Cognitive Function During the 1-Year Period Following Endarterectomy and Stenting of Patients with High-Grade Carotid Artery Stenosis. *Acta neurochirurgica*, Vol. 151, No. 12, (Epub 2009/06/18), pp. 1593-1600, ISSN 0942-0940 (Electronic) 0001-6268 (Linking).

Takaiwa, A., Kuwayama, N., Hayashi, N., Kubo, M., Matsui, M. & Endo, S. (2006). [Cognitive Function in Patients with Severe Carotid Stenosis--Evaluation of Rbans, Wais-R and Nart before Treatment of Carotid Revascularization--]. *No to shinkei = Brain and nerve*, Vol. 58, No. 8, (Epub 2006/09/22), pp. 681-686, ISSN 0006-8969 (Print) 0006-8969 (Linking).

Tavares, A., Caldas, J.G., Castro, C.C., Puglia, P., Jr., Frudit, M.E. & Barbosa, L.A. (2010). Changes in Perfusion-Weighted Magnetic Resonance Imaging after Carotid Angioplasty with Stent. *Interventional neuroradiology : journal of peritherapeutic neuroradiology, surgical procedures and related neurosciences*, Vol. 16, No. 2, (Epub 2010/07/21), pp. 161-169, ISSN 1591-0199 (Print) 1591-0199 (Linking).

Teng, M.M., Cheng, H.C., Kao, Y.H., Hsu, L.C., Yeh, T.C., Hung, C.S., Wong, W.J., Hu, H.H., Chiang, J.H. & Chang, C.Y. (2001). Mr Perfusion Studies of Brain for Patients with Unilateral Carotid Stenosis or Occlusion: Evaluation of Maps of "Time to Peak" and "Percentage of Baseline at Peak". *Journal of computer assisted tomography*, Vol. 25, No. 1, (Epub 2001/02/15), pp. 121-125, ISSN 0363-8715 (Print) 0363-8715 (Linking).

Theron, J.G., Payelle, G.G., Coskun, O., Huet, H.F. & Guimaraens, L. (1996). Carotid Artery Stenosis: Treatment with Protected Balloon Angioplasty and Stent Placement. *Radiology*, Vol. 201, No. 3, (Epub 1996/12/01), pp. 627-636, ISSN 0033-8419 (Print) 0033-8419 (Linking).

Thurley, P.D., Altaf, N., Dineen, R., MacSweeney, S. & Auer, D.P. (2009). Pial Vasodilation and Moderate Hyperaemia Following Carotid Endarterectomy: New MRI Diagnostic Signs in Hyperperfusion/Reperfusion Syndrome? *Neuroradiology*, Vol. 51, No. 6, (Epub 2009/04/25), pp. 427-428, ISSN 1432-1920 (Electronic) 0028-3940 (Linking).

Turk, A.S., Chaudry, I., Haughton, V.M., Hermann, B.P., Rowley, H.A., Pulfer, K., Aagaard-Kienitz, B., Niemann, D.B., Turski, P.A., Levine, R.L. & Strother, C.M. (2008). Effect of Carotid Artery Stenting on Cognitive Function in Patients with Carotid Artery Stenosis: Preliminary Results. *AJNR. American journal of neuroradiology*, Vol. 29, No. 2, (Epub 2007/11/09), pp. 265-268, ISSN 1936-959X (Electronic) 0195-6108 (Linking).

Van Laar, P.J., Hendrikse, J., Mali, W.P., Moll, F.L., van der Worp, H.B., van Osch, M.J. & van der Grond, J. (2007). Altered Flow Territories after Carotid Stenting and Carotid Endarterectomy. *Journal of vascular surgery : official publication, the Society for Vascular Surgery [and] International Society for Cardiovascular Surgery, North American Chapter*, Vol. 45, No. 6, (Epub 2007/06/05), pp. 1155-1161, ISSN 0741-5214 (Print) 0741-5214 (Linking).

van Laar, P.J., van der Grond, J., Moll, F.L., Mali, W.P. & Hendrikse, J. (2006). Hemodynamic Effect of Carotid Stenting and Carotid Endarterectomy. *Journal of vascular surgery : official publication, the Society for Vascular Surgery [and] International Society for Cardiovascular Surgery, North American Chapter*, Vol. 44, No. 1, (Epub 2006/05/30), pp. 73-78, ISSN 0741-5214 (Print) 0741-5214 (Linking).

van Rooij, W.J., de Gast, A., Sluzewski, M., Nijssen, P.C. & Beute, G.N. (2006). Coiling of Truly Incidental Intracranial Aneurysms. *AJNR. American journal of neuroradiology*, Vol. 27, No. 2, (Epub 2006/02/18), pp. 293-296, ISSN 0195-6108 (Print) 0195-6108 (Linking).

van Rooij, W.J. & Sluzewski, M. (2006). Procedural Morbidity and Mortality of Elective Coil Treatment of Unruptured Intracranial Aneurysms. *AJNR. American journal of neuroradiology*, Vol. 27, No. 8, (Epub 2006/09/15), pp. 1678-1680, ISSN 0195-6108 (Print) 0195-6108 (Linking).

Vitek, J.J., Roubin, G.S., Al-Mubarek, N., New, G. & Iyer, S.S. (2000). Carotid Artery Stenting: Technical Considerations. *AJNR. American journal of neuroradiology*, Vol. 21, No. 9, (Epub 2000/10/20), pp. 1736-1743, ISSN 0195-6108 (Print) 0195-6108 (Linking).

Vlak, M.H., Rinkel, G.J., Greebe, P., van der Bom, J.G. & Algra, A. (2011). Trigger Factors and Their Attributable Risk for Rupture of Intracranial Aneurysms: A Case-Crossover Study. *Stroke; a journal of cerebral circulation*, Vol. 42, No. 7,

(Epub 2011/05/07), pp. 1878-1882, ISSN 1524-4628 (Electronic) 0039-2499 (Linking).

Vos, J.A., van den Berg, J.C., Ernst, S.M., Suttorp, M.J., Overtoom, T.T., Mauser, H.W., Vogels, O.J., van Heesewijk, H.P., Moll, F.L., van der Graaf, Y., Mali, W.P. & Ackerstaff, R.G. (2005). Carotid Angioplasty and Stent Placement: Comparison of Transcranial Doppler Us Data and Clinical Outcome with and without Filtering Cerebral Protection Devices in 509 Patients. *Radiology*, Vol. 234, No. 2, (Epub 2004/12/24), pp. 493-499, ISSN 0033-8419 (Print) 0033-8419 (Linking).

Waaijer, A., van Leeuwen, M.S., van Osch, M.J., van der Worp, B.H., Moll, F.L., Lo, R.T., Mali, W.P. & Prokop, M. (2007). Changes in Cerebral Perfusion after Revascularization of Symptomatic Carotid Artery Stenosis: Ct Measurement. *Radiology*, Vol. 245, No. 2, (Epub 2007/09/13), pp. 541-548, ISSN 0033-8419 (Print) 0033-8419 (Linking).

Wiebers, D.O., Whisnant, J.P., Huston, J., 3rd, MeISSNer, I., Brown, R.D., Jr., Piepgras, D.G., Forbes, G.S., Thielen, K., Nichols, D., O'Fallon, W.M., Peacock, J., Jaeger, L., Kassell, N.F., Kongable-Beckman, G.L. & Torner, J.C. (2003). Unruptured Intracranial Aneurysms: Natural History, Clinical Outcome, and Risks of Surgical and Endovascular Treatment. *Lancet*, Vol. 362, No. 9378, (Epub 2003/07/18), pp. 103-110, ISSN 1474-547X (Electronic) 0140-6736 (Linking).

Wilkinson, I.D., Griffiths, P.D., Hoggard, N., Cleveland, T.J., Gaines, P.A., Macdonald, S., McKevitt, F. & Venables, G.S. (2003). Short-Term Changes in Cerebral Microhemodynamics after Carotid Stenting. *AJNR. American journal of neuroradiology*, Vol. 24, No. 8, (Epub 2003/09/19), pp. 1501-1507, ISSN 0195-6108 (Print) 0195-6108 (Linking).

Wintermark, M., Flanders, A.E., Velthuis, B., Meuli, R., van Leeuwen, M., Goldsher, D., Pineda, C., Serena, J., van der Schaaf, I., Waaijer, A., Anderson, J., Nesbit, G., Gabriely, I., Medina, V., Quiles, A., Pohlman, S., Quist, M., Schnyder, P., Bogousslavsky, J., Dillon, W.P. & Pedraza, S. (2006). Perfusion-Ct Assessment of Infarct Core and Penumbra: Receiver Operating Characteristic Curve Analysis in 130 Patients Suspected of Acute Hemispheric Stroke. *Stroke; a journal of cerebral circulation*, Vol. 37, No. 4, (Epub 2006/03/04), pp. 979-985, ISSN 1524-4628 (Electronic) 0039-2499 (Linking).

Xu, K., Wang, H., Luo, Q., Li, Y. & Yu, J. (2011). Endovascular Treatment of Bilateral Carotid Artery Occlusion with Concurrent Basilar Apex Aneurysm: A Case Report and Literature Review. *International journal of medical sciences*, Vol. 8, No. 3, (Epub 2011/04/14), pp. 263-269, ISSN 1449-1907 (Electronic) 1449-1907 (Linking).

Yadav, J.S., Wholey, M.H., Kuntz, R.E., Fayad, P., Katzen, B.T., Mishkel, G.J., Bajwa, T.K., Whitlow, P., Strickman, N.E., Jaff, M.R., Popma, J.J., Snead, D.B., Cutlip, D.E., Firth, B.G. & Ouriel, K. (2004). Protected Carotid-Artery Stenting Versus Endarterectomy in High-Risk Patients. *The New England journal of medicine*, Vol. 351, No. 15, (Epub 2004/10/08), pp. 1493-1501, ISSN 1533-4406 (Electronic) 0028-4793 (Linking).

Yamada, K., Kawasaki, M., Yoshimura, S., Enomoto, Y., Asano, T., Minatoguchi, S. & Iwama, T. (2010). Prediction of Silent Ischemic Lesions after Carotid Artery Stenting Using Integrated Backscatter Ultrasound and Magnetic Resonance Imaging. *Atherosclerosis*, Vol. 208, No. 1, (Epub 2009/07/21), pp. 161-166, ISSN 1879-1484 (Electronic) 0021-9150 (Linking).

Part 4

Treatment of Intracranial Hypertension

12

Medical and Surgical Management of Intracranial Hypertension

James Scozzafava[1,2], Muhammad Shazam Hussain[3] and Seby John[4]
[1]*Department of Adult Critical Care Medicine, Saskatoon Health Region, Saskatoon, Saskatchewan,*
[2]*Division of Stroke Neurology, Neurosciences Program, Department of Medicine, University of Alberta, Edmonton, AB,*
[3]*Medical Director, Primary Stroke Center, Vascular Neurology and Endovascular Surgical Neuroradiology, Cerebrovascular Section, Cleveland Clinic, Cleveland, OH*
[4]*Department of Neurology, Cleveland Clinic, Cleveland, OH*
[1,2]*Canada*
[3,4]*USA*

1. Introduction

Elevated intracranial pressure (ICP) is a frequently encountered problem in neurological and neurosurgical patients in the intensive care unit (ICU). It is most often seen in the setting of acute head trauma, however can also result from such causes as tumor, stroke, intracranial hemorrhage, or infection. Regardless of the cause, the degree and duration of ICP elevation has a direct and inverse relationship with morbidity and mortality. Consequently, despite the severity of the initial injury to the brain, a great deal of attention must often be focused on the monitoring and the management of ICP in acute neurological patients in the ICU setting.

2. Physiology

The brain is enclosed within inelastic container (the skull) or "closed box", and the sum of volumes of intracranial contents is constant. The intracranial contents include blood, cerebrospinal fluid (CSF) and brain and an increase in the volume of one must be offset by a decrease in one or more of the other contents or an increase in ICP will result. Elevations in ICP are the result of an increase in cranial volume and a decrease in cranial compliance. This is best described by the Monro-Kellie hypothesis. (Figure 1) CSF is the most accommodating of the intracranial contents, however the compensatory capabilities of CSF are limited and once exhausted, small increases in intracranial volume result in large increases in ICP.

The most common cause for raised ICP is severe head trauma. Regardless of etiology, the pathophysiology underlying the secondary injury from raised ICP is similar for all patients with brain injury and it results from such issues as swelling, edema, and neuronal cytotoxicity. Brain edema is a frequent occurrence in neurological diseases and can accumulate intracellular or extracellular. Intracellular edema is usully the result of cytotoxic edema damaging the cell membranes often destroying the sodium/potassium exchange pump. This leads to unregulated passage of sodium and water into neuronal cells. Extracellular edema often results from capillary injuries at specific areas. This leads to breakdown of the blood brain barrier and leakage of protein and fluids into the extracellular space affected. A third category of edema, interstitial edema, is sometimes seen in acute obstructive hydrocephalus (See Table 1)

Type of Edema	Pattern	Mechanism	Differential Diagnosis	Potential Therapies
Vasogenic	Extracellular spaces in white matter.	Blood brain barrier breakbown secondary to capillary injury near focal lesions.	Tumors. Hemorrhage. Infection. Inflammation.	Surgery Antibiotics Steroid
Hydrostatic	Extracellular spaces in white matter and grey matter. Often diffuse. Can favor posterior circulation in PRES. Can be unilateral in CHS.	Increased cerebral capillary water influx across blood brain barrier because of elevated pressure states.	Hypertensive Emergencies. Hyperperfusion Syndromes. Hepatic Encephalopathy	Antihypertensives
Interstitial	Prefers Periventricular white matter, especially frontal and occipital lobes.	Transependymal flow of cerebrospinal fluid in hydrocephalus.	Acute Obstructive Hydrocephalus.	Shunt or EVD
Cytotoxic	Often intracellular. Prefers grey matter	Na/K Pump breakdown and damaged cell membranes leading to water entry into cells and cellular swelling.	Infarct. Trauma. Toxins. Hepatic Encephalopathy. Hypoxic Ischemic Encephalopathy	Mannitol Hypertonic Saline

Table 1. Types of Cerebral Edema

When ICP remains severely and persistently elevated it is sometimes referred to as malignant intracranial hypertension and secondary brain injury is often the far more severe and deadly issue for the patient. ICP monitoring should be considered in any traumatic head injury presenting with a Glasgow coma score (GCS) of less than 9 and an abnormal computed tomography (CT) of head. Even in the absence of an abnormal CT scan, evidence suggests severe head trauma patients are still at significant risk of elevated ICP, particularly those greater than 40 years of age, showing signs of motor posturing, or having a systolic blood pressure less 90 mmHg or less.[1] A GCS of less than 9 and at least two of these three risk factors warrants consideration for continuous ICP monitoring, regardless of CT findings. ICP monitoring may also be warranted in cases where CSF flow appears disrupted or potentially disrupted. This is because of the potential for obstruction of drainage of flow of CSF and in the setting of acute brain injury, obstructive hydrocephalus can have an accelerating effect on ICP. This is often seen in the context of tumors, infections or hemorrhage either within the ventricular system or just outside of it. Whether the result of direct intraventricular obstruction or extraventricular mass effect with secondary ventricular system obstruction, these situations can lead to rapidly rising ICP and risk of deadly brain herniation.

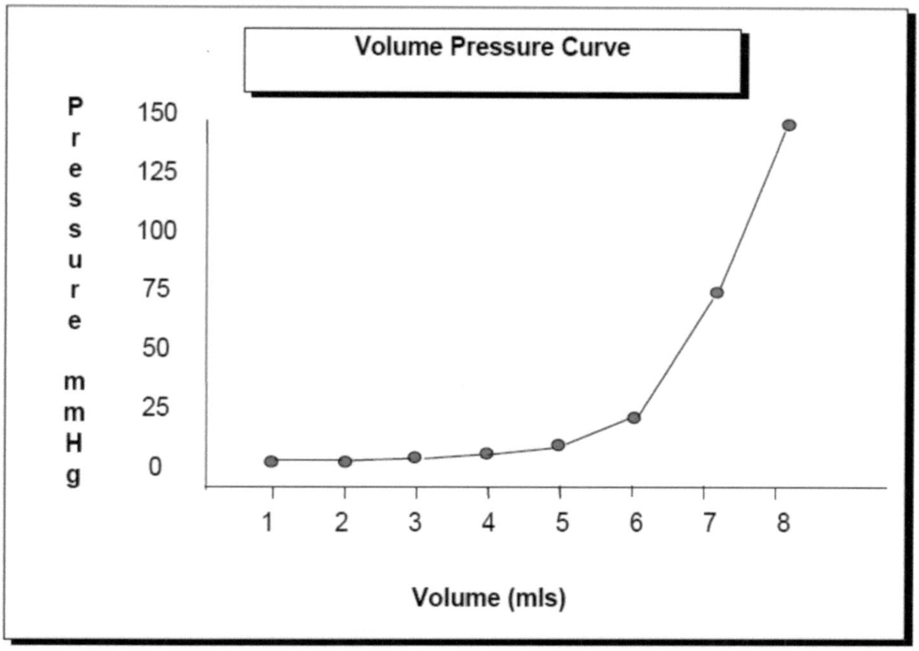

Fig. 1. ICP – Volume Curve showing exponential increases in ICP as skull compensatory mechanisms exhaust in the face of increasing intracranial volume.

3. Monitoring

Often by the time of clinical deterioration, ICP has been elevated for some time causing significant neuronal damage through progressive tissue swelling, ischemia, and hypoperfusion. Without emergent intervention, the process of deadly brain herniation may have already begun. Unfortunately, only the early stages are reversible and the clinical changes can be sudden. As a result, non-clinical modes of assessment such as continuous ICP monitoring have proven invaluable in head injury patients. Although there are several different forms of ICP monitoring devices, the external ventricular drain or catheter (EVD) has become the gold standard for accurate measurement and monitoring of ICP in the ICU. As its name implies, the EVD not only provides ICP monitoring, but allows for CSF drainage in response to ICP levels. Although associated with small risks of infection and bleeding, the ability to recalibrate over time for sustained accuracy and the ability to drain CSF in response to ICP make the EVD superior to other ICP monitoring devices.

The goal of ICP management is to prevent secondary brain injury, particularly from herniation or ischemia and infarction. Brain herniations occur as a result of pressure gradients created in the skull by either localized or generalized mass effect. Although, herniation is more likely to occur with progressive increases in ICP, there is no specific ICP known to be a threshold for herniation. Furthermore, certain herniations can occur without raised ICP, resulting solely from a localized pressure gradient such as in tentorial (uncal) herniation from a temporal lobe mass.[2] Brain herniation is only reversible if treated early with aggressive medical and surgical intervention. Delays in treatment and inadequate treatment are associated with progressive herniation and severe and irreversible brain injury often leading to death. Special attention to the signs of brain herniation must be paid to all patients with raised ICP and to all patients with temporal lobe masses.

Marked elevations in ICP lead to compromise in cerebral perfusion either locally or diffusely which in turn leads to ischemia and eventually infarction. CBF is closely correlated to cerebral perfusion pressure (CPP) through cerebral vascular resistance (CVR). Although, measures of CBF and CVR and not easily accessible, CPP can be calculated as the difference between mean arterial pressure (MAP) and ICP and CPP is often used as a surrogate for CBF. Although there is some conflicting evidence and opinion regarding the optimal CPP for head injury patients with raised ICP, the largest group of patients studied is within the traumatic brain injury population. It suggests that a CPP between 60 to 70 mmHg and ICP less than 20 mmHg may be associated with improved outcome and fewer complications from therapy in patients with brain injury and elevated ICP.[3]

4. Managing ICP

The medical management of ICP starts with principles of physiologic homeostasis including optimizing cerebral venous return. Evidence supports the practice of elevating the head of the bed to 30° or more for patients with raised ICP, however, in patients who are hypovolemic, this change in position may cause decreases in CPP by decreasing the MAP more than the ICP. Attention to CPP and ICP should be paid to ensure CPP is not compromised. Facilitating physiologic homeostasis also includes early treatment of factors that can worsen ICP such as fever, hypercarbia and seizures.

Other medical management aimed at treating elevated ICP has been investigated for several years with limited success. Although several strategies have been shown to reduce ICP for short periods of time, very few have proven to be effective long term therapies or have a significant impact on patient outcome. One of the more commonly accepted approaches is to minimize agitation and excess muscle activity which can worsen ICP. This usually involves the use of analgesics and sedatives. The agents most commonly used are narcotics such as morphine and fentanyl for analgesics and benzodiazepines such as midazolam and lorazepam for sedation. In many cases, barbituates and anesthetics such as propofol, can be used in small or large doses in an attempt to decrease cerebral metabolism and cerebral oxygen requirements. This has been shown to reduce ICP and the decrease in cerebral oxygen requirements may result in a decrease in the absolute amount of cerebral blood flow needed to prevent further brain injury. The use of "drug induced coma" to decrease cerebral metabolic rate remains controversial. This is largely because of the lack of convincing evidence that this approach changes outcome and the significant systemic and hemodynamic complications associated with these medications at high doses for long periods of time.[5,6] Regardless, both barbituates and anesthetics such as propofol have been shown to be effective in decreasing ICP in a dose related fashion and assuming a patient is hemodynamically stable, these agents can often be used safely for extended periods with favorable responses on ICP.

Occasionally, neuromuscular blockers are required, however routine use of neuromuscular blockers has not been shown to improve outcome and is associated with increased complications including pneumonia and sepsis. In the setting where paralysis is considered to control excess movements in association with persistently elevated ICP, encephalography (EEG) be considered to rule out ongoing seizures. Continuous EEG is used in some centres to monitor brain wave activity and to observe for the potential of non-clinical seizures. Although standard 20 minute EEG has not been shown to correlate with a high detection rate of seizures, any concern regarding the possibility of non-clinical seizures warrants assessment with EEG, if not continuous then intermittent.[5]

Other medications focus on diuresing fluid, presumably edema, from the brain with osmotic diuretics or hypertonic solutions. The most extensively used agent is mannitol and various clinical and experimental studies have shown reductions in ICP with the use of mannitol.[5] The effect of mannitol on ICP involves several mechanisms including decreasing blood viscosity, decreasing CSF production, and fluid shift from brain tissue to intravascular compartments. The effects are usually not sustained and repeated doses over a prolonged period are often required. Earlier studies have also suggested the effects of mannitol on ICP could be prolonged with the concomitant use of furosomide. Whether used alone or in combination with another diuretic, prolonged mannitol use can lead to problems with dehydration, elevated serum osmolality, and renal impairment. Furthermore, the use of mannitol is less effective and potentially aggravating in cases of localized lesions with significant vasogenic edema, such as hemispheric ischemic strokes. In such cases, localized pressure gradient or a compartmentalized ICP is a more likely cause of neurological deterioration rather than a generalized increase in ICP.[7] The effects of mannitol in these cases will be seen maximally in normal and non affected areas of the brain which can potential worsen the pressure gradient further.[8] Kauffman et al investigated mannitol in the treatment of vasogenic cerebral edema following ischemic stroke and found that with multiple mannitol injections, non-infarcted hemisphere shrunk more compared to the infarcted hemisphere. He concluded that prolonged mannitol use may actually cause a reversal of the osmotic gradient between edematous brain and plasma, and worsen localized vasogenic edema in cases of large ischemic strokes with vasogenic edema.[9] Similarly, several stroke specific trials found that if hyperosmolar therapy was used continuously for greater than 48 hours, it was associated with worse outcome and increased risk of herniation. The effects are felt to be secondary to increasing ICP differentials between infarcted brain and healthy brain and thereby increasing the chances of herniation, as the osmotic agents may collect in the infarcted tissue and then induce increasing edema. Consequently, mannitol

---

[5] Scozzafava J, Hussain MS, Brindley PG, Jacka MJ, Gross D. The role of 20 minute EEG in the comatose patient. Journal of Clinical Neurosciences. Jan 2010;17(1):64-8.

should be used cautiously in any patient with hemispheric masses or lesions and in cases where repeated dose are to be used for a prolonged period of time. In more recent years, hypertonic saline has also been used and like mannitol, hypertonic saline acts by pulling water from brain tissue to the intravascular space. Hypertonic saline is associated with slightly less concerns regarding electrolyte disturbance, dehydration and renal injury, however similar precautions should be taken in regards to hemispheric lesions, particularly ischemic strokes with significant vasogenic edema. At the present time, there are no established guidelines for its use, however most studies have used 3% or 7.5% at rates of 20 to 40 cc/hr. Prolonged infusions should be run only in an ICU setting and discontinuation should be tapered slowly to prevent rebound hyponatremia.

Several other measures should be considered second tier in the management of ICP because of their lack of proven benefit and/or the significant risks associated with their use. These include hyperventilation and hypothermia. For years, hyperventilation was thought to be a safe and effective method of lowering ICP. Its action is through generalized vasoconstriction of small cerebral arteries in response to low carbon dioxide levels in the blood. It causes a relatively quick decrease in ICP at the expense of a corresponding decrease in CBF and at the risk of resultant cerebral ischemia. Further evidence has shown hyperventilation to be an ineffective method of managing ICP and possibly detrimental in several subgroups of patients.[13] Its use should be avoided in most circumstances.

The concept of hypothermia as a possible neuroprotectant has been explored for several years. Evidence exists that induced hypothermia lowers ICP, likely through decreasing cerebral metabolic rate, however it has not been shown to correlate to a decreased rate of secondary brain injury. Although increasing evidence suggests a role for therapeutic hypothermia to protect the brain from secondary injury following cardiac arrest, similar evidence has not developed in other situations, particularly primary brain injury such as trauma and stroke. A randomized controlled trial in severe closed head injury failed to show any benefit of induced hypothermia on outcome.[14] Further trials in various subgroups of patients are underway.

In many cases, ICP proves resistant to aggressive medical management and surgical intervention is required beyond the EVD. This is frequently seen in association with mass lesions. Mass lesions can include tumors, abscesses or hemorrhages which are all frequently prone to edema which further aggravates ICP and poses an increasing risk of herniation. In the case of tumors, resection often results in resolution of ICP issues. In the case of abscess, drainage not only serves to reduce ICP, but treats the infection and prevents further spread.

The recommended management of hematoma depends on location, size of hematoma and extent of neurological injury as well as ICP. Subdural and epidural hematomas causing increased ICP generally should be considered for surgical evacuation. Subarachnoid hemorrhage (SAH) is often the result of a ruptured cerebral artery aneurysm or arteriovenous malformation. Definitive management of the hemorrhage and the ICP must involve securing the ruptured vessel(s). In cases of parenchymal intracerebral hemorrhage (ICH) or hemorrhagic stroke, several randomized trials have failed to demonstrate benefit associated with surgical evacuation of supratentorial ICH, regardless of ICP.[15]

However, there is one sub-group of hemorrhagic stroke that has shown to benefit from aggressive and early surgical intervention. Although, there are no randomized trials looking specifically at surgical evacuation of cerebellar hematomas, a non-randomized case series has shown urgent surgical evacuation of cerebellar hemorrhage improves outcome compared to medical management alone.[16] It included patients with GCS < 13, and hematoma > 40 mm. 45 patients were treated medically and 30 patients were treated with decompressive surgery. Good outcome occurred 58% with surgery while only 18% with conservative medical therapy. A large reason for this may be because of the reasonable potential for recovery regardless of stroke burden in cerebellum. The factor which is often more critical is the acute and progressing edema and/or hemorrhage which may prove to compromise the nearby fourth ventricle of the ventricular system and lead to obstructive hydrocephalus. As a result, a cerebellar hemorrhage which would otherwise have potentially good outcome, could rapidly progress to malignant intracranial hypertension and death. Our recommendation is for the prophylactic insertion of EVD or suboccipital decompressive craniectomy for patients with cerebellar hemorrhage and hematoma > 3 cm in diameter or obstructive hydrocephalus.

Medically intractable ICP is also frequently seen with severe head trauma. Unlike mass lesions and vascular anomalies, an underlying structural abnormality cannot be removed unless associated with significant secondary hematoma. Surgical treatment of ICP has focused on brain decompression in these patients. Both hemispheric and bifrontal craniectomy have both been shown to be effective in managing medically intractable ICP following head trauma and should be considered when medical measures have failed.[17-22]

Decompressive craniectomy for prevention of fatal brain herniation has been for almost 100 years. The rationale for surgery is to change the inelastic container or 'closed box' and provide a mechanical outlet for the edematous brain to stretch beyond the skull thereby preventing herniation. As a consequence, secondary benefits include rapid reduction of intractable ICP and restoration of cerebral perfusion.

5. Considerations specific to ischemic stroke

Issues regarding raised ICP are less often encountered in ischemic stroke compared to hemorrhagic stroke. Similarly, the management of malignant intracranial hypertension is far less seen is ischemic stroke patients compared to hemorrhagic stroke or other neurology and neurosurgery patients presenting with mass lesions. This is because in the case of ischemic stroke, a new mass or volume is not immediately introduced into our "closed box" model of intracranial contents like with other causes of brain injury like hemorrhage or tumor. However, immediately after the onset of ischemia changes begin to occur including cerebral edema.

It has been estimated that anywhere from 1-10% of supratentorial ischemic strokes can cause rapid neurological deterioration from space-occupying cerebral edema. Although ischemic strokes display some vasogenic edema as part of the inflammatory phase, the primary swelling is the result of cytotoxic edema. Cytotoxic edema is the result of damaged cell membranes during ischemia. The result is that neuronal cells fill with plasma ultrafiltrate. Although this usually occurs between the second and fifth day after stroke, it can occur as early as 24 hours. Such a presentation following an ischemic stroke involving the entire middle cerebral artery (MCA) territory is called 'malignant MCA infarction' (MMI). This is often a consequence of occlusion of the internal carotid artery or the proximal portion of the middle cerebral artery. The prognosis of MMI is poor and mortality is as high as 80%, with most deaths occurring during the first week from cerebral edema and brain herniation.[23]

Multiple non-randomized studies have shown that decompressive surgery, consisting of a hemicraniectomy and duraplasty reduces mortality in patients with MMI.[24-26] However, its popularity decreased because clinicians were concerned as to whether survival was at the expense of poor functional outcome. In the midst of uncertainty regarding functional outcome, three European trials have addressed the role of decompressive hemicraniectomy on functional outcome since 2000: the French DECIMAL (decompressive craniectomy in malignant middle cerebral artery infarcts) trial; the German DESTINY (decompressive surgery for the treatment of malignant infarction of the middle cerebral artery) trial; and the Dutch HAMLET trial (hemicraniectomy after middle cerebral artery infarction with life-threatening edema trial). A pooled analysis of all three trials confirmed suggestions from earlier non-randomized trials. Decompressive hemicraniectomy undertaken within 48 hours of onset of MMI reduces mortality and increases the number of patients with a favorable functional outcome.[27-30]

The trials were both praised and criticized on many points. Analysis of all the trials was assumed possible due to similar design and use of the modified Rankin Scale (mRS) as a common primary outcome measure. Significant difference between the trials included imaging modalities and longer treatment window allowed in HAMLET, which allowed patients to be treated up to 96 hours after onset of stroke symptoms. However, the pooled analysis included only the patients randomized and treated within 48 hours. The primary outcome measure for the pooled analysis was the mRS dichotomized between "favorable" (defined as mRS 0-4) and "unfavorable" (mRS 5-6). There was clearly a difference between the two treatment arms, with 75% achieving a favorable outcome in the hemicraniectomy group as compared to 24% in the medical treatment alone arm ($p < 0.01$). The most robust effect was seen on survival, which increased from 28% to nearly 80% with DCH. Perhaps the most significant result was the proportion of patients who were independent with disability (mRS 2) which increased more than five times with DCH from 2.5% to 14%. Forty three percent of patients had a good clinical outcome with mRS 2-3 after a DCH compared to 21.5% of patients who received conservative therapy. However, the proportion of patients surviving with moderate-to-severe disability was increased more than 12-fold (31% vs. 2.5%) but the rates of very severe disability was not increased after DCH (4%vs. 5%).

Despite the evidence provided by these trials, the most crucial question remains the selection of patients for surgery because not all middle cerebral infarctions lead to MMI and no single prognostic factor has been identified as a predictor of fatal outcome in MMI. Neuroimaging combined with clinical examination does provide valuable information to identify patients at risk. Early ischemic changes (less than 6 hours) on CT scan that involves greater than 50% of the MCA territory have been associated with fatal outcomes. This included such early CT changes as localized cerebral edema causing sulcal effacement or compression of the lateral ventricle.[31] The DECIMAL trial used a critical stroke volume of 145 mL on diffusion-weighted MRI and confirmed this cut off value suggested by previous authors. Analysis showed 78% mortality in strokes with >145 mL volume and no deaths when the stroke volume was less than 145mL. NIHSS $\geq$ 20 (dominant hemisphere) or $\geq$ 15 (non-dominant hemisphere) within 6 hours of symptom onset along with CT findings of hypodensity >50% were also associated with high risk for developing malignant cerebral edema.[32]

Age is another important consideration in this population following MMI. The upper age limit in the above randomized trials was 60-years and therefore the results are not easily transferable to individuals older than 60 years. As we have seen with many interventions, with older patient populations come more medically complex and often fragile patients. DESTINY II is underway to evaluate the benefit of surgery in this older patient population, which may be more indicative of the stroke population seen in most hospitals. However, non-randomized trials which investigated this question showed that mortality and poor outcome were significantly higher in patients older than 60 years of age.[33,34] Regardless of the results of DESTINY II, the decision to operate on patients older than 60 years of age should be individualized and must take into consideration patients declared wishes and social support. It must be kept in mind that outcomes in this population depend on several factors like admission functional status, cognitive ability and presence of social support.

The timing of surgery is also significant. From the individual results and pooled analysis of DECIMAL and DESTINY, patients who were surgically treated with decompressive hemicraniectomy within 48 hours did better when compared to the HAMLET trial in which there was no improvement in functional outcome despite decrease in mortality in patients who received delayed surgical treatment (up to 96 hours after symptoms onset). Early decompressive hemicraniectomy (<48hours) is recommended for all patients with MMI and impending herniation who are felt to have good potential for recovery from the initial ischemic injury of the stroke.

In contrast to supratentorial ischemic strokes, the question of surgical intervention in large cerebellar strokes is less controversial. Fatal space occupying edema can develop in 17% - 54% of cerebellar strokes resulting in obstructive hydrocephalus, transforaminal or transtentorial herniation and brain stem compression. Although lacking evidence from randomized clinical trials, it is widely accepted that surgical intervention with suboccipital decompressive craniectomy or insertion of an external ventricular drainage is lifesaving in malignant cerebellar infarctions and with the potential for good clinical outcomes. The long term outcome in survivors is also good, especially where there is no associated brainstem infarction.[35,36] Similar to cerebellar hematoma, these patients should be considered for early decompression.

6. ICP in aneurysmal subarachnoid hemorrhage

Case report and case series evidence in aneurysmal SAH continues to emerge describing beneficial effects of decompressive craniectomy on elevated ICP and on outcome. The factors affecting ICP in aneurysmal SAH can include ongoing hemorrhage, edema and vasospasm with secondary ischemia. In cases of high grade SAH with severely and persistently elevated ICP, decompressive craniectomy should be considered in conjunction with early securing of aneurysm in an effort to maximize the chances of good outcome.[37,38] Location of SAH may also play a factor in both outcome and surgical approach. Case series evidence has described dramatic recoveries in patients with high grade SAH from anterior communicating artery aneurysm ruptures and malignant ICP managed with bifrontal decompressive craniectomy.[39] The evidence may be related to the observation that high grade SAH from anterior communicating artery aneurysm ruptures often present without significant initial brain tissue injury but with significant subarachnoid hematoma and dramatic bilateral cerebral edema. The prominent edema associated with these SAH's often precipitates uncontrollable ICP levels and surgical management of the ICP with craniectomy may offer the opportunity for hematoma resolution and often dramatic recovery. A randomized controlled trial has been initiated to investigate this further.

7. Conclusion

The management of ICP is critical in patients with brain injury, whether from trauma, stroke or hemorrhage. Although recovery is inevitably limited by the degree of initial injury incurred, morbidity and mortality rise dramatically in the face of uncontrollable ICP. The management of elevated ICP has evolved significantly over the past century. Despite extensive research for many years, evidence for most medical therapies is lacking. In recent years, redirection toward early surgical intervention has re-emerged, particularly with subsets of trauma, stroke and subarachnoid hemorrhage patients. Evidence from case reports and small trials will need to be supported by larger trials. Even evidence from large trials needs to be scrutinized and challenged again with further trials to determine which patients may benefit the most from which therapies, particularly decompressive surgery. This may lead to more focused and more successful management of all patients with ICP issues. Inevitably all therapies have to be individualized to each individual case. In the mean time, courage and determination are needed by healthcare providers dealing with these emergencies since delays in recognition and intervention are often fatal.

8. References

[1] Narayan RK, Kishore PR, Becker DP et al. Intracranial pressure: to monitor or not to monitor? A review of our experience with acute head injury. J Neurosurg 56:650-659, 1982.

[2] Kincaid MS, Lam AM, Monitoring and Managing Intracranial Pressure. Continuum: Lifelong Learning in Neurology. 2006; 12(1): 93-108.

[3] Robertson, C. S., Valadka, A. B., Hannay, H. J., Contant, C. F., Gopinath, S. P. Cormio, M., Uzura, M., Grossman, R. G. (1999). Prevention of secondary ischemic insults after severe head injury. Critical Care Medicine, 27, 2086-2095.

[4] Fan JY, Effect of backrest position on intracranial pressure and cerebral perfusion pressure in individuals with brain injury: a systematic review. J Neurosci Nurs. 2004 Oct;36(5):278-88.

[5] Piek J: Barbiturate coma in patients with severe head injuries: long-term outcome in 79 patients. Adv Neurosurg 21:178--183, 1993

[6] Eisenberg H, Frankowski R, Contant C et al. Comprehensive central nervous system trauma centers: High dose barbiturate control of elevated intracranial pressure in patients with severe head injury. J Neurosurg 1988;69: 15-23

[7] Weaver DD, Winn HR, Jane JA. Differential intracranial pressure in patients with unilateral mass lesion. J Neurosurg 1982:56;660-5.

[8] Kalita J, Ranjan P, Misra UK. Current status of osmotherapy in intracerebral hemorrhage. Neurology India 2003; 51(1): 104-109.

[9] Kaufmann AM, Cardaso ER. Aggravation of vasogenic cerebral edema by multiple dose of mannitol. J Neurosurg. 1992;77:584-9.

[10] Juttler E, Schellinger P, Aschoff A, et al. Clinical review: therapy for refractory intracranial hypertension in ischaemic stroke. Crit Care. 2007;11:231–44.
[11] Bereczki D, Liu M, Fernandes do Prado G, et al. Mannitol for acute stroke. Cochrane Database Syst Rev. 2007; Issue 3. Art No: CD001153.
[12] Qureshi AI, Suarez JI, Bhardwaj A, et al. Use of hypertonic (3%) saline/acetate infusion in the treatement of cerebral edema: effect on intracranial pressure and lateral displacement of the brain. Crit Care Med. 1998;26:440–6.
[13] Muizelaar JP, Marmarou A, Ward JD, et al. Adverse effects of prolonged hyperventilation in patients with severe head injury: a randomized clinical trial. J Neurosurg 1991;75:731–9.
[14] Clifton GL, Miller ER, Choi SC, et al. Lack of effect of induction of hypothermia after acute brain injury. N Engl J Med 2001;344:556–63.
[15] Fernandes HM, Gregson B, Siddique S, et al. Surgery in intracerebral hemorrhage: the uncertainty continues. Stroke. 2000 Oct;31(10):2511-6.
[16] Kobayaski S, Miyata A, Serizawa T, et al. Treatment of cerebellar hemorrhage—surgical or conservative. Stroke. 1990; 21(8) Suppl: I-62.
[17] Kjellberg RN, Prieto A. Bifrontal decompressive craniotomy for massive cerebral edema. J Neurosurg. 1971;34:488-493.
[18] Kleist-Welch Guerra W, Gaab MR, Dietz H, et al. Surgical decompression for traumatic brain swelling: indications and results. J Neurosurg 1999;90:187–96.
[19] Polin RS, Shaffrey ME, Bogaev CA, et al. Decompressive bifrontal craniectomy in the treatment of severe refractory posttraumatic cerebral edema. Neurosurgery 1997;41:84–92.
[20] Guerra WK, Gaab MR, Dietz H, Mueller JU, Piek J, Fritsch MJ. Surgical decompression for traumatic brain swelling: indications and results. J Neurosurg. 1999 Feb;90(2):187-96.
[21] Ziai WC, Port JD, Cowan JA, Garonzik IM, Bhardwaj A, Rigamonti D. Decompressive craniectomy for intractable cerebral edema: experience of a single center. J Neurosurg Anesthesiol. 2003 Jan;15(1):25-32.
[22] RS, Shaffrey ME, Bogaev CA, et al. Decompressive bifrontal craniectomy in the treatment of severe refractory posttraumatic cerebral edema. Neurosurgery 1997;41:84–92.Surg Gyn Obstet 1905, 1:297-314.
[23] Hacke W, Schwab S, Horn M, et al. Malignant middle cerebral artery territory infarction: clinical course and prognostic signs. *Arch Neurol.* 1996;53:309–315.
[24] Robertson SC, Lennarson P, Hasan DM, et al. Clinical course and surgical management of massive cerebral infarction. Neurosurgery. 2004;55:55- 61;discussion 61–2.
[25] Walz B, Zimmermann C, Bo¨ttger S, et al. Prognosis of patients after hemicraniectomy in malignant middle cerebral artery infarction. J Neurol. 2002;249:1183–90.
[26] Schwab S, Steiner T, Aschoff A, et al. Early hemicraniectomy in patients with complete middle cerebral artery infarction. Stroke. 1998;29:1888 –1893.
[27] Juttler E, Schwab S, Schmiedek P, et al. Decompressive surgery for the treatment of malignant infarction of the middle cerebral artery (DESTINY): a randomized, controlled trial. Stroke. 2007;38:2518 –2525.
[28] Vahedi K, Vicaut E, Mateo J, et al. Sequential-design, multicenter, randomized, controlled trial of early decompressive craniectomy in malignant middle cerebral artery infarction (DECIMAL Trial). Stroke. 2007;38:2506 –2517.

[29] Hofmeijer J, Kappelle LJ, Algra A , et al. Surgical decompression for space-occupying cerebral infarction (the Hemicraniectomy After Middle Cerebral Artery infarction with Life-threatening Edema Trial [HAMLET]): a multicentre, open, randomised trial. Lancet Neurol. 2009 Apr;8(4):326-33.

[30] Vahedi K, Hofmeijer J, Juettler E, et al Early decompressive surgery in malignant middle cerebral artery infarction: a pooled analysis of three randomized controlled trials. Lancet Neurol. 2007;6:215–222.

[31] von Kummer R, Meyding-Lamade U, Forsting M, et al. Sensitivity and prognostic value of early CT in occlusion of the middle cerebral artery trunk. Am J Neuroradiol. 1994;15:9 –15; discussion 16–8.

[32] von Kummer R, Meyding-Lamade U, Forsting M, et al. Sensitivity and prognostic value of early CT in occlusion of the middle cerebral artery trunk. Am J Neuroradiol. 1994;15:9 –15; discussion 16–8.

[33] Holtkamp M, Buchheim K, Unterberg A, et al Hemicraniectomy in elderly patients with space occupying media infarction: improved survival but poor functional outcome. *J Neurol Neurosurg Psych.* 2001;70:226 –228.

[34] Van der Worp HB, Kappelle LJ. Early decompressive hemicraniectomy in older patients with nondominant hemispheric infarction does not improve outcome. *Stroke.* 2011;42:845- 846.

[35] Adams HP Jr, del ZG, Alberts MJ, et al Guidelines for the early management of adults with ischemic stroke: a guideline from the American Heart Association/American Stroke Association Stroke Council, Clinical Cardiology Council, Cardiovascular Radiology and Intervention Council, and the Atherosclerotic Peripheral Vascular Disease and Quality of Care Outcomes in Research Interdisciplinary Working Groups: The American Academy of Neurology affirms the value of this guideline as an educational tool for neurologists. *Circulation.* 2007;115: e478–e534.

[36] Thomas Pfefferkorn, Ursula Eppinger; Jennifer Linn, et al Long-Term Outcome After Suboccipital Decompressive Craniectomy for Malignant Cerebellar Infarction *Stroke.* 2009;40:3045-3050.)

[37] Smith ER, Carter BS, Ogilvy CS. Proposed use of prophylactic decompressive craniectomy in poor-grade aneurysmal subarachnoid hemorrhage patients presenting with associated large sylvian hematomas. Neurosurgery. 2002 Jul;51(1):117-24; discussion 124.

[38] Fisher CM, Ojemann RG. Bilateral decompressive craniectomy for worsening coma in acute subarachnoid hemorrhage. Observations in support of the procedure. Surg Neurol 1994;41:65–74.

[39] Scozzafava J, Brindley PG, Mehta V, Findlay JM. Decompressive Bifrontal Craniectomy for Malignant Intracranial Pressure Following Anterior Communicating Artery Aneurysm Rupture: two case reports. Neurocritical Care. February, 2007; 6(1): 49-53.

13

An Innovative Technique of Decompressive Craniectomy for Acute Ischemic Stroke

Marcelo M. Valença[1], Carolina Martins[2], Joacil Carlos da Silva[1], Caio Max Félix Mendonça[1], Patrícia B. Ambrosi[1] and Luciana P.A. Andrade-Valença[1,3]
[1]Neurology and Neurosurgery Unit, Federal University of Pernambuco,
[1]Neurosurgery Unit, Hospital Esperança,
[2]Medical School of Pernambuco Imip,
[3]Neurology Unit, Medical School of the University of Pernambuco,
Recife,
Brazil

1. Introduction

Decompressive hemicraniectomy (DH) is a lifesaving surgical procedure to arrest a rapidly rising intracranial pressure, not controllable with other methods. This procedure decreases the intracranial mass effect, reducing the risk of transtentorial herniation, therefore preventing secondary brain injury, brainstem compression and death.

Currently, malignant hemispheric infarction is one of the main indications for DH (Chen et al., 2006; Delashaw et al., 1990; Rengachary et al., 1981; Schwab et al., 1998; Merenda & DeGeorgia, 2010) either when the intracranial hypertension does not respond to conservative therapies (*i.e.*, osmotic therapy, hyperventilation, etc.); or when there is an important brain shift of the midline structures, or both (Gerriets et al., 2001).

It is estimated that about 10% to 20% of the patients with infarction in the territory of the middle cerebral artery (MCA) or of the internal carotid artery (ICA) develop hemispheric cerebral edema/swelling, presenting signs of uncal and cingulate herniations and neurological deterioration (Paciaroni et al., 2011; Minnerup et al., 2011; Thomalla et al., 2010; Paciaroni et al., 2011). Clinical deterioration of patients with massive MCA infarction (with or without additional involvement of anterior cerebral artery or posterior cerebral artery) is the result of brain edema, which peaks on days 3 to 5, followed afterwards by progressive edema reduction within the following two weeks (Shaw et al., 1959; Merenda & DeGeorgia, 2010; Ng & Nimmannitya, 1970). This form of malignant cerebral infarction - often irresponsive to medical treatment - is associated with very high mortality rates (i.e. 70% to 90% of cases) (Jourdan et al., 1993; Rieke et al., 1995; Ropper, 1986; Moulin et al., 1985).

Decompressive surgery comprehends the surgical measures taken to arrest a rising intracranial pressure and can be divided into external and internal decompression – at times combined into a single procedure. External decompression involves removal of parts of calvaria and duraplasty. Internal decompression is the term used when removal of affected parts of the brain or cerebellum is performed to achieve stabilization of intracranial pressure.

Decompressive surgery has been used in a number of emergency settings as brain trauma (Britt & Hamilton, 1978; Ransonhoff et al., 1971; Venes & Collins, 1975), *pseudotumor cerebri* (Martin & Corbett, 1996), encephalitis (Taferner et al., 2001), cerebral venous thrombosis (Valença et al., 2009), severe lead encephalopathy (McLaurin & Nichols, 1957), and Reye´s syndrome (Ausman et al., 1976). The use of DH has also been reported in the treatment of acute rupture of cerebral aneurysms (Fisher CM, Ojemann, 1994; Mitka, 2001).

2. Decompressive craniectomy – historical note

The idea of opening the skull to relieve pressure is an old concept. It was exemplified in the Greek myth of Athena, goddess of wisdom, according to Hesiod in the poem *Theogony* (circa 700 BC). Zeus, the supreme god of Olympus, after swallowing his first wife Metis, had a severe and progressively worsening headache attributed to the growth of his daughter Athena inside his skull. Hephaestus, the god of fire and metalwork, struck Zeus´ head with an axe splitting his cranium in order to alleviate his pain, and Athena was born through this primitive craniotomy (Brasiliense et al., 2010).

Trepanation has been performed since the Mesolithic Era, before the development of metal instruments and written language. There is widespread archaeological evidence from South America, Europe, Asia and Africa documenting this ritualistic skull opening as long as 10.000 B.C. (Marino & Gonzales-Portillo, 2000; Martin, 1995, Missios, 2007).

The ancient Egyptians performed trephinations with local anesthesia, prepared by mixing ground marble with vinegar. The success of these procedures is evidenced by the discovery of healed wounds in skulls unearthed from tombs dating back 6000 years. On the tomb of Bany Hassan, there are paintings that depict a surgeon performing cranial surgery with the patient seated (El Gindi, 2002).

The pre-conquest South America nations were extremely skilled in the art of opening the skull. Operations were apparently performed for trauma, fractures, diseases of the cranium, scalp and cranial infections, epilepsy, headaches, mental disease, and some thaumaturgic rituals. The main archaeological findings related to trephination are associated with excavations in the territories of the Paracas, Nazca, Huari, and Içá cultures. They were less popular in the more recent Inca culture and were apparently forbidden in the Inca Empire long before the Hispanic conquest (Marino & Gonzales-Portillo, 2000).

The written description and codification of trepanation began in Greece during the fifth century B.C. Hippocrates, in his work *On Injuries of the Head*, described the different types of skull fractures and provided specific instructions as well as warnings on the use and dangers of trepanation (Martin, 2000).

Galen of Pergamon (A.D. 129–200) lived in an era when trepanation was better known and more widely accepted by practicing physicians as well as the public. He advocated use of the procedure for relief of intracranial pressure and drainage of phlegmatous lesions (Missios, 2007).

After the collapse of Roman Empire, medical knowledge was preserved by Middle Eastern cultures. The most famous representative author or this period was Ibn Sina or Avicenna (A.D. 980-1037); although he was well versed in intracranial anatomy, his contributions on cranial surgery are mostly speculative (Sarrafzadeh et al., 2001).

The practice of surgery in the Middle Ages was well described by a less well-known author called Serefeddin Sanbuncuoglu (A.D. 1385–1468?), the author of *Cerrahiyyetü'l-Haniyye* (Imperial Surgery), which was written in Turkish in 1465. The book contains wonderful color illustrations of craniotomy procedures (Elmaci, 2000).

Berengario da Carpi (1460-1530) wrote the most important work on craniocerebral surgery of the Renaissance Period, the *Tractatus de Fractura Calvae sive Cranei* (Treatise on Fractures of the Calvaria or Cranium), in which he described an entire set of surgical instruments to be used for cranial operations to treat head traumas, which became a reference for a later generation of physicians (Di Ieva et al., 2011).

The modern concept of decompressive surgery was born with the statement of Kocher, in 1901: *"If there's no CSF pressure, but brain pressure exists, then pressure relief must be achieved by opening the skull."* This author introduced the concept of elevated intracranial pressure as a global phenomenon requiring a large skull opening to be treated (Gautschi & Hildebrandt, 2009).

Harvey Cushing designed a standard subtemporal decompressive craniectomy largely used during his huge neurosurgical series with highly impressive results (Cushing, 1905).

Despite its sporadic use for trauma and other conditions characterized by elevated ICP, decompressive surgery was not routinely employed until the 1990s. Probably the initial poor results when used in patients without Intensive Care Unit (ICU) support led some authors to abandon the procedure. As stated by Clarke in 1968: "The reason for reporting this experience is to warn others from doing similar surgery" (Clark et al., 1968).

The landmark work of Guerra and colleagues (1999) described their results using decompressive craniectomy for traumatic brain swelling. The good results from this series established the procedure as an appropriate therapy for refractory intracranial pressure of any origin (Guerra et al., 1999).

Decompressive craniectomy for cerebral ischemia was seldom indicated. Ivamoto, in 1974, reported the first clinical series, but with no great repercussions (Ivamoto et al., 1974). However, after the definition of the "malignant" MCA territorial infarction, which courses with dramatic intracranial hypertension, there has been a number of experimental and clinical studies on the benefits of early decompression of the brain during ischemic events (Hacke et al., 1996).

Finally the scientific evidence of mortality reduction demonstrated by three randomized trials has made decompressive surgery part of the modern arsenal for treatment of ischemic stroke (Vahedi et al., 2007).

3. Decompressive hemicraniectomy for cerebral hemispheric ischemic lesions

The better results obtained after DH in patients with hemispheric lesions (e.g. stroke), when compared with patients with severe brain trauma, could be explained by the fact that lesions in unconscious patients after trauma are usually diffuse rather than restricted, as is the case of an unilateral cerebral hemispheric lesion, such as MCA ischemia.

3.1 When we should perform a decompressive hemicraniectomy?

Decompressive hemicraniectomy in cerebral hemispheric ischemia is a lifesaving procedure that should be used without delay when the intracranial hypertension does not respond to conservative therapies (*i.e.*, osmotic therapy, hyperventilation, etc.); or when there is an important brain shift of the midline structures, or both, in a symptomatic patient with motor deficit and impairment of consciousness. In a previous healthy patient with no severe cognitive deficit the age of the patient (e.g. >75 years of age) or side of the lesion (left sided lesion in a right handed patient) does not contraindicate the procedure (Merenda & DeGeorgia, 2010).

In such a setting, the immediate removal of the bone corrects the cerebral displacement, relieving the pressure exerted on the rostral midbrain structures. In theory, DH may also improve perfusion of the collateral leptomeningeal vessels, improve retrograde perfusion of MCA, optimize perfusion of the penumbra, and, consequently reduce the size of the infarct and neurological deficit, even in the absence of a brainstem deformation or uncal compression. Uncal and cingulate herniations may cause secondary vascular compressions. During the development of a cingulate (subfalcine) herniation, the cingulate gyrus shifts across and under the falx cerebri and it may compress the anterior cerebral artery, cutting off the blood supply to normal brain areas. Uncal herniation, likewise, causes cistern obliteration and may compress the posterior cerebral artery. The neurological repercussions caused by an uncal herniation (a subtype of transtentorial herniation) due to a MCA territory infarction depends on the degree of brain and brainstem atrophy, intracranial volume reserve (subaracnoid space/CSF) and the type of variation of the tentorial aperture.

Since the works of Sunderland (1958) and Corsellis (1958) a relationship between the notch size/shape and patterns of uncal herniations was recognized. In this regard, Adler and Milhorat (2002) studied 100 human autopsy cases (23 female cadavers) and derived from data related to the maximum notch width and notch length, they classified the specimen tentorial apertures as: narrow (15%), wide (12%), short (8%), long (15%), typical (24%), large (9%), small (10%), and mixed (7%). Accordingly, patients with a wide tentorial notch (i.e. maximum notch width >32 mm)(Adler & Milhorat, 2002) and a considerable space between the brainstem and the free tentorial edge have a lower chance of brainstem compression by cerebral tissue herniation.

Furthermore, the territory irrigated by the MCA is variable and the volume of ischemic area after occlusion of M1 segment of the MCA is unpredictable, it depends on the extension of collateral circulations (mainly those supplied by the anterior cerebral artery and posterior cerebral artery).

There are predictive signs of MCA malignant infarction observed on CT-scan (performed within 12 hours of ictus), such as: (1) hypodensity in more than 50% of the MCA territory, (2) hyperdense MCA, and (3) diffuse attenuation between the cortex and the white matter. In addition, MRI findings with diffusion weighted imaging signals occupying >89 ml volume is 90% sensitive and 96% specific in predict malignant cerebral hemispheric infarction (Arenillas et al., 2002).

In a prospective cohort of 1,010 patients with acute ischemic stroke, Paciaroni and associates (2011) observed an incidence of hyperdense MCA and/or ICA arteries of about 15%, which was associated with a final infarct involving more than one third of the MCA territory and poor functional outcome at 3 months. Overall, 163 patients (16.1%) had a final infarct covering more than one third of the MCA territory. 52.7% of the individuals with hyperdense MCA and/or ICA had an infarct involving more than one third of the MCA territory compared to only 9.9% of the patients without artery hyperdensity. At 3 months, 18 patients were lost to follow-up, 325 patients (32.8%) were disabled and 165 died (16.5%). The following factors were associated with unfavorable outcome: age, NIH Stroke Scale score for 1 added point on admission, stroke due to atherosclerosis, hemorrhagic transformation of the ischemic lesion, and hyperdense MCA and/or ICA (OR 2.0, 95% CI 1.0-4.0).

1. Initial NIH Stroke Scale (NIHSS) score at least 20 for dominant stroke

2. NIHSS score at least 15 for nondominant stroke

3. Younger age

4. Early hypodensity >50% of the MCA territory, including the basal ganglia

5. Additional involvement of ipsilateral ACA or PCA territories

6. CT-documented anteroseptal shift >5 mm

7. CT-documented pineal shift >2 mm

8. Volume of ≥145 ml on DWI MRI

Fig. 1. Clinical and neuroradiological features predictive of a malignant course after a middle cerebral artery (MCA) infarction. ACA, anterior cerebral artery; PCA, posterior cerebral artery.

Recent study (Minnerup et al., 2011), enrolling 52 ischemic stroke patients with carotid-T or MCA main stem occlusion and ischemia (reduced cerebral blood volume) on perfusion CT performed on admission, showed that 26 (50%) of them developed malignant MCA infarction. Two subgroups were separated according to the development of malignant MCA infarction defined by clinical signs of herniation. Age, a decreased level of consciousness on admission, cerebral blood volume (CBV) lesion volume, CSF volume, and the ratio of CBV lesion volume to CSF volume were significantly different between malignant and nonmalignant subgroups. The best predictor of a malignant course was the ratio of CBV lesion volume to CSF volume.

Merenda and DeGeorgia (2010) cited a number of clinical and neuroradiological features that appear to be predictive of a malignant course after MCA infarction, as illustrated in Figure 1. Although, a significant number of patients presenting predictors of malignant course may not develop fatal brain swelling (Merenda & DeGeorgia, 2010).

3.2 The classic decompressive craniectomy

The technical recommendations for the performance of DH stress that craniectomy should be performed in the frontotempoparietal region, reaching the base of the frontal bone and sparing the calvarium ≈ 1 cm from the midline to prevent injury to bridging veins and additional bleeding (Wagner et al., 2001). A diameter of around or above 12 cm is desirable, as it has been shown that doubling the diameter from 6 to 12 cm results in an increase in decompressive volume from 9 to 86 cm (Wirtz et al., 1997; Wagner et al., 2001). The bone removal towards the occipital squama adds little to the decompressive procedure itself, as the falx prevents the medial incursion of cerebrum in this region and has a constant length of approximately 6 cm (Rhoton Jr & Ono, 1996), besides, it might produce problems of stability and patient positioning (Wagner et al., 2001).

3.3 Duraplasty

It is estimated that only 15% reduction is achieved with craniectomy alone and duraplasty may decrease intracranial pressure by an additional 55% (Chen et al., 2006). Consequently, a generous duraplasty would cause significant intracranial pressure reduction. Usually a planed, large skin incision, allows for sufficient quantities of pericranium to be used as a dural graft and this has resulted in lower deep wound infection rates (5%) than when foreign material was used (e.g. neuro-patch, 15%) (Malliti et al., 2004), Additionally, CSF leaks were more frequent in the neuro-patch group of patients (13% vs. 1.6%). Synthetic dural grafts should therefore be reserved for situations when autologous grafts are not sufficient or unavailable.

3.4 Disadvantages and secondary problems following decompressive craniectomy

As HD increased in popularity among surgeons as a life-saving treatment of malignant intracranial hypertension following a vascular insult or brain trauma, reports on its technical execution and potential complications have been appearing in the literature. Among them are hemorrhagic complications, overindication, problems with flap storage and sterilization, metabolic changes associated with lack of bony covering, syndrome of the trephined, hydrocephalus and others (Chen et al., 2006; Valença et al., 2009; Faleiro et al., 2008).

3.5 Hemorragic complications

Parenchymal hemorrhages and infarcts, secondary to the initial site of ischemia, and associated with the hemicraniectomy occurred with high frequency rates (Wagner et al., 2001). An increasing risk of mortality is also linked to hemicraniectomy-associated bleeding.

This hemicraniectomy-associated bleeding is related to the size of the craniectomy performed, that is to say, the smaller the hemicraniectomy, the more often lesions will occur. The mechanism leading to hemicraniectomy-associated lesions involves the shear forces at the edges of the trephination. Small craniectomies cause a higher shear stress on the swollen brain, particularly if the craniectomy edges are sharp.

Creation of vascular channels to reduce the risk of vascular congestion in the herniated tissue, thereby decreasing the secondary injury caused by the craniectomy was a procedure designed by Csókay and coworks (2002) to decrease the brain lesion induced by the classic decompressive hemicraniectomy.

3.6 Increased brain edema

After vascular occlusion brain water content increases, reaching its maximum 2-4 days later (O´Brien & Waltz, 1973) and significant changes in blood brain barrier may occur for 20 days (O´Brien et al., 1974). Thus, we can subdivided the brain swelling in stroke in three phases: (1) an immediate increase (within first minute) in brain volume due to vasoparalysis (tissue acidosis); (2) a metabolic (cytotoxic) type of brain edema; and (3) a vasogenic brain edema which follows the metabolic type of edema after 4-6 hours, reaching its maximum after a few days. In about 3-4 days after cerebral arterial occlusion it is detected the maximal anatomic brain deformity, and it is only after this period of time that a gradual reabsortion of the excessive parenchymal fluid takes place (Hossmann & Schuier, 1979; Ayata & Ropper, 2002; Valença et al., 2009b; Shaw et al., 1959).

In addition, the combination of reperfusion and hemicraniectomy experimentally caused an increase in brain edema and breakdown of the blood-brain barrier, although a decrease in the infarction size was observed (Engelhorn et al., 2003). Comparable results were obtained by Cooper and colleagues (1979) in edema induced by cold in dogs, with edema volume seven times larger in craniectomized animals. Since the driving force in edema pathophysiology is the pressure gradient across the injured capillaries, cerebral decompression may decrease interstitial fluid pressure and, as a result, increase edema formation. In this regard, craniectomy caused a significant decline in tissue pressure in the cortical gray matter in cat models (Hatashita et al., 1985), a phenomenon that probably occurs after decompressive craniectomy in a patient with brain infarction caused by traumatic vessel occlusion, who had a poor outcome (Venes & Collins, 1975). This increased edema can lead to higher shear forces and contribute to hemicraniectomy-associated lesions.

3.7 Flap storage and sterilization

The gradual acceptance of decompressive craniectomy as a therapeutic tool in a selected, but significant number of patients with stroke or intracranial hypertension will result in an increased demand for cranioplastic procedures with its inherent costs.

During classic decompressive hemicraniectomy procedure, the bony flap is removed and stored under the abdominal fat or in a freezer. The replacement of the bone or cranioplasty is performed after the resolution of the hemispheric swelling, which occurs six to 20 weeks later, in surviving patients. The storage of the a large bony flap under the abdominal fat is time-consuming procedure and very probably is a cause of moderate to severe abdominal pain, not infrequently associated with sizeable hematomas. Infection may occur, as well. It is well-know that increase in the intracranial pressure is the result of pain and agitation, thus this is another disadvantage of storing the bone in the abdominal subcutaneous tissue. On the other hand, bone resorption may be high (up to 50%) after cryopreservation (Grant et al., 2004).

Autoclaving the bone flap and sterilization with ethylene oxide were also tried. Missori and coworkers (2003) repositioned the bone after an average period of 4.3 months in 16 patients in whom the bone flap was removed for decompressive craniectomy, washed in hydrogen peroxide and then placed in hermetically-sealed bags and sterilized using ethylene oxide. The authors reported that, after an average follow-up of 20 months, the esthetic and functional results were good in 15 patients (Missori et al., 2003).

Autoclaving the bone flap is another option to be used (Jankowitz & Kondziolka, 2006; Osawa et al., 2000, Vanaclocha et al., 1997). Biopsies of the tumor invaded bone flaps autoclaved for 20 minutes at 134 degrees C and 1 kg/cm^2 and re-implanted showed newly formed bone partly re-populated by osteocytes. After studied 6 cases the authors concluded that autoclaved calvarial bone, when replaced with direct contact with living bone, will be gradually repopulated with osteocytes (Vanaclocha et al., 1997). In this regard, we believe that an alternative that may increase the re-population rate of bone flaps that underwent autoclave sterilization is to reintroduce bone cells (e.g. osteoclasts and osteoblasts), using fresh bone powder or stem cells, into perforations made in the craniotomy flap to be replaced.

3.8 Overindication

In 1995, Rieke and colleagues reported the first prospective study with the use of DH in space-occupying supratentorial infarction, obtaining a significant benefit with the combination of bone removal and dural patch enlargement. Subsequently, Schwab and associates (1998) demonstrated that if the patients were submitted to decompressive surgery within the first 24 hours (even before any significant brain shift of the midline structures), a further reduction in mortality rates to 16% occur. The remaining question is: do all patients with early (predictive) clinical/neuroimaging signs of malignant cerebral infarction require DH for the correction of intracranial hypertension?

False-positive classification as malignant infarction is of major concern as it may result in unnecessary surgical treatment, while a high threshold for indicating this procedure may severely compromise results.

Although this question does not have an answer at this point, it is our feeling that performance of a decompressive surgery, using a technique that allows the bony flap to open gradually, according to the intracranial pressure (Valença et al., 2009), would be a less harmful alternative to some patients, since, in the absence of a gradual mass effect, the flap would stay in a normal anatomic position, while absent flap problems and metabolic changes due to lack of bone covering would be prevented. Moreover, a decompressive surgery may improve perfusion of the collateral leptomeningeal vessels, optimizing perfusion of the penumbra regions.

3.9 Syndrome of trephined

As a larger-sized hemicraniectomy is desirable to prevent complications related to insufficient decompression (Stiver, 2009), the number of problems related to a sinking defect might be expected to increase.

The "syndrome of the trephined" or "sinking skin flap syndrome" (Stiver, 2009; Winkler et al., 2000) involves the occurrence of severe headache, convulsion, dizziness, undue fatigability, poor memory, irritability, and intolerance to vibration. The syndrome of the sinking skin flap has been related to the action of atmospheric pressure and changes in hydrostatic pressure of cerebrospinal fluid (CSF), and those have been shown to resolve after cranioplasty.

It is known that skull defects are associated with a blood flow decrease related to postural changes, suggesting that DH has an impact on dynamic cerebral blood flow regulation. Cranioplasty affects postural blood flow regulation, cerebrovascular reserve capacity and cerebral glucose metabolism (Winkler et al., 2000). These effects were associated with a clinical improvement in all patients except in those whose cranioplasty was delayed for a long time following the HD. Therefore early cranioplasty or a decompressive surgery which uses a technique that allows the bony flap to be kept in place and open gradually, according to the intracranial pressure is desirable to facilitate rehabilitation in patients after decompressive surgery (e.g. "window-like"craniotomy) (Valença et al., 2009).

3.10 Hydrocephalus

Hydrocephalus is also a common complication after DH, especially in trauma (Stiver, 2009). Its causes are multiple, but absence of a bone flap causes major disturbance to the intracranial pressure regulatory mechanisms. As a result of the lower intracranial pressure, CSF absorption may be deficient, since a pressure gradient between CSF and the venous blood is required for this to happen.

Again, a decompressive surgery which uses a technique that allows the bony flap to be kept in place and open gradually, according to the intracranial pressure would result in sooner reestablishment of intracranial biomechanical conditions. If such an assumption is correct, a lower incidence of hydrocephalus would be expected with the use of our proposed "window-like"craniotomy procedure (Valença et al., 2009).

4. "Window-like"craniotomy *versus* classic decompressive craniectomy

Although DH has led to a substantial decline in mortality rates, several problems remained to be solved. Many of them resulted from the lack of bony covering and the changes associated with it. A decompressive surgery that used a technique allowing the bony flaps to be open gradually depending on the intracranial pressure would theoretically sort out many of the problems related to DH (Valença et al., 2009).

The idea of performing an "window-like" craniotomy as an alternative to substitute decompressive hemicraniectomy was reported in 2003 (Valença & Valença, 2003). In 2009 we published our results with a series of cases (Valença et al., 2009), emphasizing the advantages of the new procedure over the classic decompressive hemicraniectomy (Figure 2).

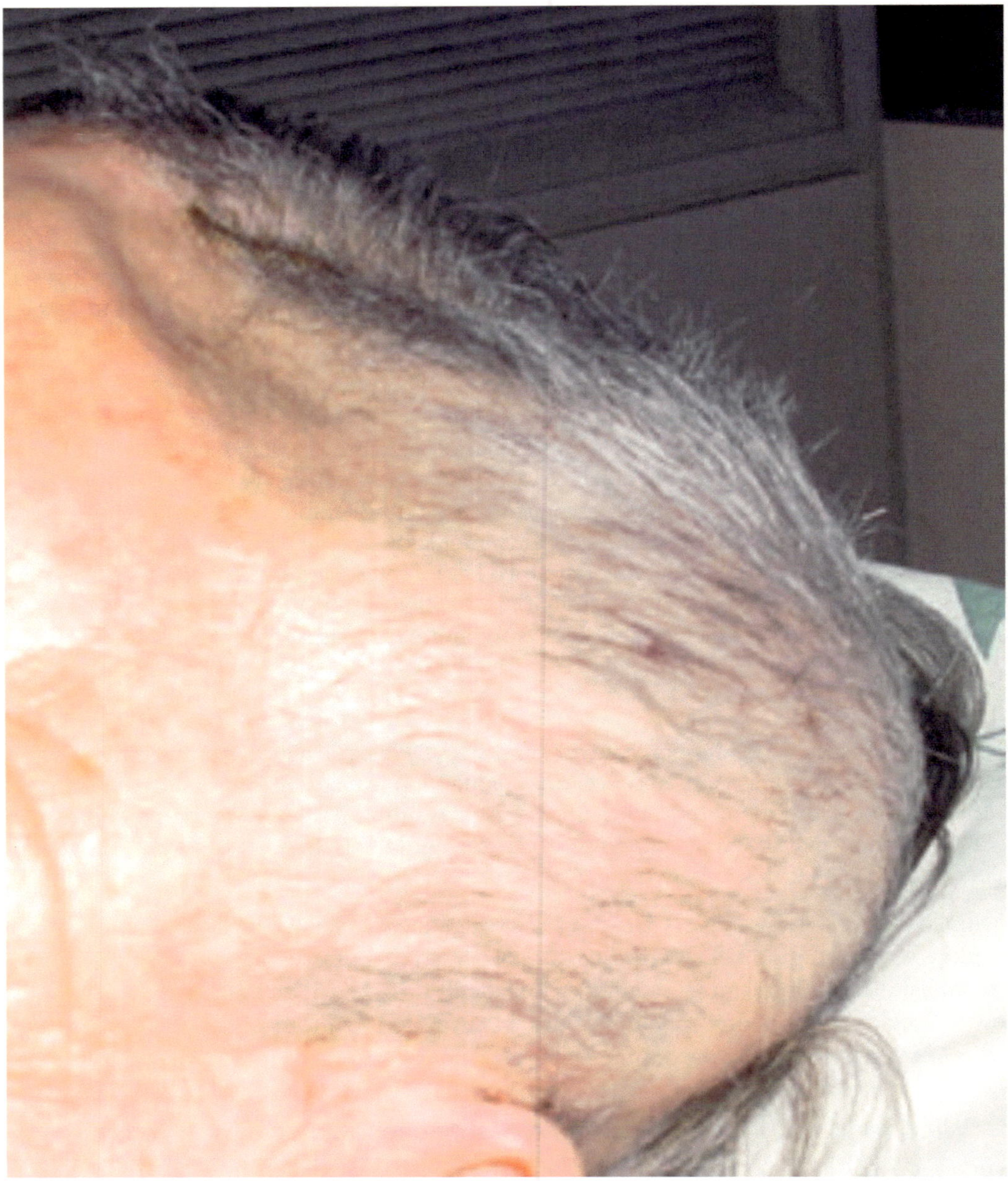

Fig. 2. The classic decompressive craniectomy, which causes trephine syndrome.

5. "Window-like" craniotomy: Surgical technique

The surgical procedure involves a large semicircular skin incision starting in the proximities of the midline and extending to the posterior parietal area, ending at the level of the *tragus*. The scalp must be elevated from the underlying pericranium to ensure greater elasticity and a looser skin flap to allow the stitching up of the skin at the end of the "window-like"operation, covering again the bone with the skin.

An extensive, near rectangular-shaped craniotomy is performed involving frontal, temporal, and parietal bones, and part of the occipital squama (diameter 12-15 cm). The angle of the bone cut must be beveled outwards to allow the upper part of the craniotomy bone flap to rest on the adjacent skull and prevent penetration into the intracranial cavity. An anterior temporal craniectomy (subtemporal decompression) is added to relieve temporal lobe pressure (Figure 3). Dural incisions are performed and the dura is fixed at the bone border to prevent epidural bleeding (Valença et al., 2009).

Using a vertical cut, the bone flap is divided into two similarly sized pieces, which will be the opening of the "window lids". The outer frontal and parietooccipital sides of the flap are each one tied to the skull at two points (see arrows in Figure 4) using a synthetic nonabsorbable suture (i.e. polypropylene) to function as a hinge joint that allows opening of the window but prevents downward movement of the bone inside the skull. We advise inserting sutures at the edges of the bone flap in such a way that the lateral portion of the window is divided into three parts (Figures 3 and 4).

The recommended beveling of the craniotomy edges is important in the upper part of the craniotomy. What may prevent the bone from sinking in, besides the above-mentioned aspect, are the following: 1. the underlying attached dura mater; therefore, the dura-bone stitches should be placed 0.5 cm from the bone edge. 2. a good alignment of the bone flaps with the calvarium; 3. the intracranial pressure; and 4. the fact that the close of the bone "window" and the cranial vault reconstruction follow the principles used in the construction of vaults (a semicircular arch-shaped structure) where many segments are held in place by lateral thrust (representing the calvarium where the bone flaps will be fixated, Figure 3). The anterior and posterior parts of the craniotomy - the suture placement - should be cut vertically in order to permit a good alignment of the bone flaps with the calvarium. Thus, the central, bisected portion of the bone does not "sink in" when the brain later atrophies after infarct or major trauma (Valença et al., 2009). In our experience we have seen a good aesthetic result as early as 2-3 months following the procedure, although a foreseeable defect at the edge and along the central vertical line of the craniotomy can be felt on palpation.

After 5 months of follow-up the appearance of the "window-like" craniotomy was similar to that of a normal classical "fixated" craniotomy, with no signs of movement of the partially free bone flaps.

A dura patch, using synthetic graft, lyophilized cadaver dura mater, pericardium, or homologous pericranium, fascia lata or temporal fascia is placed in the incision (~ 13-20 cm in length and 4-8 cm wide). The bone flap is divided into two similar-size pieces by a vertical cut, which will be the opening of the "window lids". The outer frontal and parieto-occipital sides of the flap are tied to the skull at 2 points, to function as a hinge joint that allows the opening of the window but prevents the bone from moving downward inside the skull.

A number of difficulties are encountered when performing the duraplasty using the stellar durotomy. In order to facilitate the duraplasty we have developed a new form of dural closure resembling an anteroposterior bridge between the dural edges (Valença et al., 2009). The dura mater is opened in a way that resembles two semicircular "ears" with the bases facing the upper and lower bone boundaries, respectively. On the anterior and posterior limits of the craniotomy 1 cm of dura mater is left in order to permit a subsequent suture of the rectangular dural graft.

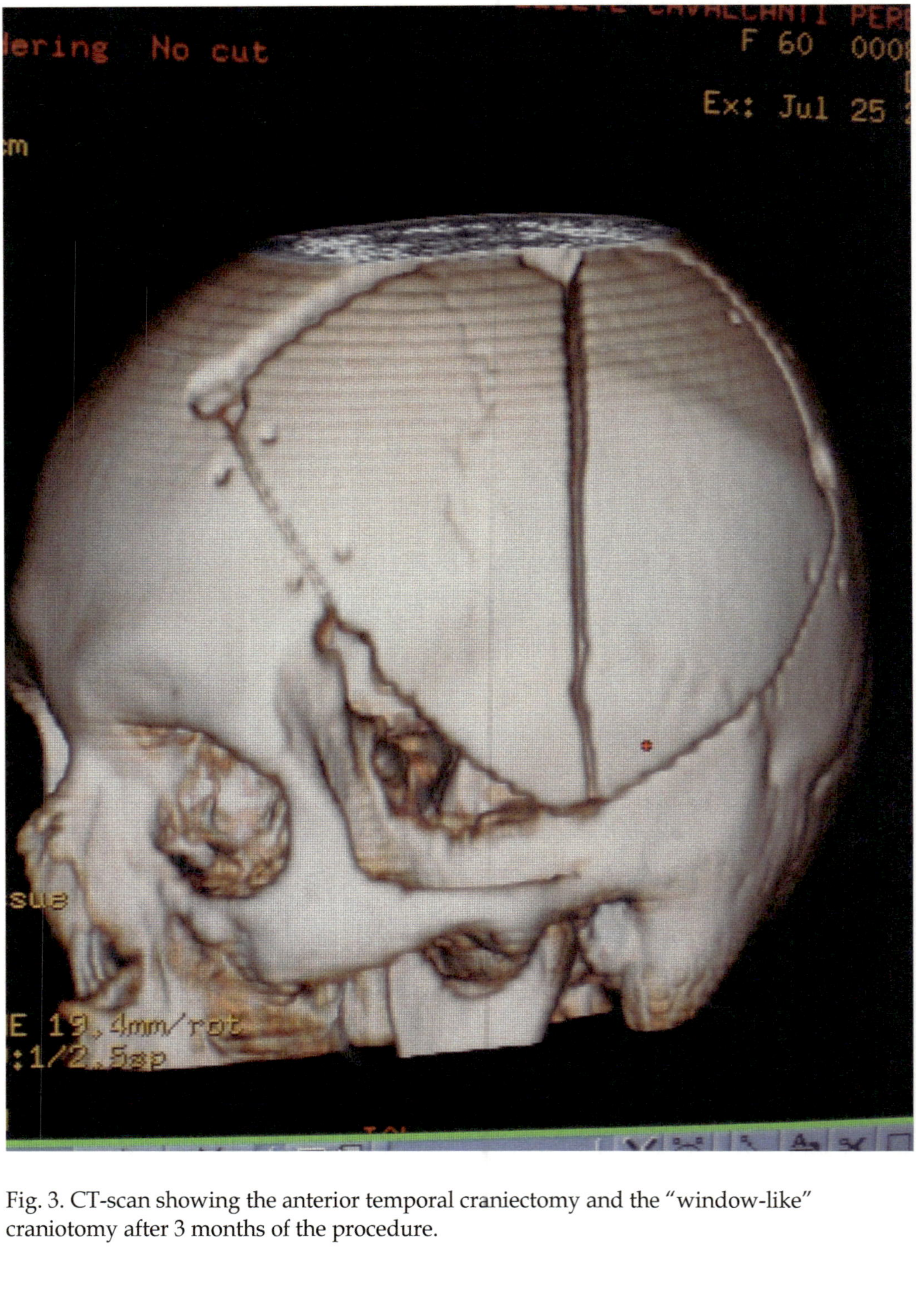

Fig. 3. CT-scan showing the anterior temporal craniectomy and the "window-like" craniotomy after 3 months of the procedure.

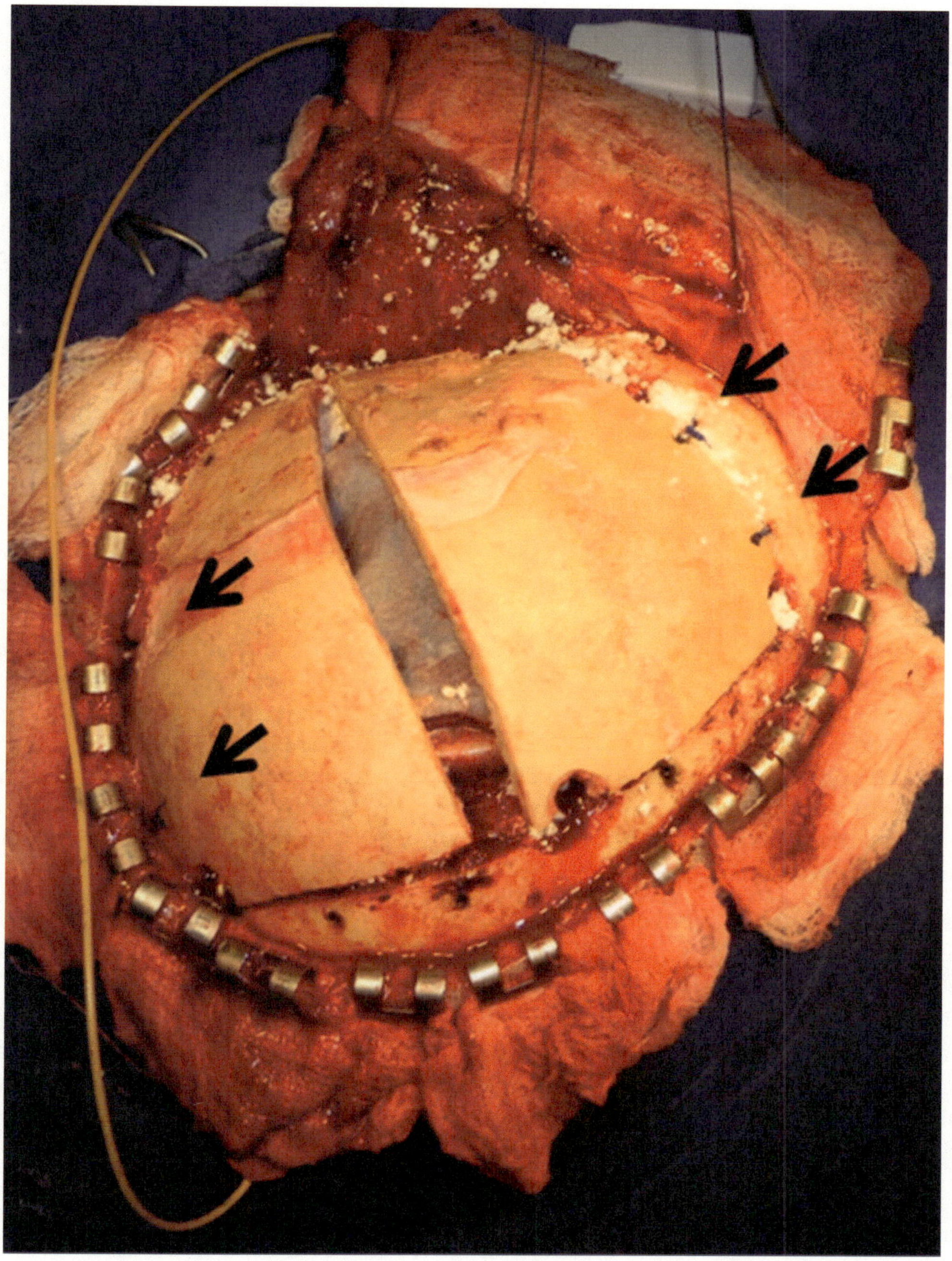

Fig. 4. Left side "window-like"craniotomy. The outer frontal and parietooccipital sides of the flap are each one tied to the skull at two points (see arrows) using a synthetic nonabsorbable suture to function as a hinge joint that allows opening of the window but prevents downward movement of the bone inside the skull. We advise inserting sutures at the edges of the bone flap in such a way that the lateral portion of the window is divided into three parts.

6. Advantages of decompressive surgery, using a technique that allows the bony flap to open gradually, according to the intracranial pressure

The proposal of decompressive surgery using a craniotomy in an "window-like" fashion presents the following advantages: (i) it allows the edematous cerebral parenchyma to herniate with a gradual opening, simultaneously relieving the elevated intracranial pressure; (ii) it is an anatomic option that preserves the brain-skull biomechanics and avoids the use of expensive bio-prosthetic materials; (iii) it may avoid the development of sinking skin flap symptoms; (iv) it obviates the need for a second bone-replacement surgical procedure at a later stage, which may have an impact on the overall treatment cost; (v) it is not necessary the storage of a large bony flap under the abdominal fat, a procedure that is time-consuming and is cause of moderate to severe abdominal pain (pain causes agitation thereby increasing intra-abdominal and intracranial pressures), sometime associated with sizeable hematomas and infection; (vi) probably there is an attenuation in the formation of brain edema because the procedure may allow pressure at the herniated tissues close to normal levels with the presence of the bone flap.

The "window-like" procedure allows the performance of a large hemicraniectomy to adequately decompress the ischemic brain and avoid hemicraniectomy-associated lesions. At the same time, it offers an anatomic, inexpensive solution that allows the gradual accommodation of the herniated brain tissue, with a decrease in the intracranial pressure after brain insult. This surgical treatment may avoid edge trauma and ischemia, since the transcalvarial (external) herniation may be directed more uniformly towards the center of the craniotomy by the angle formed between the skull and the partially opened bone flaps. Again, the principle of the window-like craniotomy prescribed here mimics the opening of the sutures observed in child hydrocephalus. On the other hand, by recomposing a resistant barrier between the brain and the environment, it possibly prevents the sinking flap symptoms that detract from the functional outcome. A comparable procedure - hinge craniotomy - was recently described (Schmidt et al., 2007).

As to whether a young survivor could resume sports after the "window-like" procedure, we still lack sufficient evidence to confidently state that after one year the bone has healed sufficiently to allow sporting activities, we believe that such activities can be recommended without the need for further surgery to stabilize the bone. However, to be certain of this, further studies are required to clearly demonstrate bone resistance sufficient to sustain significant impacts, not only in the case of the procedure presented here but also after the subsequent repositioning of the bone to correct the defect left by a conventional decompressive hemicraniectomy, since bone resorption might occur (Stiver, 2009, Faleiro et al., 2008).

The "window-like" craniotomy was used by us in eight patients with malignant ischemic stroke with no mortality. None of these patients develop hydrocephalus and the aesthetic result was good.

7. Future perspectives

Present results indicate that, despite adequately addressing several drawbacks of the conventional DH, "window-like" techniques for DH will benefit further from observations of clinical practice. From a technical stand-point the technique might be improved by the development of two devices: (a) a hinge and (b) a synthetic dura mater substitute in the form of an ellipsoid "bag" rather than the commercial available rectangular graft (Valença et al., 2009). Although this second device could save time during duroplasty, it can only be acceptable if it presents complication rates similar or smaller than using pericranium grafts.

We reported (Valenca et al., 2009) the possibility of using metal hinges to replace the synthetic nonabsorbable sutures. To this end, the hinges (2 on each side of the craniotomy) should be fixed into the calvaria and to the bone flaps by using screws. The metallic implants must be compatible with the use of MR imaging. The opening must be situated toward the outer calvaria, a kind of hinge designed to lock the bone flaps in a natural anatomical position (horizontal plane), thus preventing the central portion of the bone from sinking in when the brain later atrophies.

Using the same rational to develop the idea of the "window-like" craniotomy, we recently use a similar approach to substitute the bifrontal decompressive craniectomy. The one-piece bifrontal craniotomy is tied to the skull posteriorly at two points, allowing the opening of the bone flap in a fashion similar to a "car hoop" (car hood decompressive craniotomy). We used this procedure with success in two occasions: a man with severe brain trauma with diffuse bifrontal contusion and a patient with a large fronto-basal meningeoma with significant brain edema. The "car hood" procedure may be useful in patients with bilateral infarction involving the anterior cerebral arteries, particularly in those with occlusion of a dominant A1.

8. Conclusion

Decompressive craniectomy in cerebral ischemia is a lifesaving procedure. The history of the development of this treatment, its indications, the techniques available to perform the procedure, its expected results and complications were discussed. Alternative procedures to the classic decompressive hemicraniectomy, including "window-like" craniotomy, were presented. Decompressive surgery, using a technique that allows the bony flap to open gradually, according to the intracranial pressure, adds the advantage of avoiding a second surgical procedure to close the bone defect, therefore preventing the metabolic cerebral impairment associated with lack of the overlying skull.

In conclusion, the present proposal of decompressive surgery using a craniotomy in an "window-like" fashion presents the following advantages: (i) it allows the edematous cerebral parenchyma to herniate with a gradual opening, simultaneously relieving the elevated intracranial pressure; (ii) it is an anatomic option that preserves the brain-skull biomechanics and avoids the use of expensive bio-prosthetic materials; (iii) it may avoid the development of sinking skin flap symptoms; (iv) it obviates the need for a second bone-replacement surgical procedure at a later stage, which may have an impact on the overall treatment cost.

9. References

Adler DE, Milhorat TH (2002) The tentorial notch: anatomical variation, morphometric analysis, and classification in 100 human autopsy cases. Journal of Neurosurgery 96:1103-1112

Arenillas JF, Rovira A, Molina CA, Grivé E, Montaner J, Alvarez-Sabín J. Prediction of early neurological deterioration using diffusion- and perfusion-weighted imaging in hyperacute middle cerebral artery ischemic stroke. Stroke. 2002 Sep;33(9):2197-203.

Ausman JI, Rogers C, Sharp HL: Decompressive craniectomy for the encephalopathy of Reye´s syndrome. Surg Neurol 6:97-99, 1976

Ayata C, Ropper AH. Ischaemic brain oedema. J Clin Neurosci 2002;9:113-124.

Brasiliense LB, Safavi-Abbasi S, Crawford NR, Spetzler RF, Theodore N. The legacy of Hephaestus: the first craniotomy. Neurosurgery. 2010 Oct;67(4):881-4; discussion 4.

Britt RH, Hamilton RD. Large decompressive craniotomy in treatment of acute subdural hematoma. Neurosurgery 2:195-200, 1978

Chen C, Smith ER, Ogilvy CS, Carter BS. Decompressive craniectomy: physiologic rationale, clinical indications, and surgical considerations. In: Schmidek & Sweet Operative Neurosurgical Techniques: Indications, Methods, and Results. Elsevier, Philadelphia, 70-80, 2006

Clark K, Nash TM, Hutchison GC. The failure of circumferential craniotomy in acute traumatic cerebral swelling. J Neurosurg. 1968 Oct;29(4):367-71.

Corsellis JAN. Individual variation in the size of the tentorial opening. J Neurol Neurosurg Psychiatry 21:279-283, 1958

Csókay A, Együd L, Nagy L, Pataki G: Vascular tunnel creation to improve the efficacy of decompressive craniotomy in post-traumatic cerebral edema and ischemic stroke. Surg Neurol 57:126-129, 2002

Cooper PR, Hagler H, Clark WK, Barnett P. Enhancement of experimental cerebral edema after decompressive craniectomy: implications for the management of severe head injuries. Neurosurgery. 1979;4:296-300

Cushing H. The stablishment of cerebral hernia as a decompressive measure for inaccessible brain tumor; with the description of intramuscular methods of making the bone defect in temporal and occipital regions. Surg Gynecol Obstet. 1905;1:297-314.

Delashaw JB, Broaddus WC, Kassel NF, Haley EC, Pendleton GA, Vollmer DG, Maggio WW: Treatment of right hemispheric cerebral infarction by hemicraniectomy. Stroke 21:874-881, 1990

Di Ieva A, Gaetani P, Matula C, Sherif C, Skopec M, Tschabitscher M. Berengario da Carpi: a pioneer in neurotraumatology. J Neurosurg. 2011 May;114(5):1461-70.

El Gindi S. Neurosurgery in Egypt: past, present, and future-from pyramids to radiosurgery. Neurosurgery. 2002 Sep;51(3):789-95; discussion 95-6.

Elmaci I. Color illustrations and neurosurgical techniques of Serefeddin Sabuncuoglu in the 15th century. Neurosurgery. 2000 Oct;47(4):951-4; discussion 4-5.

Engelhorn T, Doerfler A, de Crespigny A, Beaulieu C, Forsting M, Moseley ME. Multilocal magnetic resonance perfusion mapping comparing the cerebral hemodynamic effects of decompressive craniectomy versus reperfusion in experimental acute hemispheric stroke in rats. Neurosci Lett. 2003;344:127-31

Faleiro RM, Faleiro LC, Caetano E, Gomide I, Pita C, Coelho G, Brás E, Carvalho B, Gusmão SN. Decompressive craniotomy: prognostic factors and complications in 89 patients. Arq Neuropsiquiatr. 2008 Jun;66(2B):369-73.

Gautschi OP, Hildebrandt G. Emil Theodor Kocher (25/8/1841-27/7/1917)--A Swiss neurosurgeon and Nobel Prize winner. Br J Neurosurg. 2009 Jun;23(3):234-6.

Gerriets T, Stolz E, König S, Babacan S, Fiss I, Jauss M, Kaps M: Sonographic monitoring of midline shift in space-occupying stroke. Stroke 32:442-, 2001

Grant GA, Jolley M, Ellenbogen RG, Roberts TS, Gruss JR, Loeser JD. Failure of autologous bone-assisted cranioplasty following decompressive craniectomy in children and adolescents. J Neurosurg. 2004 Feb;100(2 Suppl Pediatrics):163-8.

Guerra WK, Gaab MR, Dietz H, Mueller JU, Piek J, Fritsch MJ. Surgical decompression for traumatic brain swelling: indications and results. J Neurosurg. 1999 Feb;90(2):187-96.

Fisher CM, Ojemann RG: Bilateral decompressive craniectomy for worsening coma in acute subarachnoid hemorrhage. Observations in support of the procedure. Surg Neurol 41:65-74, 1994

Hossmann K-A, Schuier FJ. Metabolic (cytotoxic) type of brain edema following middle cerebral artery occlusion in cats. In Price TR, Nelson E (eds). Cerebrovascular diseases. New York: Raven Press 1979:141-165.

Jankowitz BT, Kondziolka DS. When the bone flap hits the floor.Neurosurgery. 2006 Sep;59(3):585-90; discussion 585-90.

Jourdan C, Convert J, Mottolese C, Bachour E, Gharbi S, Artu F: Evaluation of the clinical benefit of decompression hemicraniectomy in intracranial hypertension not controlled by medical treatment. Neurochirurgie 39:304-310, 1993

Ivamoto HS, Numoto M, Donaghy RM. Surgical decompression for cerebral and cerebellar infarcts. Stroke. 1974 May-Jun;5(3):365-70.

Hacke W, Schwab S, Horn M, Spranger M, De Georgia M, von Kummer R. 'Malignant' middle cerebral artery territory infarction: clinical course and prognostic signs. Arch Neurol. 1996 Apr;53(4):309-15.

Hatashita S, Koike J, Sonokawa T, Ishii S. Cerebral edema associated with craniectomy and arterial hypertension. Stroke 16:661-668, 1985

Malliti M, Page P, Gury C, Chomette E, Nataf F, Roux FX. Comparison of deep wound infection rates using a synthetic dural substitute (neuro-patch) or pericranium graft for dural closure: a clinical review of 1 year. Neurosurgery 54:599-603; discussion 603-604, 2004

Marino R, Jr., Gonzales-Portillo M. Preconquest Peruvian neurosurgeons: a study of Inca and pre-Columbian trephination and the art of medicine in ancient Peru. Neurosurgery. 2000 Oct;47(4):940-50.

Martin G. Trepanation in the South pacific. J Clin Neurosci. 1995 Jul;2(3):257-64.

Martin G. Was Hippocrates a beginner at trepanning and where did he learn? J Clin Neurosci. 2000 Nov;7(6):500-2.

Martin TJ, Corbett JJ: Pseudotumor cerebri, in Youmans JR (ed): Neurological Surgery, 4th ed . Philadelphia: WB Saunders, vol 4, pp 2980-2997, 1996

McLaurin RL, Nichols JB Jr: Extensive cranial decompression in the treatment of severe lead encephalopathy. Pediatrics 20:653-667, 1957

Minnerup J, Wersching H, Ringelstein EB, Heindel W, Niederstadt T, Schilling M, Schäbitz WR, Kemmling A. Prediction of Malignant Middle Cerebral Artery Infarction

Using Computed Tomography-Based Intracranial Volume Reserve Measurements. Stroke. 2011 Sep 8. [Epub ahead of print]

Merenda A, DeGeorgia M. (2010). Craniectomy for acute ischemic stroke: how to apply the data to the bedside. Current Opinion in Neurology, vol 23, pp. 53-58

Missios S. Hippocrates, Galen, and the uses of trepanation in the ancient classical world. Neurosurg Focus. 2007;23(1):E11.

Missori P, Polli FM, Rastelli E, Baiocchi P, Artizzu S, Rocchi G, Salvati M, Paolini S, Delfini R. Ethylene oxide sterilization of autologous bone flaps following decompressive craniectomy. Acta Neurochir (Wien). 2003 Oct;145(10):899-902; discussion 902-3.

Mitka M. Hemicraniectomy improves outcomes for patients with ruptured brain aneurysms. JAMA 2001; 286:2084

Moulin DE, Lo R, Chiang J, Barnett HJM: Prognosis in middle cerebral artery occlusion. Stroke 16:282-284, 1985

Ng L, Nimmannitya J: Massive cerebral infarction with severe brain swelling: A clinicopathological study. Stroke 1:158-163, 1970

O´Brien MD, Waltz AG. Transorbital approach for occluding the middle cerebral artery without craniectomy. Stroke 1973;4:201-206.

O´Brien MD, Jordan MM, Waltz AG. Ischemic cerebral edema and the blood-brain barrier. Distributions of pertechnetate, albumin, sodium and antipyrine in brains of cats after occlusion of the middle cerebral artery. Arch Neurol 1974;30:461-465.

Osawa M, Hara H, Ichinose Y, Koyama T, Kobayashi S, Sugita Y. Cranioplasty with a frozen and autoclaved bone flap. Acta Neurochir (Wien). 1990;102(1-2):38-41.

Paciaroni M, Agnelli G, Floridi P, Alberti A, Acciarresi M, Venti M, Alagia MG, Fiacca A, Gallina MC, Guercini G, Pantaleoni R, Leone F, Pieroni A, Caso V. Hyperdense Middle Cerebral and/or Internal Carotid Arteries in Acute Ischemic Stroke: Rate, Predictive Factors and Influence on Clinical Outcome. Cerebrovasc Dis. 2011 Aug 23;32(3):239-245. [Epub ahead of print]

Rengachary SS, Batnistsky S, Morantz RA, Arjunan K, Jeffries B. Hemicraniectomy for acute massive cerebral infarction. Neurosurgery 8:321-328, 1981

Ransonhoff J, Benjamin MV, Gage EL Jr, Epstein F. Hemicraniectomy in the management of acute subdural hematoma. J Neurosurgery 34:70-76, 1971

Rieke K, Schwab S, Krieger D, von Kummer R, Aschoff A, Schuchardt V, Hacke W: Decompressive surgery in space-occupying hemispheric infarction: Results of an open, prospective trial. Critical Care Medicine 23:1576-1587, 1995

Rhoton Jr AL, Ono M. Microsurgical anatomy of the region of the tentorial incisura. In: Wilkins Rh, Rengachary SS (ed): Neurosurgery. 2nd Ed, Vol. 1, New York, McGraw-Hill, 897-915, 1996

Ropper AH. Lateral displacement of the brain and level of consciousness in patients with an acute hemispheral mass. N Engl J Med 314:953-958, 1986

Sanderland S. The tentorial notch and complication produced by herniations of the brain through that aperture. Br J Surg 45: 422-438, 1958

Sarrafzadeh AS, Sarafian N, von Gladiss A, Unterberg AW, Lanksch WR. Ibn Sina (Avicenna). Historical note. Neurosurg Focus. 2001;11(2):E5.41

Schmidt JH 3rd, Reyes BJ, Fischer R, Flaherty SK. Use of hinge craniotomy for cerebral decompression. Technical note. J Neurosurg107(3):678-82, 2007

Schwab S, Steiner T, Aschoff A, Schwarz S, Steiner HH, Jansen O, Hacke W. Early hemicraniectomy in patients with complete middle cerebral artery infarction. Stroke 29:1888-1893, 1998

Shaw CM, Alvord EC, Berry GR: Swelling of the brain following ischemic infarction with arterial occlusion. Arch Neurol 1:161-177, 1959

Stiver SI. Complications of decompressive craniectomy for traumatic brain injury. Neurosurg Focus 26 (6):E7, 2009

Taferner E, Pfausler B, Kofler A, Spiss H, Engelhardt K, Kampfl A, Schutzhard E: Craniectomy in severe, life-threatening encephalitis: a report on outcome and long-term prognosis of four cases. Intensive Care Med 27:1426-1428, 2001

Thomalla G, Hartmann F, Juettler E, Singer OC, Lehnhardt FG, Köhrmann M, Kersten JF, Krützelmann A, Humpich MC, Sobesky J, Gerloff C, Villringer A, Fiehler J, Neumann-Haefelin T, Schellinger PD, Röther J; Clinical Trial Net of the German Competence Network Stroke. Prediction of malignant middle cerebral artery infarction by magnetic resonance imaging within 6 hours of symptom onset: A prospective multicenter observational study. Ann Neurol. 68(4):435-45, 2010

Turney TM, Garraway WM, Whisnant JP: The natural history of hemispheric and brain stem infarctions in Rochester, Minnesota. Stroke 15:790-794, 1984

Vahedi K, Hofmeijer J, Juettler E, Vicaut E, George B, Algra A, et al. Early decompressive surgery in malignant infarction of the middle cerebral artery: a pooled analysis of three randomised controlled trials. Lancet Neurol. 2007 Mar;6(3):215-22.

Valença MM, Martins C, da Silva JC. "Window-like" craniotomy and "bridgelike" duraplasty: an alternative to decompressive hemicraniectomy. J Neurosurg. 2010 Nov;113(5):982-9. Epub 2009 Dec 11.

Valença MM, Valenca LPAA, Gonçalves da Silva G, Antunes-Rodrigues J, Leite JP. Post-ischemic stroke first-time seizures: influence of gender and age of the patient on the latent period. Neurobiologia (Recife) 72:27-32, 2009b

Valença MM, Valença LPAA. (2003) Doenças Cerebrovasculares, In *Neurologia Clínica*, edited by Gilson Edmar Gonçalves e Silva; Marco Otávio Saraiva Valença, 340-369. Ed. Universitária da UFPE, Recife

Vanaclocha V, Sáiz-Sapena N, García-Casasola C, De Alava E. Cranioplasty with autogenous autoclaved calvarial bone flap in the cases of tumoural invasion. Acta Neurochir (Wien). 1997;139(10):970-6.

Venes JL, Collins WF: Bifrontal decompressive craniectomy in the management of head trauma. J Neurosurgery 42:429-433, 1975

Wagner S, Schnippering H, Aschoff A, Koziol J, Schwab S, Steiner T. Suboptimum hemicraniectomy as a cause of additional cerebral lesions in patients with malignant infarction of the middle cerebral artery. J Neurosurgery 94:693-696, 2001

Winkler PA, Stummer W, Linke R, Krishnan KG, Tatsch K. Influence of cranioplasty on postural blood flow regulation, cerebrovascular reserve capacity, and cerebral glucose metabolism. J Neurosurg 93:53-61, 2000

Wirtz CR, Steiner T, Aschoff A, et al. Hemicraniectomy with dural augmentation in medically uncontrollable hemispheric infarction. Neurosurgical Focus 2 (5):Article 3, 1997